THE ANTEBELLUM PRESS

The Antebellum Press: Setting the Stage for Civil War reveals the critical role of journalism in the years leading up to America's deadliest conflict by exploring the events that foreshadowed and, in some ways, contributed directly to the outbreak of war.

This collection of scholarly essays traces how the national press influenced and shaped America's path towards warfare. Major challenges faced by American newspapers prior to secession and war are explored, including: the economic development of the press; technology and its influence on the press; major editors and reporters (North and South) and the role of partisanship; and the central debate over slavery in the future of an expanding nation. A clear narrative of institutional, political, and cultural tensions between 1820 and 1861 is presented through the contributors' use of primary sources. In this way, the reader is offered contemporary perspectives that provide unique insights into which local or national issues were pivotal to the writers whose words informed and influenced the people of the time.

As a scholarly work written by educators, this volume is an essential text for both upper-level undergraduates and postgraduates who study the American Civil War, journalism, print and media culture, and mass communication history.

David B. Sachsman holds the West Chair of Excellence in Communication and Public Affairs. He came to the University of Tennessee at Chattanooga from California State University, Fullerton, where he served as dean and professor of the School of Communications. Previously, he was chair of the Department of Journalism and Mass Media at Rutgers University. Dr. Sachsman is the director of the annual Symposium on the 19th Century Press, the Civil War, and Free Expression. His previous works include *The Civil War and the Press* (2000), *Sensationalism* (2013), *A Press Divided* (2014), and *After the War* (Routledge, 2017).

Gregory A. Borchard, a professor in the Hank Greenspun School of Journalism and Media Studies at the University of Nevada, Las Vegas, has written numerous works on journalism history, including *A Narrative History of the American Press* (Routledge, 2019). Together with David W. Bulla, he is the author of *Lincoln Mediated* (2015), and *Journalism in the Civil War Era* (2010). He is also the author of *Abraham Lincoln and Horace Greeley* (2011) and editor of *Journalism History*, a quarterly journal published by the Association for Education in Journalism and Mass Communication's History Division.

THE ANTEBELLUM PRESS

Setting the Stage for Civil War

Edited by David B. Sachsman and Gregory A. Borchard

With Dea Lisica

Routledge
Taylor & Francis Group

NEW YORK AND LONDON

First published 2019
by Routledge
52 Vanderbilt Avenue, New York, NY 10017

and by Routledge
2 Park Square, Milton Park, Abingdon, Oxon, OX14 4RN

Routledge is an imprint of the Taylor & Francis Group, an informa business

© 2019 Taylor & Francis

The right of David B. Sachsman and Gregory A. Borchard to be identified as the authors of the editorial material, and of the authors for their individual chapters, has been asserted in accordance with sections 77 and 78 of the Copyright, Designs and Patents Act 1988.

Library of Congress Cataloging-in-Publication Data
Names: Sachsman, David B., editor. | Borchard, Gregory A., editor. |
Lisica, Dea, editor.
Title: The antebellum press : setting the stage for Civil War /
edited by David B. Sachsman and Gregory A. Borchard with Dea Lisica.
Description: New York, NY : Routledge, 2019. | Includes index.
Identifiers: LCCN 2019005762 | ISBN 9780367196806 (hardback) |
ISBN 9780367196820 (paperback) | ISBN 9780429242588 (ebook)
Subjects: LCSH: United States–History–1815–1861–Press coverage. |
American newspapers–History–19th century. |
Press and politics–History–19th century.
Classification: LCC E302.1 .A58 2019 | DDC 070.4/4932097309034–dc23
LC record available at https://lccn.loc.gov/2019005762

ISBN: 978-0-367-19680-6 (hbk)
ISBN: 978-0-367-19682-0 (pbk)
ISBN: 978-0-429-24258-8 (ebk)

Typeset in Bembo
by Newgen Publishing UK

CONTENTS

List of Figures   *viii*

Preface   *xi*
    David B. Sachsman

Introduction   1
    Gregory A. Borchard

1 Newspapers, Agenda Setting, and a Nation Under Stress   14
    Donald L. Shaw and Thomas C. Terry, with Milad Minooie

2 The "Irrepressible Conflict" and the Press in the Late
    Antebellum Period   23
    Debra Reddin van Tuyll

PART I
Nullification, Abolition, and Division   **37**

3 Nat Turner's Revolt Spurs Southern Fears and Sparks
    Public Debate over Slavery   39
    James Scythes

4 Disunion or Submission? Southern Editors and the
    Nullification Crisis, 1830–1833   49
    Erika Pribanic-Smith

5 Abolitionist Editors: Pushing the Boundaries of
   Freedom's Forum   61
   David W. Bulla

6 When the Pen Gives Way to the Sword: Editorial Violence
   in the Nineteenth Century   70
   Abigail G. Mullen

7 An Editorial House Divided: The Texas Press Response
   to the Compromise of 1850   77
   Mary M. Cronin

8 "The Good Old Cause": The Fugitive Slave Law and
   Revolutionary Rhetoric in *The Boston Daily Commonwealth*   91
   Nicole C. Livengood

9 Franklin Pierce and the Failure of Compromise: Newspaper
   Coverage of the Compromise Candidate, the "Nebraska Act,"
   and the Midterm Elections of 1854   101
   Katrina J. Quinn

10 Abolitionism, the Kansas-Nebraska Act, and the End of
   Compromise   114
   Dianne M. Bragg

11 "Like so many black skeletons": The Slave Trade through
   American and British Newspapers, 1808–1865   125
   Thomas C. Terry and Donald L. Shaw

PART II
The Election of 1856, Dred Scott, and the
Lincoln-Douglas Debates   **139**

12 1856: A Year of Volatile Political Reckoning   141
   Dianne M. Bragg

13 Doughface Democrats, James Buchanan, and Manliness in
   Northern Print and Political Culture   155
   Brie Swenson Arnold

14 "Free Men, Free Speech, Free Press, Free Territory, and Frémont" 167
   Gregory A. Borchard

15 Newspaper Coverage of Dred Scott Inflames a
Divided Nation   179
William E. Huntzicker

16 "More than a Skirmish": Press Coverage of the
Lincoln–Douglas Debates   188
David W. Bulla

PART III
The Election of 1860 and the Crisis of Secession   **199**

17 The Democrats Divide: Newspaper Coverage of the 1860
Presidential Conventions   201
Brian Gabrial

18 Fanning the Flames: Extremist Rhetoric in the
Antebellum Press   212
Phillip Lingle

19 The Fire-Eating *Charleston Mercury*: Stoking the Flames
of Secession and Civil War   223
Debra Reddin van Tuyll

20 "Our all is at stake": The Anti-Secession Newspapers
of Mississippi   234
Nancy McKenzie Dupont

21 Exchange Articles Carried by the New York *Evening Post*,
December 13–31, 1860   245
Erika Thrubis

22 War of Words: Border State Editorials During the
Secession Period   251
Melony Shemberger

About the Editors   *257*
Contributors   *260*
Index   *265*

FIGURES

Many of the illustrations in this book come from prints that were created and sold independently by Currier & Ives and other publishers. This was a thriving business from the 1850s on and was the principle way of distributing political cartoons at the time, as most newspapers were not publishing illustrations through much of the period.

Book Cover

"Democratic Platform Illustrated," lithograph published by James G. Varney, 1856. This illustration is an attack on the 1856 Democratic Party as pro-South and proslavery. On the right, two slaves are pictured chained to the Democratic platform and crying, "Is this Democracy?" The illustration on the left appears to represent the caning of abolitionist Senator Charles Sumner by Democratic Representative Preston Brooks. (From Library of Congress Prints and Photographs Division Washington DC.)

0.1 "Democratic Platform Illustrated," lithograph published by James G. Varney, 1856. This illustration is an attack on the 1856 Democratic Party as pro-South and proslavery. (Library of Congress)   10

2.1 "The Republican Party Going to the Right House," published by Currier & Ives, 1860. This satirical cartoon depicts Abraham Lincoln's 1860 presidential campaign supporters as a motley crew of fanatics and oddballs. (Library of Congress)   32

3.1 "Horrid massacre in Virginia," by Samuel Warner, 1831. This illustration depicts events from Nat Turner's Rebellion, as recounted by white witnesses. (Library of Congress)   41

4.1 "Jackson and the nullifiers," 1832. This anti-nullification verse to
the tune of "Yankee Doodle Dandy" was printed as a broadside
and sold in New York City. (Library of Congress)   54

7.1 "Congressional scales, a true balance," lithograph published by
Currier & Ives, 1850. This satirical cartoon shows President
Zachary Taylor striving to balance opposing viewpoints on slavery
between the North and South and depicts the struggle that
occurred in Congress to meet an agreement on key issues
relating to the expansion of slavery into Western territory.
(Library of Congress)   80

8.1 "Anthony Burns," printed by R. M. Edwards in Boston, 1855.
A portrait of fugitive slave Anthony Burns ringed by drawings
of important events, such as his arrest and imprisonment.
(Library of Congress)   97

9.1 "Ornithology," lithograph published by John Childs, 1852. This
political cartoon depicts Whig presidential candidate Winfield
Scott as a turkey and Democratic candidate Franklin Pierce as a
gamecock on either side of the "Mason & Dixson's Line."
(Library of Congress)   104

12.1 "Arguments of the chivalry," lithograph by Winslow Homer, 1856.
This image illustrates the May 22, 1856, attack on Massachusetts
Senator Charles Sumner by South Carolina Representative
Preston S. Brooks. (Library of Congress)   145

12.2 "Forcing Slavery Down the Throat of a Freesoiler," lithograph
by John L. Magee, 1856. This anti-Democratic political cartoon
focuses on the "Crime Against Kansas," a major issue in the 1856
presidential election. (Library of Congress)   149

13.1 "Letting the Last 'Democratic Drop,'" lithograph by unknown
author and publisher, 1856. In this political cartoon, James
Buchanan, wearing female attire, is bled by Republican
candidate John C. Frémont and an obscure figure.
(Library of Congress)   161

14.1 "The Great Republican Reform Party, Calling on their
candidate," lithograph published by Nathaniel Currier, 1856.
This political cartoon presents a negative view of Republican
presidential nominee John C. Frémont during the 1856 election,
depicting him standing before an assorted group of constituents
(including a Catholic priest and a free black) and responding to
their demands. (Library of Congress)   171

14.2 "The 'Mustang' Team," lithograph published by Nathaniel Currier,
1856. This cartoon mocks the abolitionist Republican Party,
and jabs at the newspaper editors who supported Republican
presidential candidate John C. Frémont. (Library of Congress)   173

15.1 "Visit to Dred Scott—his family—incidents of his life—decision
of the Supreme Court," *Frank Leslie's Illustrated Newspaper,* June
27, 1857. Woodcut portraits of Dred Scott, his wife, Harriet, and
their children, printed on the front page of *Frank Leslie's Illustrated
Newspaper* following the final court appearance and emancipation
of Dred Scott. (Library of Congress)   184

15.2 "The Political Quadrille. Music by Dred Scott," lithograph by
unknown artist, 1860. This cartoon, depicting Scott playing
the violin as the four 1860 presidential contenders dance with
stereotypical caricatures of their constituents, illustrates the
importance and impact of the Dred Scott decision on the
presidential campaign. (Library of Congress)   186

17.1 "The Great Exhibition of 1860," lithograph published by Currier
& Ives, 1860. This cartoon satirizes the Republicans' attempts
to deemphasize their antislavery position in the 1860 election.
(Library of Congress)   205

17.2 "National," lithograph by unknown author and publisher, 1860.
This political cartoon, also called "Dividing the national map,"
shows clear understanding of the sectional nature of the 1860
political election by depicting Lincoln, Douglas, Breckinridge, and
John Bell fighting and tearing apart a map of the nation.
(Library of Congress)   208

18.1 "Harper's Ferry Insurrection—Burying the Dead Insurgents,"
Frank Leslie's Illustrated Newspaper, November 5, 1859. This
front-page illustration from *Frank Leslie's Illustrated Newspaper*
introduces its November 5, 1859, follow-up story on John
Brown's Raid. (Library of Congress)   218

18.2 "An Heir to the Throne, or the Next Republican Candidate,"
lithograph published by Currier & Ives, 1860. This openly racist
political cartoon argues that the Republicans' next presidential
candidate will be a black man. (Library of Congress)   220

20.1 "The 'Secession Movement,'" lithograph published by Currier &
Ives, 1861. This illustration depicts secessionist states as being led
over a cliff to their doom. (Library of Congress)   240

PREFACE

David B. Sachsman

Freedom of speech and of the press is fundamental to American democracy. The press has played a critical role in politics since colonial times and all the nation's political and social discussions and debates have taken place in the pages of its newspapers.

But how powerful is the press? Does it change our attitudes and beliefs, or simply provide us with information and ideas to fuel our discussions and set the agenda for our debates and deliberations? How powerful is the press today, and how powerful was it during the great turning points in American affairs?

The Civil War is the watershed event in American history. The war changed the culture, the politics, the industry, the economics, and the very freedom of the American people. The Civil War turned the United States into what would become the greatest industrial nation in the world—a nation crisscrossed by railroads and telegraph lines. It created what we think of now as modern America.

The cause or causes of the Civil War have been the subject for debate since before the war began. Southerners argued for and against secession for at least thirty years before the election of Abraham Lincoln precipitated the event. Slavery was always the essential issue—the point of the spear—but the nation had been separated by culture, politics, and economics from its very beginnings.

Was the Civil War inevitable? This question also continues to be open to debate. The South, or Southern interests, held control of American politics for most of the period from 1800 to 1860. If Southern interests had been able to continue controlling the presidency, the Congress, and the courts throughout the nineteenth century, the calls for secession might have remained in check. If changes in the culture and economics of the South had made slavery unprofitable—had replaced slavery with wage slavery—the spear might have lost its point.

But this was not to be. By 1860, the North and West had enough Electoral College votes to elect the president of the United States, and it was clear to Southerners, Northerners, and Westerners that the political power of the Northern and Western states would continue to grow with western expansion.

When the Republicans developed as a distinctly Northern and Western regional power and took control of the presidency, it was much more than a shift of political power. It was seen by Southern secessionists as a fundamental, cultural, and societal change. In the months from the election of Abraham Lincoln to the day in March 1861 when he took office, these secessionists were able to push seven states into leaving the Union. When President Lincoln, after Fort Sumter, required that the states choose sides, Virginia, Arkansas, North Carolina, and Tennessee seceded as well.

All of the events that would lead to the Civil War can be found in the pages of the nation's newspapers—from the Missouri Compromise of 1820, to Nat Turner's Revolt, to the Nullification Crisis of 1832–1833, to the Compromise of 1850, to the Kansas-Nebraska Act of 1854, to the caning of Senator Charles Sumner and the political year of 1856, to the Dred Scott decision of the Supreme Court, to the Lincoln-Douglas Debates, to John Brown's Raid, and the election of 1860. The American press from 1800 to 1860 was highly partisan, and the arguments that both shaped the nation and separated its people were given voice by the press.

For more than twenty-five years, the Symposium on the 19th Century Press, the Civil War, and Free Expression has explored the role of the nation's newspapers during the pivotal years leading up to the events of 1860 and 1861. The Symposium has consistently featured cutting-edge research, including the work of Donald L. Shaw, the creator of the agenda-setting theory of the press, who has spent some two decades focusing his research for the Symposium on agenda setting in the historical context of the Antebellum and Civil War period. This volume begins with the conclusions and questions raised by Shaw and Thomas C. Terry (with Milad Minooie) in "Newspapers, Agenda Setting, and a Nation Under Stress," a work they developed specifically for this book and presented at the conference. It also begins with the answers provided by Debra Reddin van Tuyll, the nation's leading expert on the Southern Antebellum and Civil War press, in "The 'Irrepressible Conflict' and the Press in the Late Antebellum Period" (which also was written to introduce this volume and was presented at the conference). The question of whether the Civil War was inevitable runs through this volume as its authors explore the American newspapers of the Antebellum period and their coverage of a "nation under stress." What role did fire-eating, Southern newspaper editors play in the secession movement? What role did the abolitionist press play in changing the nature of the debate?

The Antebellum Press describes the involvement of the nation's newspapers through all the crucial moments and discussions leading up to the breakdown in the very fabric of America. It offers seasoned research and new insights as it tells the story of how the American press set the stage for the Civil War.

This is a work of history, a major focus of which was the racial hatred that existed in the United States. This racial hatred was openly reflected in the language of the newspapers of the time. That language, when provided in direct quotations, is generally retained in this book in order to accurately reflect the historical context.

It has been my honor to serve for more than twenty-five years as the holder of the George R. West, Jr. Chair of Excellence in Communication and Public Affairs at the University of Tennessee at Chattanooga, and as the founder and director of the Symposium on the 19th Century Press, the Civil War, and Free Expression. It has been my pleasure to work on this volume with Gregory Borchard of the University of Nevada, Las Vegas, and with Dea Lisica, who serves as assistant director of the Symposium and curator of the Symposium Library. I am also grateful to the University of Tennessee at Chattanooga for its support of these efforts. This book is first and foremost the creation of its authors, many of whom I have worked with from the time they were graduate students. It has been their goal to provide answers to the most important questions of the Antebellum period, and we believe they have made great progress towards accomplishing this task.

INTRODUCTION

Gregory A. Borchard

Newspapers have played a direct role in the development of the nation. This was true during the American Revolution and during the creation of the Constitutional Republic. The presidential election of 1800 was fought in the pages of rabidly partisan newspapers, and from this time on the partisan press would separate the nation, North and South, East and West, free and slave. The Antebellum press foreshadowed and in some ways contributed directly to the outbreak of America's bloodiest conflict. Indeed, the press set the stage for civil war.

The nation's founding documents reflected sentiments popular in some of the most widely read journalistic pieces of the day. Phrases from the Preamble of the Declaration of Independence mirror those in Thomas Paine's *Common Sense*. Aided by the institution of the American press itself, which demonstrated the democratizing effects of the Enlightenment philosophy of natural law, newspapers fueled a government of the people, by the people, and for the people, designed to promote life, liberty, and the pursuit of happiness.

The goals of the Revolution suggested a hopeful trajectory for the United States that would sustain its people through an unknown future. However, one idea, the concept that "all men are created equal," meant different things to different people. In time, the irreconcilable nature of an America in which free people coexisted with a system of slavery reached levels beyond any peaceful remedy. The Three-Fifths Compromise found in the Constitution—an agreement between slaveholding and non-slaveholding representatives—had counted the slave population as a portion of a state's total population in determining representation for legislative and taxation purposes. On the one hand, slaves were property. On the other, they were three-fifths human. In time, the nation could not have it both ways. The declared beliefs in equality among humans could not exist side-by-side with a system of oppression.

From the Founding to the Era of Good Feeling to the Age of Jackson

At the beginning of the War for Independence, dozens of weekly newspapers operated throughout the colonies with readership reaching an audience in the hundreds of thousands. The words of the Declaration of Independence, "that all men are created equal," were widely circulated among the Revolutionary press. Much later, James Madison explained that the Founding Fathers understood a free press to be vital to the integrity of the Republic. "A Popular government without information, or the means of acquiring it, is but a prologue to a farce or tragedy; or perhaps both," Madison wrote. "Knowledge will forever govern ignorance; and a people who mean to be their own governors must arm themselves with the power which knowledge gives."[1]

Efforts to implement the ideals of liberty in the new Republic emerged along partisan lines that more often than not represented the economic and cultural interests of the United States in forms of regional representation—almost immediately between North and South, or, in economic terms, industrial and agrarian. While George Washington, the first president to serve under this newly formed constitutional republic, subsequently received praise for maintaining a non-partisan style of administration, opposing political parties emerged even before he voluntarily stepped down from office.

Two members of Washington's cabinet—Treasury Secretary Alexander Hamilton and Secretary of State Thomas Jefferson—provided highly competitive models for organizing subsequent administrations. Although both sought to maximize liberty, they wrestled with a vision of the United States maintained by a strong central government, as promoted by Hamilton, versus a collection of interdependent states that would, according to Jefferson, secure individual rights. For Hamilton, the strength of the United States would emerge from its commerce, as determined by its businesses, its entrepreneurs, and its banks and financiers, with the federal government doing its job best in assisting competition globally. Hamilton and fellow Federalists James Madison and John Jay had expressed these ideas directly through the press in the classic series of articles titled "The Federalist," which they published in New York's *Independent Journal*, *New-York Packet*, and *Daily Advertiser* between 1787 and 1788.

Jefferson disagreed with the Federalists about the nature of the American people, believing financiers would exploit the labor of the virtuous and independent agrarian-based citizenry. Instead of assisting business, Jefferson argued, the government did best in protecting the rights and liberties of its citizens, a belief he had helped articulate in the Bill of Rights, the list of guarantees prohibiting government interference with personal liberties. Among the rights advocated by Jefferson was a free press. "Were it left to me to decide whether we should have a government without newspapers, or newspapers without a government," Jefferson wrote, "I should not hesitate a moment to prefer the latter."[2]

The Compromise of 1790 included an agreement between Hamilton and Jefferson and legislators from the North and South to relocate the nation's capital from Philadelphia to the area now recognized as the District of Columbia. The move carried the implication that the North would not raise serious objections to the institution of slavery, as the capital would be located in two slave states, Maryland and Virginia.

The balance between interests, however, was short lived, as Hamilton and Jefferson proceeded to finance newspapers as a way to persuade new followers to support each respective party. For Hamilton and the Federalists, this newspaper was *The Gazette of the United States*, edited by John Fenno; for Jefferson and the Republicans, it was *The National Gazette*, edited by Philip Freneau. These partisan newspapers often contained vitriolic personal attacks on members of the opposition. The struggle grew so intense that President John Adams used the Alien and Sedition Acts (1798) to silence opposition to Federalist policies.

The number of newspapers increased during this period, as they would throughout the nineteenth century. The Postal Service Act of 1792 resulted in a vast increase in the number of post offices and the extent of mail delivery, and in a decrease in the cost of mailing newspapers. The number of post offices routing mail and newspapers grew from fewer than 100 in 1788 to nearly 5,000 in the early Antebellum era. In 1800, publishers produced between 150 and 200 newspapers in the United States. By 1810, this number had grown to 366. Historians estimate that the annual circulation of all newspapers between 1828 and 1840 doubled from 68 million to 148 million copies, and that at the outbreak of the Civil War, the number of individual newspapers in print had grown to 4,000.[3]

The effect of this information explosion on Americans is reflected in an anecdote from the journal of Alexander Bölöni Farkas describing the phenomenon. While traveling through the United States in the 1830s, the Transylvanian reformer marveled as the stagecoach driver hurled out settlers' newspapers right and left as they passed remote cabins along the road, through the whole day's travel. "No matter how poor a settler may be, nor how far in the wilderness he may be from the civilized world," Farkas wrote, "he will read a newspaper."[4] Frenchman Alexis de Tocqueville agreed, writing after his tour of the United States that he saw "scarcely a hamlet that has not its newspaper."[5]

Meanwhile, the number of slaves and the extent of slavery also increased. During the Antebellum period, the nation's African American population would increase from 400,000 to 4.4 million, of whom 3.9 million were slaves. (In 1860, slaves composed 57 percent of South Carolina's population, with Mississippi following at 55 percent, then Louisiana at 47 percent, Alabama at 45 percent, and Florida and Georgia, both at 44 percent.)[6]

In an effort to maintain control over the role slavery would play in national affairs, Congress had passed the Missouri Compromise on May 8, 1820, which sought a balance of power between North and South. The Compromise, which legislators at the time believed provided a permanent solution to the slavery question, allowed the admission of Maine as a free state along with Missouri as a slave state and prohibited slavery north of the 36°30' parallel (excluding Missouri). The press, politicians, and the people celebrated the legislation as securing the stability of the United States, and as a crowning achievement for the Era of Good Feelings, a term which reflected the national mood under James Monroe, a relatively non-partisan (though Southern) president. This period of compromise—essentially a transition between 1817 and 1825—included at least a momentary dissolution of national political identities, as the discredited Federalist Party succumbed to a succession of Democratic administrations that blurred previously defined lines of division. (In time, different political parties would evolve—in the 1830s, the Federalists reorganized as Whigs, and shortly before the Civil War, the Whigs took the name Republicans.)

The ensuing era—the Age of Andrew Jackson—celebrated the "common man," even though the tension over the disparity between free people and people who were slaves had reached the point where the state of South Carolina would attempt to nullify its membership in the United States. This disparity between free men and slaves found further articulation in the ideas of the Whig Party, namely those promoted by leading spokesperson Henry Clay of Kentucky. Having devoted his life to politics and known as the "Great Compromiser," Clay's political ideals promoted a form of nationalism he dubbed the American System, which organized state and local governments with ties to the federal system, linking manufacturing in the Northeast with grain production in the West and the cotton and tobacco crops of the South. The key to the success of Clay's model included individual contributions from a particular kind of citizen, which he described (in helping to coin the phrase) as a "self-made."[7] This model also influenced subsequent members of both politics and the press, including Abraham Lincoln and his match in the press, Horace Greeley of *The New York Tribune*.[8]

The Press and the Antebellum Era

In the 1830s, cities that grew with rapid urbanization supported multiple daily newspapers, and big-city dailies emerged as sophisticated enterprises that included mechanized printing plants, large staffs, and circulations that would reach between 10,000 and 50,000 readers. Weeklies in small towns enjoyed growth likewise, utilizing hand-operated presses, smaller staffs, and modest-but-healthy circulation counts. The Postal Act of 1792 had permitted newspaper publishers to exchange their papers with other publishers without any postage at all. This exchange system gave small and large newspapers alike free access to news from nearly everywhere. According to the 1840 census, more than 1,500 newspapers circulated in the United States, and by 1850, the number increased to more than 2,500, with a total annual circulation of 500 million copies for a population of less than 23 million people.[9]

During the 1830s and 1840s, the press expanded content to reach a wider audience. The sales and business model most clearly defining the press during this period has been called the "penny press" for the simple reason that newspapers increasingly lowered their prices, oftentimes selling for as little as one cent. While this trend was a national one, New York newspapers in particular, such as the New York *Sun*, *Herald*, and *Tribune*, typified the changes in content and marketing strategies of publishers. While other penny press newspapers had their own unique approach to targeting particular audiences, they all shared a motivation to reach as many readers as possible, putting them in contrast with the preceding era's partisan press. The content of each newspaper sometimes enlightened and educated large groups of readers, but more commonly—in efforts to grab advertising dollars to supplement the low cost of the newspaper—it entertained them, enticing readers with sensational stories of scandal, sex, and crime.

The 1830s also saw the rise of the most influential abolitionist newspaper in the Antebellum era. William Lloyd Garrison and Isaac Knapp's *The Liberator* (1831–1865) expressed a view of freedom and slavery that was not open to compromise of any kind. In 1844, Garrison, in "No Union with Slaveholders," actually advocated for secession, calling for "the existing national compact" to be "instantly dissolved."[10] At the same time, Frederick Douglass championed one of the most remarkable stories of both press history and of American history. Though born into slavery, Douglass learned to read and write. His autobiography *Narrative of the Life of Frederick Douglass, an American Slave, Written by Himself* (1845)—and later, the abolitionist newspapers he published—demonstrated his talents as a journalist.[11]

Meanwhile, Southern fire-eaters used the press to advance proslavery ideals, which at an extreme called for separation of the Southern states from the Union and the formation of a new nation. Robert Barnwell Rhett, one of the most outspoken proponents of this movement, even used the *Charleston* (SC) *Mercury* to press for the re-opening of the international slave trade. Rhett pleaded with Southerners to maintain their sovereignty and—appealing to racist impulses—to resist alleged efforts from the North to extinguish their way of life. "Pent up and confined within compulsory limits, the labor of the two races may become valueless," he said in a speech. "We are of the dominant Caucasian race, and will perform our part in civilizing the world, and bring into beneficent subjection and cultivation its most productive regions of sun and beauty."[12] Calls for secession from a system Rhett described as oppressive resulted gradually in hardening the nation's sectionalism. The antagonistic views of the fire-eaters of the South and the abolitionists of the North, although initially considered radical, became increasingly salient in the years leading to war.

The Telegraph and the Press

The development of the telegraph in the 1840s had a profound effect on mass communication in America. The telegraph made speed the key trait of information transmission and reception from the late 1840s onward. Newspaper editors in New York were among the first to realize the potential of the new invention. Along with a daily newspaper in Philadelphia, they formed the Associated Press (AP) wire service to share information over long distances, forever changing the way readers would receive news. Prior to the telegraph, people saw themselves as parts of isolated geographic regions, having little knowledge of immediate news from other parts of the world. The telegraph, however, permitted a message to travel almost instantaneously. As communication scholar James Carey explained, "The simplest and most important point about the telegraph is that it marked the decisive separation of 'transportation' and 'communication.' Until the telegraph, these words were synonymous. The telegraph ended that identity and allowed symbols to move independently of geography and independently and faster than transport."[13]

The concurrent expansion of the railroads at a rate of more than 1,000 new miles each year in the 1850s also allowed publishers means for selling newspapers over vast distances, with the amount of track jumping from 8,589 miles in 1850 to 30,591 in 1861. Furthermore, in the 1850s, the Post Office typically delivered more than 100 million journals in a single year, with the bulk of those being newspapers.[14] The advent of new technologies—from increasingly faster steam-driven printing presses, to machine-made paper and railroads—meant that newspapers could appear more frequently and carry more news.

The Mexican-American War, the Compromise of 1850, and the Kansas-Nebraska Act

The context for these developments included political, social, and cultural issues that extended well beyond the publishing world, and although the press played a major role in reporting news on these events, editors, reporters, and publishers often found themselves simply reacting to external events instead of anticipating them. The Mexican-American War (1846–1848) provided Americans with a voluminous supply of news. The war symbolized the power behind the concept of Manifest Destiny, exacerbating the slavery issue and making previous compromises obsolete by upsetting the delicate balance between slave and free states.

Congress passed a package of five separate bills, dubbed the Compromise of 1850, designed to defuse political confrontations between slave and free states on the status of territories acquired during the Mexican-American War. While the compromise temporarily averted sectional conflict, controversy arose over a provision regarding fugitive slaves. The Fugitive Slave Act required that all escaped slaves were, upon capture, to be returned to their masters, and it required officials and citizens of free states to cooperate in enforcing this law.

Meantime, although the Whig Party had succeeded in the election of two presidents (William Henry Harrison in 1840 and Zachary Taylor in 1848), both chief executives died in office, facilitating the collapse of the party in 1852. The creation of the Republican Party in 1854 would transform social institutions and the press itself. At the same time, any sense of immediate relief from the passage of the Compromise of 1850 quickly deteriorated as settlers of the territories in the West organized in attempts to join the Union as states.

In legislation drafted by Democratic Senator Stephen A. Douglas of Illinois, the highly controversial Kansas–Nebraska Act of 1854 created the territories of Kansas and Nebraska with the goal of opening up thousands of new farms in the West and facilitating creation of the Transcontinental Railroad. The popular sovereignty clause of the law in effect repealed the Missouri Compromise of 1820, which had prohibited slavery north of the 36°30' parallel (with the exception of Missouri), and led pro- and antislavery elements to flood into Kansas with the goal of voting slavery up or down. The conflict between these settlers erupted into fighting and created the conditions that would be called Bleeding Kansas.

In 1856, the Republican Party nominated explorer John C. Frémont as its first presidential nominee. Frémont represented the pioneer spirit in settlement of the West. He received support from free-soilers who advocated an end to the spread of slavery into territories. Democrat James Buchanan won the election, but his administration did nothing to quell the growing national crisis about the role slavery would or would not play in the West.

Meanwhile, the Dred Scott case was winding its way through the courts. Scott, a slave who had been taken to free states and territories by his owners, was attempting to sue for his freedom. The Supreme Court ruled on *Dred Scott v. Sandford* on March 6, 1857. It soon became clear that Chief Justice Roger B. Taney and the majority intended to settle the slavery issue once and for all, ruling that "a negro, whose ancestors were imported into [the US], and sold as slaves," whether enslaved or free, could not be an American citizen and therefore had no standing to sue in federal court.[15] Moreover, the Court ruled that the federal government had no power to regulate slavery in the federal territories acquired after the creation of the United States. Rather than settling the slavery issue, this decision inflamed sentiments on both sides, causing Southerners to toughen their stand on the absolute right of slavery everywhere, and causing outraged abolitionists to reject the ruling.

The events of the next four years would ignite these sentiments into war. In October 1859, John Brown and his followers captured Harpers Ferry, Virginia, intending to incite a slave insurrection. The raid triggered fear in the South, a fear that multiplied with the realization that Republican Party presidential candidate Abraham Lincoln could win the 1860 presidential election. In turn, South Carolina, Mississippi, Florida, Alabama, Georgia, Louisiana, and Texas all seceded from the Union before Lincoln assumed office on March 4, 1861. As Lincoln said in his Second Inaugural Address, "and the war came."[16]

Setting the Stage for Civil War

Was the Civil War inevitable? Did Antebellum press coverage of the divisions of the time allow readers to see that war was coming? The opening chapters of this book reach different conclusions regarding these questions as they discuss the role of the press as either a reflection of or an actor in the national consciousness of the era. Donald L. Shaw and Thomas C. Terry, with Milad Minooie, in "Newspapers,

Agenda Setting, and a Nation under Stress" find that American newspapers in the Antebellum period reflected the nation as a whole and at the same time represented conflicting sectional interests. While stories about the Missouri Compromise, the 1831 Nat Turner revolt, the Mexican War, the Kansas-Nebraska Act, the Dred Scott decision, and John Brown's raid reinforced the importance of divisive issues in the minds of Americans, Shaw, Terry, and Minooie suggest this newspaper content did not predict the eventuality of civil war.

Debra Reddin van Tuyll, in "The 'Irrepressible Conflict' and the Press in the Late Antebellum Period," argues that the press played a much more active role in determining the course of events. The antecedents of the Civil War, van Tuyll contends, occurred as early as—if not before—Missouri's admission to the Union as a slave state in 1820, and subsequent developments in politics and the press simply brought the irreconcilable differences between regions to inevitable blows. She suggests that the press may not have had a direct role in creating hostilities, but newspapers at least reflected the underlying tensions of the nation, and as part of these antagonisms, enabled those in power to act upon divisive rhetoric.

As presented in these opening chapters, the differing perspectives on whether or not the war was avoidable reflect to a large extent the continuing and conflicting views about the Civil War to this day.[17] While most historians agree that a combination of slavery and western expansion created a climate that contributed directly to the outbreak of war, the authors presented in this volume provide new and valuable evidence as to the role played by the press in setting the stage for conflict.

Nullification, Abolition, and Division

The first full section of this book provides an overview of the core issues preceding the war, as well as those that contributed to defining the era. James Scythes, in "Nat Turner's Revolt Spurs Southern Fears and Sparks Public Debate over Slavery," describes the confusion and fear that swept the South in the weeks following the slave uprising, and notes that many of the news stories contained inaccuracies and exaggerations, enflaming the episode.

In 1832, a South Carolina state convention adopted an Ordinance of Nullification nullifying the tariffs of 1828 and 1832 and proclaiming that the state could decide which federal laws were enforceable in South Carolina. Erika Pribanic-Smith's account of these events in "Disunion or Submission? Southern Editors and the Nullification Crisis, 1830–1833" describes the ways newspapers addressed the question of nullification. Pribanic-Smith details the range of reactions from editors who debated the possible dissolution of ties from the Union (secession) as a remedy for federal measures seen as encroaching on state's rights.

On the other side, the abolitionist movement and William Lloyd Garrison's *The Liberator* fought for the end of slavery. David W. Bulla's "Abolitionist Editors: Pushing the Boundaries of Freedom's Forum" looks at editorials by Garrison, Frederick Douglass, and Gamaliel Bailey between 1847 and 1857. Bulla shows that these leading abolitionists focused content on policy issues ranging from the Mexican War to the Dred Scott Supreme Court decision.

Editors throughout the nation did not react uniformly to outside pressures, but in some respects, they encountered the same overriding cultural influences on their work. Reflecting on the pressures these issues put on the press in general, Abigail G. Mullen's "When the Pen Gives Way to the Sword: Editorial Violence in the Nineteenth Century" provides an overview of the ways in which editors responded to job pressures and describes their commitment to their political and journalistic ideals.

The Mexican-American War reopened the question of slavery in the Western territories that had been discussed in the Missouri Compromise of 1820. Mary M. Cronin's "An Editorial House Divided: The Texas Press Response to the Compromise of 1850" outlines the partisan perspectives in the press regarding Congress's attempts to determine the boundaries of land and of slavery in the areas taken from Mexico. Cronin provides a glimpse into the sectionalism that affected both newspapers and readers of newspapers throughout the period.

Also regarding the Compromise of 1850, Nicole C. Livengood's "'The Good Old Cause': The Fugitive Slave Law and Revolutionary Rhetoric in *The Boston Daily Commonwealth*" shows how one abolitionist newspaper sought to unify readers in rejection of the Fugitive Slave Law of 1850—a controversial element of the compromise—by harnessing symbols from the American Revolution. For abolitionists, the Fugitive Slave Law was the ultimate contradiction of the Declaration of Independence's proclamation that "all men are created equal."

The next two chapters of this book detail the collapse of compromise regarding slavery and western expansion. Katrina J. Quinn's "Franklin Pierce and the Failure of Compromise: Newspaper Coverage of the Compromise Candidate, the 'Nebraska Act,' and the Midterm Elections of 1854" examines newspaper reporting and editorial content from the 1852 Democratic National Convention to the midterm elections of 1854 to provide a profile of the Pierce administration's failed attempt to resolve the volatile issue of slavery in the West. Dianne M. Bragg's "Abolitionism, the Kansas-Nebraska Act, and the End of Compromise" suggests newspaper coverage of the Kansas-Nebraska Act fueled the expansion of antislavery politics more than almost any other Antebellum event.

Closing the first section of the book, Thomas C. Terry and Donald L. Shaw, in "'Like so many black skeletons': The Slave Trade through American and British Newspapers, 1808–1865," examine news reports that demonstrate that the transatlantic slave trade continued long after its congressional ban in 1808.

The Election of 1856, Dred Scott, and the Lincoln-Douglas Debates

The second section of this book examines the years immediately preceding the Civil War, describing moments of political and legal crisis and how the

FIGURE 0.1 "Democratic Platform Illustrated," lithograph published by James G. Varney, 1856. This illustration is an attack on the 1856 Democratic Party as pro-South and proslavery. On the right, two slaves are pictured chained to the Democratic platform and crying, "Is this Democracy?" The illustration on the left appears to represent the caning of abolitionist Senator Charles Sumner by Democratic Representative Preston Brooks. (From Library of Congress Prints and Photographs Division Washington DC. <www.loc.gov/item/2008661581/> Reproduction Number: LC-USZ62-13211 (b&w film copy neg.). James G. Varney, "Democratic Platform Illustrated.")

press responded to them. Dianne M. Bragg's "1856: A Year of Volatile Political Reckoning" begins with the caning of Senator Charles Sumner by Preston Brooks and ends with the election of Democrat James Buchanan as president. Bragg's analysis of the press coverage of the crucial political events of 1856 indicates the increasingly hardened sectional positions being taken in the country's divided political arena.

The next two chapters continue the discussion of press coverage of the fateful election of 1856. At that time, Northern Democrats who were soft on the issue of slavery were attacked as unmanly "doughfaces." Brie Swenson Arnold's "Doughface Democrats, James Buchanan, and Manliness in Northern Print and Political Culture" examines the prevalence of gendered and sexualized depictions of politicians—namely allegations that the Democratic candidate did not have the masculinity necessary for the presidency. She reveals ways the Antebellum press played an important role in shaping Northern political sentiment during the decisive elections and party realignments that precipitated the Civil War. Gregory A. Borchard's "'Free Men, Free Speech, Free Press, Free Territory, and Frémont,'" meanwhile, profiles the presidential bid of John C. Frémont in 1856 as the first Republican candidate for the office, citing commentary from leading Republican newspapers as voices of the newly formed party. Frémont lost the election, but paved the way for Republican success in the 1860 presidential campaign.

Two days after the inauguration of Buchanan, Chief Justice Taney announced that the Supreme Court had resolved the Dred Scott case. In "Newspaper Coverage of Dred Scott Inflames a Divided Nation," William E. Huntzicker shows how the Court's attempt to end the slavery question instead divided the country into sections that could not be reconciled short of physical conflict.

"'More than a Skirmish': Press Coverage of the Lincoln-Douglas Debates" by David W. Bulla provides a close look at the 1858 contest for the Illinois Senate seat that made Lincoln a viable Republican candidate for president and removed any chance that Northerner Douglas, the advocate of popular sovereignty, could unite the Democratic Party in 1860.

The chapters in the second section of the book illustrate the ways sectional divisions hardened at mid-century. As evidenced in the pages of the press (and with the advantage of hindsight), these chapters reveal a nation moving closer—in some respects unknowingly—toward war.

The Election of 1860 and the Crisis of Secession

The third and final section of this book focuses on two events in particular that triggered the Civil War—the 1860 election and the secession crisis. As described in preceding chapters, the strains of the slavery question affected politics to the point where two political parties—first the Whigs and then the Democrats—deteriorated into chaos. While the Whigs reinvented themselves as Republicans, the Democrats split in ways that required decades to heal. Brian Gabrial, in "The Democrats Divide: Newspaper Coverage of the 1860 Presidential Conventions," presents the self-destruction of the Democratic Party (and the end of any chance of a united front against the Republican candidate) in all its gory details. The division of the Democratic Party was the first step toward the secession of the Southern states. The fire-eaters would get their way, turning political defeat into revolution.

Phillip Lingle's "Fanning the Flames: Extremist Rhetoric in the Antebellum Press" explores the roles played by extremist editors, North and South, whose rhetoric separated the nation into divisions that could no longer form a United States.

Debra Reddin van Tuyll, in "The Fire-Eating *Charleston Mercury*: Stoking the Flames of Secession and Civil War," demonstrates how a leading organ of the fire-eaters, the *Mercury*, had perpetually supported Southern rights, states' rights, and secession, and emerged in the years and months before the outbreak of war as the leading voice of disunion. However, the majority of Southern newspapers did not advocate secession before Lincoln was elected president. In "'Our All is at Stake': The Anti-Secession Newspapers of Mississippi," Nancy McKenzie Dupont shows that at least three newspapers took an editorial stance opposed to immediate secession during the secession crisis of 1860–1861.

American newspapers in the Antebellum period often covered the news from other regions of the nation by simply carrying reports from distant newspapers. Accepted custom and the Post Office encouraged this practice. Thus, Northern newspapers often carried reports written by Southern editors, including fire-eaters, and Southern newspapers carried reports from the Northern press showing the antislavery position of that region. Therefore, the extremists on both sides, who sometimes edited small newspapers with limited local circulations, often had enormous influence across the nation. Erika Thrubis's "Exchange Articles Carried by the New York *Evening Post*, December 13–31, 1860" provides a record of the extensive use of articles written by other newspapers in the *Evening Post* during a critical moment in the history of the Republic. Thrubis finds patterns in these articles, including secession sentiment, name-calling, violence, slavery, speculation, and political talk and military movements.

In the border states, sentiments remained divided well into the war. Melony Shemberger's "War of Words: Border State Editorials during the Secession Period" presents the conflicting editorials of newspapers within the slaveholding border states of Kentucky, Delaware, Maryland, and Missouri.

Antebellum editors were not expected to be independent or objective. Rather, their role was to lead their readers through the difficult issues that would shape America. Newspaper editors of the Antebellum period sought to make sense of the times, offering their readers a framed perspective. As the Antebellum period progressed, the issue of slavery in the Western territories separated the nation, North and South, East and West. By the 1850s, newspaper coverage reflected this separation in what would become a "war of words" among editors. This war of words revealed a divided nation, and to at least some degree, conflicting newspaper views played a role in heightening the tension.

The authors of this book provide extraordinary perspectives on the Antebellum press. While the authors do not always agree, the chapters as a whole, from beginning to end, recognize the ability of Antebellum newspapers to bring attention to particular issues. As Donald Shaw, one of the creators of agenda-setting theory, and Thomas Terry (with Milad Minooie) said in the first chapter of this book: "[Antebellum] newspapers reflected classic agenda-setting theory. That is, they did not tell readers what to think, but certainly what to think about—a nation under stress."

Notes

1 James Madison to W. T. Barry, August 4, 1822, in *The Writings of James Madison*, ed. Gaillard Hunt, vol. 9 (New York: G. P. Putnam's Sons, 1900–1910), 103–09.

2 Thomas Jefferson to Colonel Edward Carrington, January 16, 1787, Library of Congress, www.loc.gov/item/mtjbib002478/.

3 Frank Luther Mott, *American Journalism: A History of Newspapers in the United States Through 250 Years, 1690–1940* (New York: Macmillan, 1941), 216; Dan Schiller, *Objectivity and the News: The Public and the Rise of Commercial Journalism* (Philadelphia: University of Pennsylvania Press, 1981), 12; Ted Curtis Smythe, preface to *The Gilded Age Press, 1865–1900* (Westport, CT: Praeger, 2003), x.

4 Sándor Bölöni Farkas, *Journey in North America* (Philadelphia: American Philosophical Society, 1977), in Jeffrey L. Pasley, *The Tyranny of Printers: Newspaper Politics in the Early American Republic* (Virginia: The University Press of Virginia, 2001), 8.

5 Alexis de Tocqueville, *Democracy in America*, vol. 1 (New York: Alfred A. Knopf, 1993), 186.

6 "Census of 1860 Population-Effect on the Representation of the Free and Slave States," *New York Times*, April 5, 1860, www.nytimes.com/1860/04/05/archives/census-of-1860-populationeffect-on-the-representation-of-the-free.html.

7 Henry Clay, "The American System" (1832), in *Works*, Calvin Colton, ed., vol. 7 (New York, 1904), 464.

8 Gregory A. Borchard, *Abraham Lincoln and Horace Greeley* (Carbondale: Southern Illinois University Press, 2011), 7.

9 "The Early Nineteenth-Century Newspaper Boom," *The News Media and the Making of America, 1730–1865*, American Antiquarian Society, accessed February 12, 2018, http://americanantiquarian.org/earlyamericannewsmedia/exhibits/show/news-in-antebellum-america/the-newspaper-boom.

10 "No Union with Slaveholders," *The Liberator*, May 31, 1844.

11 Frederick Douglass, *Narrative of the Life of Frederick Douglass, an American Slave, Written by Himself* (Boston: Anti-Slavery Office, 1845), 35.

12 Robert Rhett, speech at Grahamville, SC, July 4, 1859, in Daniel Wallace, *The Political Life and Services of the Hon. R. Barnwell Rhett* (1859), 40.

13 James Carey, *Communication as Culture* (New York and London: Routledge, 1989), 213.

14 David W. Bulla and Gregory A. Borchard, *Journalism in the Civil War Era* (New York: Peter Lang, 2010), 106.

15 US Supreme Court, *Dred Scott v. Sandford*, March 6, 1857, 60 U.S. (19 How.), 393.

16 Abraham Lincoln, "Second Inaugural Address," March 4, 1865.

17 Julia Martinez, "For Civil-War Scholars, a Settled Question That Will Never Die: What Caused the War?" *Chronicle of Higher Education*, October 31, 2017, www.chronicle.com/article/For-Civil-War-Scholars-a/241627.

1

NEWSPAPERS, AGENDA SETTING, AND A NATION UNDER STRESS

Donald L. Shaw and Thomas C. Terry, with Milad Minooie

Historians understand their fields the way Søren Kierkegaard says we all understand life: We live forward, but can only understand life backward. After April 1861, when the guns opened up on Fort Sumter, everyone—suddenly and obviously—could see the war coming. And it seems, many historians claim, war was inevitable. But was it? If Abraham Lincoln had not called for volunteers after Fort Sumter, would a quiet have settled on the land until someone had found a less disruptive solution? And if Lincoln had lost the 1860 election, would there ever have been a civil war?

Was the Civil War "irrepressible," "inevitable," or "avoidable?"[1] President James Buchanan believed extremist agitators were to blame for the conflict. In his first inaugural on March 4, 1861, Lincoln observed, "in your hands, my dissatisfied [Southern] fellow-countrymen, and not in mine, is the momentous issue of civil war." Lincoln's November 1860 election had spurred South Carolina's December secession, which was soon followed by the secession of the six other Deep South states and the establishment of the Confederate States of America in February 1861. Virginia did not secede until two days after the call for volunteers. Arkansas and North Carolina followed in May, Tennessee in June.

While the North blamed Southern secession for the Civil War, Southern historians would blame the fanaticism of the Republican Party for the conflict. The idea of an avoidable Civil War gained adherents among historians in the 1920s and 1930s. Historian James G. Randall, who was born in Indiana and taught at the University of Illinois, did not perceive any differences so fundamental in the social and economic systems of the North and South as to require a war.[2] In *The Coming of the Civil War*, Iowa-born Avery Craven of the University of Chicago argued that slave laborers were not much worse off than Northern industrial workers, that the institution was already on the road to extinction, and that war could have been averted if skillful and responsible leaders had worked to produce compromise.[3] Historian David Goldfield, in *America Aflame*, contended the war was "America's Greatest Failure," because slavery was simply replaced by Jim Crow and 150 years of racial control.[4]

On the other hand, Charles and Mary Beard in their influential *The Rise of American Civilization* argued that the Civil War was a question of economic competition and that slavery was a labor system.[5] Journalist and Pulitzer Prize-winning historian Allan Nevins in his eight-volume work *The Ordeal of the Union* insisted the Civil War came about because the peoples of the North and South were becoming dramatically separate societies.[6] Historian Gary Gallagher believes there is an "Appomattox Syndrome," compelling most historians and other observers to look at "Northern victory and emancipation and read the evidence backward."[7]

There are also some historians who find the causes for disunion and civil war to be mutually exclusive, though interrelated. While disunion and disagreement and Lincoln's election led to secession, these historians maintain that the causes of the combat of the Civil War were something else entirely. Miscalculation and machismo loom large in this interpretation. James McPherson's magisterial and Pulitzer-Prize-winning *Battle Cry of Freedom: The Civil War Era* equivocates on the inevitability of the Civil War.[8] This is in the tradition of many trained historians who believe nothing in history is inevitable and that explanations occur when trying to put together an organized, logical narrative.

American newspapers in the Antebellum Period reflected the nation as a whole, while also reflecting their regions. They reflected sectional perspectives when reporting about important events, such as the 1820 Compromise, the 1831 Nat Turner revolt, the Mexican War of 1846–1848, the 1850 Kansas-Nebraska Act, the 1857 Dred Scott decision, and the John Brown raid in 1859. These newspapers were filled with controversies and tension, but until the 1860 election their coverage generally did not predict a war. Observers can find a thread of events in the Antebellum years that seemed in hindsight to be leading to war, but newspaper coverage for the most part did not demonstrate a country unraveling towards dissolution and civil war. Nevertheless, secession and civil war did occur, and so the question remains as to whether newspapers played an agenda-setting role, and when the coming of civil war could be discerned in their coverage.

Newspapers in History

Historians have long acknowledged the value of newspapers in documenting events. In 1966, political scientist Richard Merritt sampled colonial newspapers to see if he could detect the emergence of an American community in the years before the Revolution.[9] Donald Shaw's analysis of 3,000 newspaper articles from 1820 to 1860 saw an increase in "local and state news at the expense of foreign news," which declined from 28 percent to 18 percent over those 40 years. He attributed this to "an increase in interest in local communities as they grew in size" and "the emergence of an American community."[10] He wrote, "The content of … newspapers reflects the day-to-day judgments of the press at one level and the intrinsic values of a social system and culture at other levels."[11] It is to "sketch the frames of events."[12] Whatever a newspaper may *claim* is its agenda, a content analysis will lay bare what *is* that agenda and how it frames that agenda.[13]

Agenda Setting Theory

Walter Lippmann entitled a chapter in his 1922 book *Public Opinion*, "The World Outside and the Pictures in Our Heads."[14] The media placed those pictures there, Maxwell McCombs and Donald Shaw hypothesized in 1972, by "influencing the salience of attitudes toward the political issues."[15] The agenda-setting theory of McCombs and Shaw "revived Lippmann's conception" of the media's contributions to creating those "pictures in our head."[16] Bernard Cohen declared in 1963 the press "may not be successful much of the time in telling people what to think, but it is stunningly successful in telling its readers what to think about."[17] And in 1966, Kurt and Gladys Lang noted, "the mass media force attention to certain issues…[by] constantly presenting objects suggesting what individuals… should think about, know about, have feelings about."[18]

In 1993, McCombs and Shaw asserted, "agenda setting is considerably more than the classical assertion that the news [media] tells us *what to think about*."[19] The media, they emphasized, also tell us *"how to think about it."*[20] Agenda setting performs a "linking function" in democratic societies, between "citizens and policymakers," according to Stuart Shulman.[21]

Historical agenda setting is a backward approach, not only because it looks back into history, but because no explanatory theoretical model exists underpinning it. It is quite different from other agenda-setting approaches because it cannot rely on the same empirical basis. It requires counterparts for polling and survey data that did not exist much before the 1930s.

Historical scholars direct their agenda-setting research light backward into history, but it is neither a laser beam, nor is it Lippmann's "searchlight."[22] The prism of hindsight mediates the light and changes it. Pulitzer Prize-winning historian Doris Kearns Goodwin remarked, "The past is not simply the past, but a prism through which the subject filters his own changing self-image."[23]

S. Kittrell Rushing took a quantitative approach to agenda setting during the 1860–61 secession crisis preceding the Civil War.[24] He examined the sixteen Antebellum newspapers published in twenty-eight East Tennessee counties in the seven months between the 1860 presidential election and the 1861 secession referendum to determine their political slant. "A standard interpretation," he wrote, "is that after Lincoln's election Southern newspapers led the way in altering Southern attitudes toward the Union," fomenting anti-union and secessionist sentiment.[25] Rushing studied the correlation between the newspapers' political views and the results of those two elections in East Tennessee. East Tennesseans voted two-to-one against secession, bucking the statewide trend that propelled the state to "officially" secede.[26] By applying "twentieth century agenda-setting theory to 19th-century press influence," Rushing argued, "a more complete understanding may be achieved of the relationship between the antebellum press and its readership."[27] The political leanings of twelve of the newspapers Rushing looked at could be gleaned and were split evenly between the Southern wing of the Democratic Party (which supported John Breckenridge in 1860) and the regular Democratic Party (which nominated Stephen Douglas). However, both the state and East Tennessee went for the Constitutional Union candidate John Bell.[28]

While "some visible relationship...between the presence of newspapers and county election returns" was "apparent," it was not always in the direction anticipated.[29] Only one comparison "closely fit the hypothesis of a direct relationship between newspapers and voting," Rushing found.[30] And that had more to do with the "influence of a prominent, aggressive editor [Knoxville's Parson Brownlow] than the editorial content of his papers."[31]

Rushing's statistical analysis detected only a "tenuous" relationship between the press and the results of the two elections.[32] Anecdotally, he remarked that his research "seems to support the observation that media reflected the attitudes and values of the readers" in East Tennessee.[33] Although he did not state so explicitly, Rushing's conclusion dovetailed with classic agenda-setting theory; that the media may not be able to mold opinions, but it does put certain issues on the front burner.

Newspapers most notably reveal their agenda in editorials, but also in their coverage: what they emphasize and where they position stories. Placing a story on the front page is a dramatic agenda-setting decision. Civil War-era newspapers, as suggested by Rushing's Tennessee study, might be merely reacting to what the public is already interested in during times of high tension. Historian Edward Caudill considered it "a reasonable assumption that the press is more useful as a guide to public opinion during times of stress."[34]

Newspapers reveal footprints across the soft ground of history in an always-incomplete puzzle. Cultures produce many products that historians examine for insights into daily life activities. Newspapers are one of those products that reflect civic life. Numbers can be revealing. Merritt took random samples of colonial newspapers from 1735 to 1775 to determine the frequency of words such as "king," "queen," or "London" versus the frequency of words such as "governor," "Boston," or "Charleston." When did word symbols of colonial America edge out word symbols of England or Europe? In Merritt's study it was approximately 1765, a decade before the outbreak of the American Revolution.[35] Does this mean colonials had developed a sense of civic community well before the conflict, or was this newspaper coverage the result of other factors? In this period, more advanced Post Office processes and improved roads made it easier for isolated colonial newspapers to obtain news from each other. Much news in this period (and for decades after) was reprinted from other newspapers. The emergence of an

American community in Merritt's study is just as likely to be a reflection of the altered technology of information distribution as of increased political awareness by colonial Americans. Perhaps it was a little of both. These two perspectives, one cultural and one technological, are the types of challenges historians face in their work.

Shaw tested some of Merritt's ideas in his study of the American press during the Antebellum years.[36] The study drew on more than 3,000 samples of stories from newspapers in the Lower South, Upper South, border states (in the Civil War), middle states (including, for example, New York), New England, and the West (in the 1820–60 period). Each sample of a story was 150 words, and so the entire study included about a half million words from stories that could be sorted by topics, sources, and a number of other variables. Did the study reveal some deeply embedded differences in the states that would later compose the Confederacy versus those in the North or West? The answer: Only faintly, if at all.

Newspapers of the North, South, and West carried news about ten broad topics in the 1820–33, 1834–46, and 1847–60 periods. There were small differences here and there as the war approached, but the story was one of continuity rather than contrast, with few signs of impending cataclysm.[37]

Whatever historians' perspectives, slavery was at the center of the Civil War. When Shaw graphed news about slavery over time, he found evidence that the regions presented a slightly different agenda, with the South, then North, then West carrying more news that involved slavery. He found that much of the news about slavery in Southern newspapers was clipped, meaning that Southern newspapers were monitoring the news from elsewhere. This process of clipping "exchange articles" from other newspapers was common at the time (driven by the low cost of postage), and a newspaper's selection of which stories to publish constituted an important agenda-setting function. The same was also true for Western newspapers, which likewise used exchange articles concerning slavery. Many of these exchange articles came from Northern newspapers. So, while Southern newspapers carried more articles on slavery than Northern newspapers, exchange articles from Northern newspapers spread their influence nationwide, playing an important role in putting slavery on the agenda. It could be argued that news about slavery emerged in Southern newspapers from monitoring this news elsewhere, while it emerged in Northern newspapers from increasing editorial involvement. Southern editors were reacting to Northern abolitionist agendas, attempting to defend and/or justify slavery. Buried in a macro analysis is the hint of regional differences that probably reflected local community values and may have shaped the agendas differently. In the South, newspapers generally took a more passive outward approach to the topic of slavery as compared with the more aggressive, often angry coverage and editorial comments in Northern newspapers.[38]

How much of this difference in coverage of slavery resulted from the fact that the Southern press was less developed than its richer, larger, and more numerous

Northern counterparts and how much from the more aggressive interest in the topic by Northern journalists? The question poses a dilemma similar to that faced with Merritt's study of the 1735–75 colonial press: technology versus regional subcultural awareness. It is not easy to answer this question. But, for whatever reason, newspapers do suggest the hint of a splitting newspaper agenda in the South, North, and West before the outbreak of the Civil War. As Nevins suggested, the United States was splitting into two peoples. But could anyone really see that war coming, until it did?

A Nation Under Stress

History comes in like constant waves crashing and tumbling along a beach until historians organize it into single files of facts and events, adding apparent order where there is really just chaos. So it is with the American Civil War as shown in a 2017 study conducted by Shaw and Thomas Terry that found two almost equally dangerous and contentious periods between 1820 and 1860.[39] Had the first period actually led to war, it would look just as inevitable now as the early 1860s does.

Why 1860, then, and not 1830? Both Vice President John C. Calhoun and President Andrew Jackson supported states' rights to varying degrees. Calhoun argued that states could veto—nullify—any federal law, something Jackson vehemently opposed. The April 13, 1830, Jefferson Day dinner at Jesse Brown's Indian Queen Hotel was the culmination of their simmering feud. Jackson proposed a toast, declaring, "Our federal Union, it must be preserved." Calhoun replied, "The Union: next to our Liberty the most dear: may we all remember that it can only be preserved by respecting the rights of the States, and distributing equally the benefit and burden of the Union."[40] This difference of perspective lies at the heart of what would be called the Nullification Crisis of 1832–33, in which the state convention of South Carolina adopted an Ordinance of Nullification stating that South Carolina could decide which federal laws were enforceable in the state. Congress responded with a bill authorizing the president to use force to apply its laws, but also passed the Compromise Tariff of 1833, thus changing the original tariffs to which South Carolina had objected. South Carolina responded by repealing its original nullification ordinance, but then nullified the federal bill that had authorized presidential use of force. Thus, a military crisis had been avoided, although the issue of nullification would not be settled for another three decades.

Shaw and Terry, using their data set of thousands of newspaper articles from 1820 through 1861, found two dramatic increases in the mentions of "secession." Secession was mentioned in 18,078 articles in the 1830–34 period, and in 15,571 articles from 1860 to July 1, 1861. The earlier time period was longer, but there were actually fewer newspapers publishing, while the latter period ended in 1861 after the Civil War had begun. The study compared mentions of "secession," "states' rights," and "civil war." From 1820 until the first crisis period, there were few mentions of these terms. During the first crisis period, while there were

18,078 mentions of "secession," there were 3,088 mentions of "states' rights" and 2,223 mentions of "civil war." From 1835 through 1849, the terms "states' rights" and "civil war" were generally mentioned more frequently than "secession." For example, there were 1,500 mentions of "states' rights" from 1835 through 1839, 1,310 mentions of "civil war," and only 450 mentions of "secession." In the period 1850 through 1854, the term "secession" heated up to 3,986 mentions, while there were 1,846 mentions of "states' rights" and 1,746 mentions of "civil war." The term "secession" dropped to 1,400 from 1850 through 1859, while "states' rights" increased to 2,184 and "civil war" increased to 3,993. Finally, in the very short period from 1860 to July 1, 1861, there were 15,571 mentions of "secession," 1,054 mentions of "states' rights," and 5,123 mentions of "civil war."[41]

Slavery was the underlying cause of the Civil War.[42] The Shaw and Terry study also counted the mentions of the terms "slave," "slavery," and "peculiar institution." The term "slave" is only mentioned 205 times in 1820. This rises to 1,203 mentions in 1833 at the time of the first crisis. It goes up gradually from then, rising to 5,632 in 1850. From the approximate time of the Compromise of 1850, the number of mentions of "slave" rises precipitously, up to 11,539 in 1859 and 16,546 in 1860. Mentions of the term "slavery" ran parallel, with only 135 in 1820 and 15,603 in 1860. This clearly demonstrates a country under stress. (The phrase "peculiar institution" occurred far less frequently, rising from 161 in 1854 to 383 in 1860.)[43]

Agenda Building

How does an issue "grow" from small attention to permeate the general civic mind? An important agenda-setting study awaits historical exploration. The growth of abolitionist thought, which seemed extreme at first, presumably increased in importance due to the widely distributed stories from William Lloyd Garrison's *Liberator*, Frederick Douglass's *The North Star*, and other publications during the years leading to 1861. While the Civil War was at first a war to preserve the Union, it had become, in part, a war over slavery by late 1862 and early 1863 with the Emancipation Proclamation.

The Antebellum press provides an invaluable window into the nation before the war. In agenda-setting terms, slavery was a growing topic in newspapers of the South, North, and West after the 1820–32 period. The topic grew most in the newspapers of the South, then the North, then the West. The division over the future of slavery fueled the secession movement, and the increase in newspaper coverage from 1850 onwards is a reflection of and perhaps contributed to the enormous increase in political stress during the period.

Newspapers have the power to provide color to a single moment in the historical construction of time, and a picture of whole frames of time. Newspapers provide agendas that inform those who read them at the time, but they continue to inform those who read them years and decades later. This chapter argues that newspapers did not predict the coming of the Civil War (which we believe was only apparent in hindsight), but confirms that newspapers reflected classic agenda-setting theory. That is, they did not tell readers what to think, but certainly what to think about—a nation under stress.

Notes

1 The term "irrepressible conflict" is from a speech by future Secretary of State William H. Seward in Rochester, NY on October 25, 1858. He forecast the collision of the socioeconomic institutions of the North and South. Lincoln spoke along the same lines in his "House Divided" speech on June 16, 1858 at the Illinois State Capitol as he accepted the GOP's nomination for U.S. Senate. The idea can be attributed earlier to Thomas Jefferson in 1821, "The Freeman's Catechism Concerning the Irrepressible Conflict," *Vincennes* (IN) *Gazette*, December 17, 1859, 1.

2 *James G. Randall, Constitutional Problems Under Lincoln* (New York and London: D. Appleton and Co., 1926).

3 Avery Craven, *The Coming of the Civil War* (Chicago: University of Chicago Press, 1942).

4 David Goldfield, *America Aflame: How the Civil War Created a Nation* (New York and London: Bloomsbury Press, 2011).

5 Charles A. Beard and Mary Ritter Beard, *The Rise of American Civilization*, 2 vols. (New York: Macmillan and Co., 1927).

6 Allan Nevins, *The Ordeal of the Union*, 8 vols. (New York: Charles Scribner and Son, 1971).

7 Tony Horwitz, "150 Years of Misunderstanding the Civil War," *The Atlantic*, June 19, 2013. www.theatlantic.com/national/archive/2013/06/150-years-of-misunderstanding-the-civil-war/277022/.

8 James M. McPherson, *Battle Cry of Freedom: The Civil War Era* (London: Oxford University Press, 1988).

9 Richard L. Merritt, *Symbols of American Community 1735–1775* (New Haven: Yale University Press, 1966).

10 Donald L. Shaw, "At the Crossroads: Change and Continuity in American Press News 1820–1860," *Journalism History* 8:2 (1981): 38–50, 43.

11 Ibid., 39.

12 Ibid., 43.

13 See Guido H. Stempel III, "Content Analysis," in *Mass Communication Research and Theory*, ed, Stempel III, David H. Weaver, and G. C. Wilhoit (Boston: Allyn & Bacon, 2003), 209–19.

14 Walter Lippmann, *Public Opinion* (1922; reprint, New York: The Free Press, 1965).

15 Maxwell E. McCombs and Donald L. Shaw, "The Agenda-Setting Function of Mass Media," *Public Opinion Quarterly* (Summer 1972): 176–87, 177.

16 McCombs and Shaw, "The Evolution of Agenda-Setting Research: Twenty-Five Years in the Marketplace of Ideas," *Journal of Communication* (Spring 1993): 58–67.

17 Bernard C. Cohen, *The Press and Foreign Policy* (Princeton: Princeton University Press, 1963), 13.

18 Kurt Lang and Gladys Engel Lang, "The Mass Media and Voting," in *Reader in Public Opinion and Communication*, ed. Bernard Berelson and Morris Janowitz (New York: Free Press, 1966), 266.

19 McCombs and Shaw, "The Evolution of Agenda-Setting Research," 62.

20 Ibid.

21 Stuart W. Shulman, "The Origin of the Federal Farm Loan Act: Agenda-Setting in the Progressive Era Print Press" (PhD dissertation, University of Oregon, June 1999), 399.

22 Lippmann, *Public Opinion*.

23 Doris Kearns Goodwin, *Team of Rivals: The Political Genius of Abraham Lincoln* (New York: Simon & Schuster, 2006).

24 S. Kittrell Rushing, "Agenda-Setting in Antebellum East Tennessee," in *The Civil War and the Press*, ed. David B. Sachsman, S. Rushing, and Debra Reddin van Tuyll, with Ryan P. Burkholder (New Brunswick, NJ: Transaction Publishers, 2000), 147, 149.

25 Ibid.

26 Ibid.

27 Ibid.

28 Ibid., 149–50.

29 Ibid.

30 Ibid., 158.

31 Ibid.

32 Ibid.

33 Ibid.

34 Edward Caudill, "An Agenda-Setting Perspective on Historical Public Opinion," in *Communication and Democracy: Exploring the Intellectual Frontiers in Agenda-Setting Theory*, ed. Maxwell McCombs, Shaw, and David Weaver (Mahway, NJ: Lawrence Erlbaum Associates, 1997), 182.

35 Merritt, *Symbols of American Community*.

36 See Shaw, "At the Crossroads"; and Shaw, "Some Notes on Methodology: Change and Continuity in American Press News 1820–1860," *Journalism History* 8:2 (1981): 51–53, 76.

37 Shaw, *The Southern Challenge to American Cultural Union: Newspaper Symbols of Public Thought* (unpublished manuscript, Chapel Hill, NC, 1984), 262.

38 Ibid.

39 Shaw and Terry, "Long Run and Short Flare: Newspapers, Agenda Setting, and the Coming (?) Civil War" (presentation, Symposium on the 19th Century Press, the Civil War, and Free Expression, Chattanooga, TN, November 2–4, 2017).

40 "Jefferson's Dinner," *Niles' Weekly Register* (Baltimore, MD), April 24, 1830, 1. Secretary of State Martin Van Buren's toast was conciliatory: "Mutual forbearance and reciprocal concession; through the agency by the union was established. The patriotic spirit from which they emanated, will forever sustain it." Speaker of the House Andrew Stevenson of Virginia urged "unanimated moderation."

41 Shaw and Terry, "Long Run and Short Flare."

42 For newspaper coverage of the Lost Cause, see Terry and Shaw, "Rebel Yells and Idle Vaporings: The Lost Cause Rises and Dissipates in the *Chicago Tribune, Atlanta Constitution*, and *New York Times*, 1860–1914" in *After the War: the Press in a Changing America, 1865–1900*, ed. Sachsman (New Brunswick, NJ: Transaction Publishers, 2017), 3–19.

43 Shaw and Terry, "Long Run and Short Flare."

2

THE "IRREPRESSIBLE CONFLICT" AND THE PRESS IN THE LATE ANTEBELLUM PERIOD

Debra Reddin van Tuyll

The seeds of what Thomas Jefferson had predicted and William H. Seward called an "irrepressible conflict" had been planted long before 1850, but that was the year the roots truly took hold and began to blossom into full-blown sectional hostility. During the 1850s, the hostility would move from the rhetorical to the practical. Reasoned discourse and compromise would be replaced by verbal barbs punctuated by the occasional Congressional brawl or caning. In 1861, Northerners and Southerners would graduate to cannon, rifles, and small arms as they waged the bloodiest war in American history.[1]

The period from 1820 to 1850 had been one of sectional crises and national compromises. By 1850, however, compromise was starting to feel like capitulation to Southerners. It chaffed. It chaffed to the point that talk spread through the 1850s of a coming irrepressible conflict. Secession ceased to be so much an abstraction as a real possibility.

Sectional consciousness began in earnest in 1850, bringing to the fore the sectional division over slavery that "had existed at least since the Constitutional Convention of 1787." By 1850, Southerners were so raw, so scared that each new crisis evoked such a dire response that "the American political system could no longer contain the sectional conflict."[2] The more radical Southerners were reaching their breaking point, and the American press, both South and North, was center stage, helping to push the country on toward the "irrepressible conflict."[3]

"Nothing is more certainly written in the book of fate than that these people [Negro slaves] are to be free. Nor is it less certain that the two races, equally free, cannot live in the same government," Jefferson wrote in 1821 in his autobiography.[4] Seward picked up on that theme in an 1858 speech in Rochester, New York, when he declared, "It is an irrepressible conflict between opposing and enduring forces, and it means that the United States must and will, sooner or later, become either entirely a slaveholding nation or entirely a free-labor nation." Seward, in a reference perhaps to Jefferson's earlier writings, reminded his audience that day that this idea was not new: "Our forefathers knew it to be true.... They regarded the existence of the servile system in so many of the States with sorry and shame...."[5]

South Carolina was the ringleader in nineteenth-century secession thought; its renowned senator John C. Calhoun was a chief architect of the doctrine of secession, and the leading fire-eating newspapers in the South were based in the Palmetto state. During the Nullification Crisis of 1832–33 (in which the state convention of South Carolina had adopted an ordinance nullifying federal tariffs), other Southern states had resisted being pulled away from the Union, but by 1860, American politics had become so fractured, sectionalism so rampant, that South Carolina's radicalism gained widespread support in the South.[6]

Throughout the 1850s, newspapers participated in and helped lead the debate over America's future. They argued almost daily over secession and whether it was called for at that particular moment in response to yet another sectional crisis.[7] However, with the exception of the fire-eating newspapers in the South, most of the Southern press in 1850 believed the Union was too valuable to break up without some sort of direct provocation. Southern newspapers did not, however, agree on what actions would meet that threshold. By 1860, the election of a Republican president was sufficient provocation for some. For others, Lincoln's insistence on maintaining Major Robert Anderson's troops at Fort Sumter was enough. For those who had not already been infected with secession fever, it was President Lincoln's call for 75,000 volunteers to put down the rebellion. One-by-one, Southern devotion to the Union cracked.[8]

In 1850, as Americans were dealing with the series of bills that became known as the Compromise of 1850, Southerners and their newspapers took one of four positions on secession. The first position was that of the fire-eaters, the smallest group, who demanded immediate secession. This was not a new position; many in the South had been demanding secession since the Nullification Crisis. A larger number of Southerners and Southern newspapers were considered to be radical Southern Righters. Radical Southern Righters were not opposed to secession, but neither did they support immediate secession. They were more numerous than the fire-eaters but not so much so as the third category, the moderate Southern Rights adherents who supported secession *only* if the North acted directly and overtly against Southern interests. The moderate Southern Rights newspapers averred the South's need to protect itself and to prevent the election of a Republican president, but the tone of these newspapers was less aggressive than that of the radical Southern Rights press.

Finally, the Unionist press in the South would not accept the idea of secession under all but the most extreme conditions. Some newspapers, such as Benjamin Perry's *Greenville* (SC) *Patriot-Mountaineer*, were almost Jacksonian in their Unionism, but papers of that ilk were scarce. Most Unionist papers in the Antebellum South praised the Union effusively and pointed out repeatedly that the election of a Republican president was no threat to the South so long as Congress remained Democratic and no Supreme Court justices would need to be replaced.[9]

By late May 1861, when the last of the Southern states, North Carolina and Virginia, seceded, most of the region's newspapers, regardless of their initial position, had become fierce secessionists.

The 1850s: A Decade of Divisions

Sectional crises were nothing new in 1850; they had been occurring regularly for at least thirty years when Southerners and Northerners went head-to-head once again over what were essentially questions of slavery and the political power to control it. They had always found a compromise, a middle path that would keep civil war at bay. This time, though, as Americans debated whether to extend slavery into territory obtained in the recent Mexican–American War, there would be no true compromise. The Compromise of 1850 effected a legal compromise, but it papered over the organic nature of the North–South conflict; it did not achieve a true compromise in the hearts and minds of Southerners—and probably not Northerners, either. The Compromise was simply one of several in "a clear progression of events between 1840 and 1854" that resulted in both a political crisis that would lead to the dissolution of the Whig Party and the fracturing of the Democratic Party, and which stoked the crisis of fear already gripping Southern hearts and minds.[10]

This was the atmosphere in which Americans had debated the issue that led to the Compromise of 1850: what to do about California, which was seeking statehood in the wake of the 1849 gold rush, and, as well, what to do with the other lands acquired in the Mexican–American War under the Treaty of Guadalupe Hidalgo. The political outcome, the Compromise of 1850, was one of the more important events in congressional history, for it sought to resolve a question that was dividing the nation.[11]

Freshman Pennsylvania congressman David Wilmot had already inflamed sectional tensions during the Mexican–American War. In 1846, three months after the war began and long before the outcome would be known, Wilmot had introduced a proviso to keep slavery out of any lands that might be acquired as a result of the war. Reaction to Wilmot's proviso had been violent. It had the effect of polarizing congress along sectional lines. All but one Northern state told their representatives and senators to ensure that Wilmot's proviso guided the organization of any lands acquired from Mexico. The proviso forced many moderate Southerners, including John C. Calhoun, into the radical fire-eaters camp, for he, like so many others in his section, fervently believed they had a constitutional right to take their slaves into any territory they wished. Finally, some Southerners decided, it was time to take a stand for their sectional interests. Many vowed to secede rather than accept the proviso.[12]

The Wilmot Proviso never became law, but this did not silence the debate over the westward expansion of slavery. As Congress continued its deliberations on the slave question, Mississippi took the leap and called for a convention of Southern states to meet in Nashville to consider Southern response should Congress outlaw slavery in the new Western territories. Mississippi proposed Nashville as the location because it was considered to be a moderate mid-South state. Meeting in the home city of the great Unionist Andrew Jackson, they hoped, would allay Northern concerns that the convention might be a viper's nest of secessionists. The convention was not as big a story as the congressional battles over what to do about slavery in the territories, but it got plenty of coverage, nevertheless.[13]

Reaction to the call for the convention was mixed. Northern papers condemned the idea, but the states' rights papers of the South were anxious for the states to gather. South Carolina and other deep-South newspapers, in particular, gave the convention enthusiastic coverage. The Democratic *Nashville Union* proclaimed Tennesseans honored to host the convention,[14] though the Whig papers were somewhat less enthusiastic. The *Nashville Whig* declared those who supported the convention to be disunionists.[15] Only two Tennessee Whig papers, the *Memphis Enquirer* and the *Trenton Banner*, completely favored the convention.[16] The Palmetto state press had led the charge the previous year to create a United South and had already called for a Southern convention and the creation of a Southern platform.[17]

During the debate over the series of bills that constituted the Compromise of 1850, Florida newspapers focused on only one of the bills: the Fugitive Slave Act. Looking to the future, B. F. Allen of the Whig *Tallahassee Sentinel*, for example, anticipated a time when the South might be weak enough that the North would attempt to repeal the act which allowed for the capture and return of escaped slaves. That, he argued, would be sufficient provocation for secession. The same would be true if Congress abolished slavery in the national capital without the agreement of Maryland.[18]

Most Tennessee newspapers were relieved by the compromise. Henry Watterson, then of the *Nashville Union*, but who would become one of the most important Civil War correspondents with his work for the *Chattanooga Rebel*, believed Congress had done the right thing in finding a compromise in 1850.[19]

Under the Compromise, California came into the Union as a free state; Texas gave up its claim to New Mexico…in consideration of a $10 million settlement. New Mexico entered the Union without any discussion of slavery. The slave trade was abolished in Washington, DC, but the Fugitive Slave Act was strengthened. The *Nashville Union* proclaimed joyfully that the Republic had been saved.[20] But the unity Congress achieved in 1850 was not to be long lived.[21]

South Carolinians would continue to debate their options for secession. They considered separate state action verses secession by the South as a whole. Some newspapers gave virtually their entire editorial space to letters devoted to this debate.[22] Mississippians continued to ruminate on the question of secession as well. The Jackson, Mississippi *Flag of the Union* argued in May 1851 that the federal government had, indeed, usurped power from the states. Nevertheless, this was not sufficient to justify secession in the minds of editors T. Palmer and Edward Pickett.[23]

Party turmoil continued to build as well. In fact, that turmoil would completely upend and replace the party system in the mid-1850s. The Whigs would collapse under the weight of the sectional conflicts. The Republican Party would form out of the remains of the Whigs, former Free Soilers, and abolitionists, and the Democrats would fracture as the 1860 presidential election approached.[24]

The debate over slavery would be the biggest issue of the decade. Congress would refuse to admit Kansas under the proslavery Lecompton Constitution in 1854. And by the end of the decade, the House of Representatives would elect a Republican speaker and John Brown and his supporters would attack the federal arsenal in Harpers Ferry, Virginia.

Kansas and the Lecompton Constitution

Congressional debates on Kansas's application for statehood would spark one of the most important debates over the place of slavery in America. The Kansas and Nebraska Acts would toss out the old 36°30' rule from the Missouri Compromise of 1820 and allow territorial residents to decide for themselves whether they would come into the Union as free or slave. Those debates would go a long way toward convincing Southerners that they could no longer rely on politics—on Congress—to support their peculiar institution. Many Americans, Southern and Northern, began to look askance at Stephen A. Douglas, the author of the Kansas and Nebraska Acts. His authorship and support of the bills led Northerners to see the senator as weak on the rights of free states. Southerners interpreted his actions as evidence that Douglas was a closet abolitionist. Proponents argued that Douglas's proposal was the only democratic way to resolve the argument over the extension of slavery into the territories, but finding a majority was more difficult than anticipated. The *New York Herald* referred to Kansas as "the apple of discord" and opined that both North and South appeared to have taken big bites of that apple.[25] The resulting violence would be called "Bleeding Kansas." Kansans would create four constitutions before they finally got one that could make it through Congress. The third attempt, known as the Lecompton Constitution, was by far the most controversial.[26]

The issue in this controversy was not whether slavery was good or evil but whether the majority should make the decision and who and what constituted the majority. When Congress passed the Kansas-Nebraska Act, the expected outcome was that Kansas would join the Union as a slave state and Nebraska as a free state. But most Kansas residents were antislavery. Slavery proponents even suspected that the timing of the referendum in Kansas was chosen to give abolitionist leagues time to move more antislavery residents into Kansas.[27]

When Congress voted down the proslavery Lecompton Constitution in the winter of 1857–58, many Southerners and even some Northerners objected. The *New Orleans Crescent* observed during the Congressional debate that "the political winds blow Northward." Some in the press considered the constitution that finally passed, the Wyandotte Constitution, "bogus."[28]

The long-term result was division in the Democratic Party, a division that would make Stephen A. Douglas the Democratic Party's "fallen star," deprive him of the (full) Party's presidential nomination in 1860, and pave the way to Southern Democrats pulling out of the national convention and running a candidate of their own, John Breckinridge.[29] The end result was the election of Lincoln in November 1860.

The Tipping Point: John Brown

John Brown's rampage in Harpers Ferry, Virginia (October 16–18, 1859), was the tipping point for Southerners, who began to wonder how long it would be before a whole host of John Browns would head South. Initially, telegraphic reports flashed only sketchy details of a violent raid on the tiny Virginia town on the Potomac. Few details were available. Newspapers could find out nothing regarding who was responsible, what their aims had been, and what they had accomplished. Early stories speculated the violence was the result of a pay dispute. Others claimed the violence was the result of a slave revolt. All anyone knew for sure was that there had been a mass shooting, and people had been killed and wounded.[30] Only after the incident was over did newspapers have enough details to print that Kansas abolitionist John Brown and a small band of men had attacked the federal arsenal in Harpers Ferry. By October 20, the *Baltimore* (MD) *Daily Exchange* had tracked down enough information that it was able to publish not only textual content but also two graphics, one that mapped the location of Harpers Ferry and the other that showed the layout of the town.[31]

Early stories either vastly exaggerated what had happened or just got the facts outright wrong. The *New Berne* (NC) *Daily Progress* reported that 150 white men had taken part in the raid.[32] The actual number was twenty-two. One of the most accurate reports came from the *Fayetteville* (NC) *Observer*. Four days after the raid began, the *Observer* correctly reported the number of participants and the number killed and wounded.[33] While the raid made for an important news story, editorial commentary was perhaps even more important because it helped to stoke Southern fears of slave insurrection and resentment toward the North by reminding readers of the link between abolitionism and the Republican Party.[34]

The Washington, DC *National Intelligencer* noted exactly this point in an October 31 editorial. The paper's editor observed that the raid showed just how deeply sectional divisions ran and that it was time to resolve the slavery issue.[35] The *New Orleans Picayune*, referring to Brown as "a desperado in Kansas," suggested that the Brown party must have been insane and wondered, "Who made them insane? What misled them if they were misled? or [sic] inspired them with such absurd expectations and frantic hopes, if they really had thoughts of overthrowing the institutions of a great State by such feeble means, and with such wretched implements?"[36] The *Memphis* (TN) *Appeal* had an answer. That paper opined that the raid was an example of "Northern madness" and proclaimed Brown to have been rendered crazy by listening to the teachings of abolitionists.[37] The *Augusta* (GA) *Chronicle and Sentinel* referred to the events in Harpers Ferry as "the late fanatical movement."[38]

Following Brown's execution, few Southern papers called for moderation. Instead, they published bitterly anti–Northern/pro-secession editorials. The *Mobile* (AL) *Register*, on the day of Brown's execution, told readers they should have learned a lesson from Harpers Ferry, and that lesson was that they needed to begin preparing for a soon-to-come armed conflict.[39] The *Charleston* (SC) *Mercury* echoed the Register's sentiment. The editors wrote:

> As we anticipated, the affair, in its magnitude, was quite exaggerated; but it fully establishes the fact that there are at the North men ready to engage in adventures upon the peace and security of the southern people, however heinously and recklessly, and capable of planning and keeping secret their infernal designs. It is a warning profoundly symptomatic of the future of the Union with our sectional enemies.[40]

The *Richmond* (VA) *Enquirer* declared that the only way to avert war with the North was for Northerners to repudiate the Republican Party in the coming presidential election.[41]

The Brown story had a profound influence on editorial perspectives. A month after the raid, the *Staunton* (VA) *Spectator's* position was moving from Unionist to moderate Southern Rights. Its editors were reprinting articles from the *Charleston Mercury* that reported on resolutions from the South Carolina legislature seeking "the establishment of a Southern Confederacy." The *Spectator* editors themselves were not yet radicals, but they had decided it was time for the South to seek affirmative measures for protection of her issues and institutions.[42]

The Brown story did more to galvanize Southern opinion regarding the value of the Union than any of the earlier sectional crises. After Brown's raid, even relatively small issues became fodder for secession sentiment. For example, Southerners looked to the election of Pennsylvania Republican William Pennington as Speaker of the House of Representatives as a telltale sign that they and their peculiar institution were no longer safe in the Union. Further, the debate over who the speaker would be had been rancorous. House members almost brawled; Georgia senator M. J. Crawford came close to a fistfight with Pennsylvania congressman Thaddeus Stevens. In a letter to his friend Alexander Stephens, a senator from Georgia, South Carolina senator James Henry Hammond observed that such acrimony reigned in Congress that no member of the House or Senate arrived at the Capital without weapons; some carried "two revolvers and a bowie knife." He feared that if a fight broke out, the government would dissolve.[43]

The debates over such issues caused Southerners to see themselves as distinct from the rest of the nation. This notion had evolved slowly over time, beginning with the Missouri Compromise of 1820 and the Nullification Crisis of 1832–33. By the end of the 1850s, Southern political leaders, and many Southern citizens, were arguing that America was a divided nation, that the South had its own identity, institutions, and perspectives that were distinct from those of the North. Those who shared this perspective believed that the United States was a confederacy and hence possessed none of the traits of a nation. The country, they argued, was a compact between states, and the idea of an American people was a myth. Loyalty to America was a moot point. State citizenship mattered more than national.[44]

The 1860 Presidential Election

The next test of the value of the Union to the South would come in the late spring of 1860 when the Democratic National Party gathered in Charleston for its presidential nominating convention. Newspaper editors labored to make sense of all these events for their readers and to soothe the sectional malevolence the convention raised. Few Southern newspapers or citizens, though, were unconditionally committed to the Union in the lead-up to the 1860 Democratic National Convention, but even so, a number of papers urged calm. "The fire of Unionism will glow brighter and brighter and spread over every city and village and hamlet in the broad union, until every vestige of disunion shall be consumed," wrote the editors of the Raymond, Mississippi *Hinds County Gazette*.[45]

The convention convened in Charleston in May 1860. Newspapers speculated over who the candidate might be. Southern Rights papers were upfront about their perspective on the convention outcome; if they got a strong states' rights man nominated, the Union would be safe a while longer. If not, secession was the only option.[46] Many other Democratic newspapers urged calm and unity in Charleston. The *Wilmington* (NC) *Daily Journal* argued that "The whole country looks to the Democratic party as the only party that can save this Union." If the Party split, the Union would as well, the paper predicted. Hence, unity was incumbent.[47] Fire-eating journals, of course, wanted nothing less than the end of the Union.

The front-runner for the nomination was Stephen A. Douglas, and as the convention convened, his supporters and detractors began their battle immediately. Murat Halstead of the *Cincinnati Commercial* observed that it was absurd for such divided delegates to try to find common ground. It simply was not going to happen.[48] Halstead was right. Only a few days into the convention, the Southern radicals pulled out of the convention. Douglas would become the nominee of the Northern Democrats; John Breckinridge would become the nominee of the Southern. And a fourth party, the Constitutional Unionists, would nominate John Bell.

Secession became the topic of the day, and the radical Southern press considered it a foregone conclusion. The Corsicana (TX) *Navarro Express* declared the Union dead and wished "peace to its ashes."[49] The *Charleston* (SC) *Mercury* was more straightforward. "LET THE SOUTH ARM," its editors declared a week before the election. The *Montgomery* (AL) *Mail* had reached the same conclusion. "Let the boys arm. Everyone that can point a shotgun or revolver should have one... Abolizationism [sic] is at your doors, with torch and knife in hand!"[50]

As the election approached, the editorial debate in the South focused not on if or whether secession was an option, but on when and how. Unionist editors counseled waiting until the North threatened the South in some overt way. More radical journals, such as the *Raleigh* (NC) *Standard* argued that electing a Republican president would be sufficient provocation.[51] The *Charleston Courier* published a letter from William Boyce in August 1860 that declared every "self-respecting South Carolinian" ought "to recognize the compulsion of disunion if Lincoln were elected." He argued that the Republicans had their roots in the sectional competition for control of the federal government and that most believed majority rule beat out states' rights.[52] *Raleigh Standard* Editor William Woods Holden railed, "Do hostile armies await overt acts? Is a declaration of war nothing until arms are used in the field?"[53] Editors like Holden, though, were somewhat malleable in their positions. He, and others, would change his stance through the election. Fire-eaters like the *Charleston Mercury* counseled immediate secession should Lincoln be elected, and the idea grew on some of the more moderate papers.

Most Southern newspapers had hopes of defeating Lincoln up until the last minute of the campaign. Their optimism distorted their campaign coverage and misrepresented the chances any one of the other candidates had of winning.[54] As late as October, when it was clear that Lincoln was the front-runner, fire-eaters were still predicting victory for their candidate, John Breckinridge.[55] By the election, few Southerners were advocating Unionism publicly, though many still held Unionist sentiments privately.

On November 6, the unthinkable happened; Lincoln, an abolitionist sectional candidate was elected. Southerners were stunned. Lincoln's election fused a Southern unity almost unlike any known before; it was not perfect, but it was tight and widespread.[56] Southerners turned to the press to help them understand how the unthinkable had happened.[57] And South Carolina turned to secession. Within five minutes of the South Carolina convention's secession vote, the *Charleston Mercury* had a special edition on the streets.[58] The editor of the *Fayetteville North Carolinian* declared, "If we submit now to Lincoln's election, before his term of office expires, your home will be visited by one of the most fearful and horrible butcheries that has cursed the face of the globe."[59] Editors of secessionist journals realized they had to act quickly to win the day; they needed to capitalize on the shock and anger Southerners were feeling from the election. Even the staunchest Unionist editors started giving up and taking on the fire-eating approach. The

FIGURE 2.1 "The Republican Party Going to the Right House," published by Currier & Ives, 1860. In this anti–Lincoln political cartoon published by Currier & Ives, the "Right House" is the "Lunatic Asylum." This satirical cartoon depicts Abraham Lincoln's 1860 presidential campaign supporters as a motley crew of fanatics and oddballs. Lincoln, the "rail splitter," is shown atop a wooden rail carried by New York *Tribune* editor Horace Greeley, an allusion to Greeley's support to secure the Republican nomination. Lincoln's constituents include a Mormon, a free black, an aging suffragette, a ragged socialist, and three hooligans, all seeking favors from the political candidate. (From the Library of Congress Prints and Photographs Division Washington, DC. <www.loc.gov/item/2003674590/> Reproduction Number: LC-USZ62-1990 (b&w film copy neg.). Currier & Ives, "The Republican Party going to the Right House.")

Montgomery (AL) *Confederation*, for example, had called secession the equivalent of treason during the campaign. By early December 1860, its editors were looking forward to withdrawing "from such an obnoxious and oppressive government."[60] The *New Orleans Bee*, also formerly Unionist, declared that the South was doomed if it did not achieve its political independence.[61]

Some newspapers in the upper South, such as the *Alexandria* (VA) *Gazette*, would hold on to its Unionist spirit until late spring 1861. Newspapers in both North Carolina and Virginia resisted calls for secession until Lincoln called for 75,000 men to put down the rebellion, which included a requirement that North Carolina provide two regiments of men to fight against their brother Southerners.[62]

Several factors can account for the radicalization of Southern opinion and Southern newspapers after the 1860 election. First was the North's decisive vote for Lincoln. Southerners saw this as the North's disregard and growing ill will for the South and for slavery. The *Augusta* (GA) *Chronicle and Sentinel* saw the Northern vote as an example of "that spirit of intolerance, of hatred, of hostility, of fanatic anti-slavery feeling, with which we cannot live in peace."[63]

In addition, Northern politicians, clergy, and newspaper editors scoffed at the idea of giving concessions to the South and their threats to secession. Southern newspapers were dismayed by the national government's reluctance to help mediate sectional differences.[64]

Southern newspapers got swept up in the emotions of the time. Southerners fell back on the Revolution of 1776 as justification for pulling out of the Union in 1860. The interests of North and South could no longer be united under one flag, the *Charleston Mercury* declared five days after South Carolina seceded. The constitution was an experiment, and it had failed.[65] The heat of the campaign and public reaction to the outcome dragged Southern editors to secessionist perspectives. As winter approached, many Southerners had come to believe they were facing too great a power to survive and that they had to get out of the Union. Thus, no one was surprised when South Carolina seceded. And the state's press was ecstatic. The Columbia (SC) *Daily Southern Guardian* rejoiced. Its editor was so excited he could barely put his pen to paper.[66] Even the staid *Charleston Courier* celebrated.[67] It would take another four months, but by April 1860, virtually all Southern newspapers had moved to the secession perspective.[68]

Notes

1 Lorman A. Ratner and Dwight L. Teeter, Jr., *Fanatics and Fire-Eaters: Newspapers and the Coming of the Civil War* (Urbana: University of Illinois Press, 2003), 1, 3; Erika Pribanic-Smith, "'Political Demagogues and Over-zealous Partizans'": Tariff of Abominations and Secession Rhetoric in the 1828 South Carolina Press, *Atlanta Review of Journalism History* 12:1 (2015): 66; Donald J. Ratcliffe, "The Nullification Crisis, Southern Discontents, and the American Political Process," *American Nineteenth Century History*, 1:2 (2000): 2; "The Most Infamous Floor Brawl in the History of the House of Representatives," *History, Art, and Archives, House of Representatives*, accessed October 20, 2017, http://history.house.gov/Historical-Highlights/1851–1900/The-most-infamous-floor-brawl-in-the-history-of-the-U-S--House-of-Representatives/ ; "The Caning of Senator Charles Sumner," *United States Senate*, accessed October 20, 2017, www.senate.gov/artandhistory/history/minute/The_Caning_of_Senator_Charles_Sumner.htm; Donald Shaw, Randall Patnode, and Diana Knott Martinelli, "Southern vs. Northern News: A Case Study of Historical Agenda-Setting, 1820–1860," in *Words At War: The Civil War and American Journalism*, ed. David B. Sachsman, S. Kittrell Rushing, and Roy Morris Jr. (West Lafayette: Purdue University Press, 2008), 15.
2 Michael F. Holt, *The Political Crisis of the 1850s* (New York: W. W. Norton, 1978), 3; see also Steven A. Channing, *Crisis of Fear: Secession in South Carolina* (New York: W. W. Norton, 1974), and Ratcliffe, "The Nullification Crisis," 2, 3.
3 Ratcliffe, "The Nullification Crisis," 3; Thomas Jefferson, "Thomas Jefferson's Autobiography, 1743–1790," *The Avalon Project: Documents in Law, History and Diplomacy*,

Yale Law School Lillian Goldman Law Library, accessed October 1, 2017, http://avalon.law.yale.edu/19th_century/jeffauto.asp; Donald E. Reynolds, *Editors Make War* (Nashville: Vanderbilt University Press, 1966), 5.

4   Thomas Jefferson, "Thomas Jefferson, July 27, 1821, Autobiography Draft Fragment, January 6 through July 27," *Library of Congress*, accessed October 20, 2017, www.loc.gov/item/mtjbib024000/.

5   William H. Seward, *The Works of William H. Seward*, ed. George E. Baker, vol. 4 (Boston: Houghton, Mifflin and Co., 1861), 57, https://archive.org/stream/sewardwilliam04sewarich/sewardwilliam04sewarich_djvu.txt.

6   James M. Banner, "The Problem of South Carolina," in *The Hofstadter Aegis: A Memorial*, ed. Stanley Elkins and Eric McKitrick (New York: Alfred A. Knopf, 1974), 60–93.

7   Shaw, Patnode, and Martinelli, "Southern vs. Northern News," 13.

8   Reynolds, *Editors Make War*, 14.

9   Ibid., 15.

10 Holt, *The Political Crisis of the 1850s*, 102; see also Holman Hamilton, *Prologue to Conflict: The Crisis and Compromise of 1850* (Lexington: University of Kentucky Press, 2005), 7.

11 Holt, "Politics, Patronage, and Public Policy: The Compromise of 1850," in *Congress and the Crisis of the 1850s*, ed. Paul Finkleman and Donald R. Kennon (Athens: Ohio University Press, 2012), 18; St. George L. Sioussat, "Tennessee, the Compromise of 1850, and the Nashville Convention," *Mississippi Valley Historical Review* 2:3 (1915): 315; Thelma Jennings, "Tennessee and the Nashville Conventions of 1850," *Tennessee Historical Quarterly* 30:1 (Spring 1971): 70.

12 Holt, "Politics, Patronage, and Public Policy," 18–19; Richard Douglas Spence, *Andrew Jackson Donelson: Jacksonian and Unionist* (Knoxville: University of Tennessee Press, 2017), 200l; Randolph Campbell, "Texas and the Nashville Convention of 1850," *Southwestern Historical Quarterly* 72:1 (1972): 2.

13 Jennings, "Tennessee and the Nashville Conventions of 1850," 71; John Barnwell, "'In the Hands of the Compromisers': Letters of Robert W. Barnwell to James H. Hammond," *Civil War History* 29:2 (June 1983): 156; Spence, *Andrew Jackson Donelson*, 201.

14 *Nashville* (TN) *Union*, December 22, 1849; December 28, 1849; January 9, 1850; January 23, 1850.

15 *Nashville Whig*, January 28, 1850.

16 Spence, *Andrew Jackson Donelson*, 201.

17 Jennings, "Tennessee and the Nashville Conventions of 1850," 71, 73; Philip M. Hamer, *The Secession Movement in South Carolina 1847–1852* (Allentown, PA: H. R. Hass & Co., 1918), 23, 21–26.

18 *Tallahassee* (FL) *Sentinel*, November 12, 1850.

19 Jennings, "Tennessee and the Nashville Conventions of 1850," 80.

20 Spence, *Andrew Jackson Donelson*, 204; *Nashville* (TN) *Union*, September 14, 1850.

21 Campbell, "Texas and the Nashville Convention of 1850," 12–13.

22 *Camden* (SC) *Journal*, September 23, 1851.

23 *Flag of the Union* (Jackson, MS), May 30, 1851.

24 Spence, *Andrew Jackson Donelson*, 215–16.

25 New York *Herald*, December 21, 1857.

26 Ratner and Teeter, *Fanatics and Fire-Eaters*, 60–61.

27 Ibid., 62.

28 *St. Louis* (MO) *Democrat*, December 25, 1857.

29 Ibid.; *Memphis* (TN) *Appeal*, December 21, 1857; *New Orleans* (LA) *Crescent*, December 14, 1857.

30 *Valley Spirit* (Chambersburg, PA), October 26, 1859; *Staunton* (VA) *Spectator*, October 18, 1859; *Richmond* (VA) *Daily Dispatch*, October 18, 1859.

31 *Baltimore* (MD) *Daily Exchange*, October 20, 1859.

32 *New Berne* (NC) *Daily Progress*, October 18, 1859.

33 *Fayetteville* (NC) *Observer*, October 20, 1858.

34 Ratner and Teeter, *Fanatics and Fire-Eaters*, 76.

35 *National Intelligencer* (Washington, DC), October 31, 1959.

36 *New Orleans* (LA) *Picayune*, October 25, 1859.

37 Memphis (TN) Appeal, October 21, 1859.

38 *Augusta* (GA) *Chronicle and Sentinel*, October 23, 1859.

39 *Mobile* (AL) *Register*, December 2, 1959.

40 *Charleston* (SC) *Mercury*, October 21, 1859.

41 *Daily Richmond* (VA) *Enquirer*, October 25, 1859.

42 Staunton (VA) *Spectator*, December 6, 1859; December 20, 1859.

43 Reynolds, *Editors Make War*, 12–13.

44 Benjamin E. Park, "The Age of Nullification: Imagining Disunion in an Era Before Secession," *Journal of the Early Republic* 37:3 (2017): 531–33.

45 *Hinds County Gazette* (Raymond, MS), February 8, 1860.

46 *Mississippi Free Trader* (Natchez, MS), January 7, 1860.

47 *Wilmingon* (NC) *Daily Journal*, May 1, 1860.

48 Reynolds, *Editors Make War*, 41.

49 *Navarro Express* (Corsicana, TX), May 19, 1860.

50 Both articles reprinted in a run-down of Southern opinion in the *New York Times*, November 1, 1860.

51 *Raleigh* (NC) *Standard*, January 7, 1860.

52 Channing, *Crisis of Fear*, 235–36.

53 *Raleigh* (NC) *Standard*, January 7, 1860.

54 Reynolds, *Editors Make War*, 118.

55 Channing, *Crisis of Fear*, 233.

56 Ibid., 252.

57 Reynolds, *Editors Make War*, 139.

58 Perry J. Ashley, ed., *Dictionary of Literary Biography, Vol. 43: American Newspaper Journalists, 1690–1872* (Gale Research Co., 1983), 388.

59 Reynolds, *Editors Make War*, 140–41.

60 *Montgomery* (AL) *Confederation*, December 7, 1860.

61 *New Orleans* (LA) *Bee*, December 14, 1860.

62 Debra Reddin van Tuyll, Nancy McKenzie Dupont, and Joseph R. Hayden, *Journalism in the Fallen Confederacy* (New York: Palgrave, 2015).

63 *Augusta* (GA) *Chronicle and Sentinel*, November 13, 1860; November 25, 1860; December 12, 1860.

64 Reynolds, *Editors Make War*, 157, 158.

65 *Charleston* (SC) *Mercury*, December 15, 1860.

66 Qtd. in *Daily South Carolinian* (Columbia), December 21, 1860.

67 *Charleston* (SC) *Courier*, December 21, 1860.

68 Reynolds, *Editors Make War*, 158, 159, 210.

PART I

Nullification, Abolition, and Division

3

NAT TURNER'S REVOLT SPURS SOUTHERN FEARS AND SPARKS PUBLIC DEBATE OVER SLAVERY

James Scythes

The bloodiest slave revolt in United States history broke out in Southampton County, Virginia, in August 1831. Over the course of two days, seventy slaves led by slave preacher Nat Turner murdered fifty-five whites. After the revolt was put down, fear spread throughout the South. Some Southerners speculated that abolitionists were responsible for the uprising, and some even attacked and murdered blacks. Newspapers across the nation carried exaggerated accounts of the revolt coupled with reports of the murder of a number of blacks in Virginia. In the North, the accounts of the revolt and the murder of blacks enraged some in a heated critique of the South's "peculiar institution" and accelerated the growth of the Abolitionist Movement. The accusation that abolitionists had incited the slave uprising sparked a public debate in Northern and Southern newspapers between proslavery and antislavery editors that elevated the debate to a national level, driving a wedge between the North and South. By the end of 1831, the Virginia legislature had begun debating measures that would prevent another revolt, while Northern abolitionists were creating the first antislavery organization in the United States. After 1831, the South, and the United States, were never the same. Stories of the Nat Turner uprising would spur Southern white fear for decades to come.

In the early morning hours of August 22, 1831, Nat Turner and four other slaves had begun the insurrection by murdering Turner's master, Joseph Travis, and his family. Over the next day and a half, some seventy slaves moved from farm to farm killing every white person they encountered. By the night of August 23, local militia had put down the revolt, and fifty-five whites were dead. The slaves involved in the uprising were either killed or captured, but Turner eluded authorities for six weeks. Turner was captured on October 30, put on trial on November 5, and hanged on November 11, 1831.[1]

An analysis of newspaper articles related to Nat Turner's revolt provides an account of the confusion and fear in the South in the weeks following the uprising. Many of these news stories contained inaccuracies and exaggerations. This is to be expected in any breaking news story, especially in the first half of the nineteenth century, but in this case, the inaccuracies and exaggerations would exacerbate hysteria and paranoia among the Southern white population. According to the Richmond *Constitutional Whig* in September 1831, editors of other newspapers "seem to have applied themselves to the task of alarming the public mind as much as possible" by printing "false, absurd, and idle rumors."[2] Editor John Hampden Pleasants depicted himself as "a detached seeker of the 'truth,' trying to sift rumor from falsehood, legend from authentic reports" in the days following the Southampton revolt, explains historian Mary Kemp Davis.[3]

Newspaper accounts speculated on how many slaves participated in the uprising, what was the intention of the slaves, and what motivated them. On August 24, 1831, the day after local militia put down the revolt, the Richmond *Compiler* explained to readers that the early reports of the revolt "have no doubt…been exaggerated," but that it was believed the slaves were heading for the North Carolina border.[4] Another publication speculated that the slaves, who were estimated to number between 150 and 400, were trying to reach the Dismal Swamp, located twenty miles to the east of Southampton County.[5] Pleasants offered readers a motive for the revolt by stating that the slaves were attempting to get to Norfolk so they could "seize a ship and go to Africa."[6] These accounts contributed to the confusion and fear many Southerners experienced during and after Nat Turner's revolt.

As word of the revolt spread throughout the country, whites in the South began to wonder if the action of the Southampton slaves was an isolated event or part of a larger slave conspiracy. In fact, newspapers in South Carolina refused to publish any details related to Nat Turner's revolt out of fear of rousing the slaves of that state.[7] By the beginning of September 1831, Southern newspapers, such as the Richmond *Constitutional Whig*, reported that Turner's revolt was believed to have been an isolated event.[8] But rumors persisted throughout the South and fear remained of other possible slave revolts. This fear swept through North Carolina in September 1831 as a rumor spread that slaves had risen up in that state—a rumor that later proved to be false. It was reported that "the Slaves of Duplin and Sampson counties…have risen in rebellion against the whites…. The most recent account states the number of families murdered at *seventeen!*"[9] It was believed, based on the testimony of a slave named Dave, that more slaves in Sampson, Duplin, and New Hanover were preparing to revolt. The article claimed that the slaves intended to "march by two routes to Wilmington, spreading destruction and murder on their way."[10] Twenty-three slaves in Duplin and twenty-five in Sampson were arrested, and a few were executed.

In Boston, William Lloyd Garrison's antislavery newspaper, *The Liberator*, reported that the slaves in North Carolina had risen up against slaveholders. Garrison, a pacifist from Massachusetts who founded *The Liberator* in January

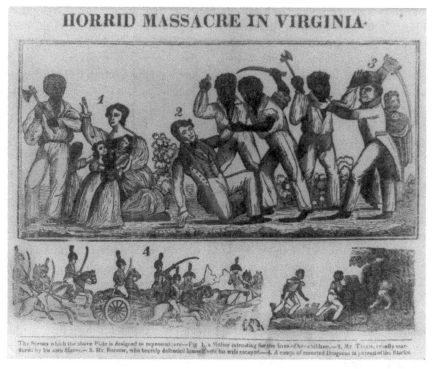

HORRID MASSACRE IN VIRGINIA.

The Scenes which the above Plate is designed to represent, are—Fig 1. a Mother intreating for the lives of her children,—2. Mr. Travis, cruelly murdered by his own Slaves.—3. Mr. Barrow, who bravely defended himself until his wife escaped.—4. A comp. of mounted Dragoons in pursuit of the Blacks.

FIGURE 3.1 "Horrid massacre in Virginia," by Samuel Warner, 1831. This woodcut appeared in a pamphlet titled "Authentic and impartial narrative of the tragical scene which was witnessed in Southampton County (Virginia) on Monday the 22d of August last when fifty-five of its inhabitants (mostly women and children) were inhumanly massacred by the blacks!" The illustration depicts events from Nat Turner's Rebellion, as recounted by white witnesses. The caption reads, "The scenes which the above Plate is designed to represent are—Fig 1. a Mother intreating for the lives of her children,—2. Mr. Travis, cruelly murdered by his own Slaves,—3. Mr. Harrow, who bravely defended himself until his wife escaped,—4. A comp. of mounted Dragoons in pursuit of the Blacks." (From the Library of Congress Rare Book and Special Collections Division Washington, DC. <www.loc.gov/pictures/item/98510363/> Reproduction Number: LC-USZ62-38902 (b&w film copy neg.). Samuel Warner, "Horrid massacre in Virginia.")

1831, was a staunch abolitionist who demanded the immediate end of slavery in the United States. In the September 24, 1831, edition of *The Liberator*, he mocked the slaveholders of the South and seemed to take pleasure in reporting that there was an insurrection in North Carolina. He wrote: "So much for oppression! so [*sic*] much for the happiness of the slaves! so [*sic*] much for the security of the South! Where now are our white boasters of liberty…. Let the blood which is now flowing rest upon the advocates of war—upon the heads of the oppressors and their apologists. Yes, God will require it at their hands. MEN MUST BE FREE!"[11] These comments surely agitated Southern slaveholders and contributed to the growing debate that began between slaveholders and abolitionists in 1831.

By the end of September, newspapers began reporting that the rumor of a slave uprising in North Carolina was false. The Democratic Richmond *Enquirer* wrote that the "accounts from our Sister State of North Carolina have been very much exaggerated.... All these reports turn out to be 'the mere coinage of the brain,' and it is now exceedingly questionable whether a single black has been under arms. It is certain that not a drop of white man's blood has been shed."[12] North Carolina governor Montfort Stokes admitted in November 1831 that there had "been no insurrection of slaves in North Carolina," but "ten or twelve negroes have been convicted for conspiracy to raise an insurrection, and most of them have been executed."[13] Governor Stokes believed news from Nat Turner's revolt caused the slaves in North Carolina to become restless and unruly, but no "concerted or extensive plan [had] been discovered."[14]

The execution of a number of blacks in North Carolina and the killing of blacks in Virginia demonstrated the hysteria that swept through the South after the revolt. On September 3, 1831, Pleasants wrote in an article titled "Southampton Affair," "How many [blacks] have...been put into death (generally by decapitation or shooting)?"[15] He attempted to answer his own question by speculating that as many as forty blacks had been killed. One man interviewed for the article claimed to have killed ten or fifteen blacks himself and believed he was justified "on the grounds of the barbarities committed on the whites."[16] Toward the end of the article Pleasants offered an apology for the people of Southampton County, believing that "human nature urged them to such extremities."[17] Pleasants was criticized for his comments, and it is believed that this criticism, which accused Pleasants of having abolitionist tendencies, was the first of a series of events that led to the 1846 duel between Pleasants and Thomas Richie, Jr., the editor of the Richmond *Enquirer*, that resulted in Pleasants's death.[18]

Similar stories were repeated in other newspaper articles over the next few weeks. The *Niles Register* of Baltimore reported that two blacks were killed in Maryland by a mob of armed whites, and there was "much *fear* and *feeling* in several of the lower counties of the state...[as] the white inhabitants seem to be in a constant excitement."[19]

Northern newspaper reports were very critical of the attacks and murders of blacks in Southampton County. *The Liberator* published Pleasants's September 3 account of the "Southampton Affair," commenting that such violent acts would backfire on the whites and lead to more violence, in which whites could "expect to be butchered without mercy."[20]

A Massachusetts abolitionist newspaper, the *Worcester Spy*, described whites from Southampton County as being savage and bloodthirsty, and believed the only event more barbarous than the scene in Virginia after Nat Turner's revolt was "Gen. Jackson's barborous [*sic*] massacre of the Indians" at the Battle of Horseshoe

Bend in 1814.[21] The editor feared that if there were further slave uprisings, the reaction might be the killing of all the blacks in the South.

The pacifist William Lloyd Garrison did not approve of the violent actions Nat Turner and his followers committed against the whites of Southampton County, but he was even more appalled by the reaction of Southern whites against blacks. Garrison believed Southern whites acted in an uncivilized manner, and he wrote that "The Indians of North America were never more savage, blood-thirsty and revengeful, than the southern slaveholders."[22] Slaves were described as having their noses and ears cut off, the flesh of their cheeks cut out, and their jaws broken. If blacks were out after dark without a pass, they would be shot. Southerners, according to Garrison, were so determined to protect their way of life that they were willing to put the entire "black population to the sword."[23] Garrison's rhetoric was extreme. He believed that there was no longer a middle ground for Southerners. Slaves were expected to be totally servile to Southern whites or there was no sense in having them. If Southerners could not be slaveholders, Garrison believed the only alternative for Southern whites would be the extermination of the black race in the South.

Not all Northern newspapers were critical of the South, however. At least two papers defended the actions taken by whites in Virginia and pledged support for the South. The editor of the antislavery New York *Journal of Commerce* believed the Virginians had the right to "shoot down without mercy not only the perpetrators, but all who are suspected of participation in the diabolical transaction," adding that at least some of the instigators "no doubt deserve to die."[24] The author of the article also pledged Northern support of the South if another slave uprising occurred. He believed a "million" men would march on short notice from the North to defend the South if necessary. "For, much as we abhor slavery; much as it is abhorred throughout the Northern and Eastern States," he wrote, "there is not a man of us who would not run to the relief of our friends in the South."[25] The editor of the Albany *Argus*, a Democratic newspaper with close political ties to Southern Democrats, stated: "we know with how much alacrity the men of the North will come to the aid of their fellow citizens of the South. The cause is a common one; and the claim upon us, although less direct, is not the less the claim of humanity and patriotism."[26]

From the moment the first reports of Nat Turner's revolt appeared in the newspapers, some Southerners had speculated that Northern abolitionists played a part in it. In a letter to South Carolina governor James Hamilton, the governor of Virginia, John Floyd, accused abolitionists from the North of circulating "inflammatory papers and pamphlets" amongst the slaves. Floyd believed Northerners were responsible for telling "the black man [he] was as good as the white; that all men were born free and equal; that they can not serve two masters; that the white people rebelled against England to obtain freedom; so have the blacks a right to do."[27] According to the Charleston *Mercury*, a group of men calling themselves the "Vigilance Association of Columbia" offered a $1,500 reward for the apprehension and conviction of any white person distributing a copy of *The Liberator* in the state of South Carolina.[28] In October 1831, the Attorney General of North Carolina went so far as obtaining an indictment against Garrison for distributing his newspaper in that state, and the Georgia legislature offered a $5,000 reward for anybody who would kidnap Garrison and drag him to Georgia for trial.[29] Some Southerners assumed that it was no coincidence that Nat Turner's revolt started eight months after Garrison published his first issue of *The Liberator*.

As the public argument between proslavery supporters and abolitionists continued in the newspapers, Governor Floyd privately expressed his feelings in his diary about Garrison and his desire for a law that would punish Garrison for urging "the slaves and free negroes in this and the other States to rebellion and to murder the men, women and children of those states."[30] The governor could not understand how a man in one state could plot treason against another state and go unpunished. He believed the federal government should correct this problem because if this was to go unchecked "it must lead to a separation of these states."[31]

Southerners were giving Garrison and the abolitionist movement too much credit. The abolitionist movement in the North was not that strong or organized in early 1831. It would have been impossible at this time for abolitionists to accomplish all of the things they were accused of doing. In 1831, Garrison was still relatively unknown and *The Liberator* was not very popular. By blaming the Northern abolitionist for inciting Nat Turner's Revolt, Southerners made Garrison a national figure.[32]

After Nat Turner's revolt, slaveholders in Virginia wanted the state legislature to take action and strengthen laws that would prevent future uprisings. They also wanted state officials to find a way to remove free blacks from Virginia. The editor of the Richmond *Enquirer*, Thomas Richie, wanted the Virginia legislature to create a law that would prohibit black men from becoming preachers. If the legislature failed to act, Richie believed the Southampton revolt would be repeated in the future in some other part of the state.[33] Less than a week later, the same paper advocated for a law "against unlawful assemblies of the colored people."[34] Later in the month, a letter appeared in the *Enquirer* from a resident of Southampton County who believed that security could only be found if the slave laws of the state were strictly enforced.[35] The editor of the Richmond *Compiler* also believed that the enforcement of existing laws, such as forbidding the education of slaves and prohibiting slaves from leaving their master's property, would improve security.[36]

By the end of 1831, a movement started to gain support in Virginia that would remove all free blacks from the state and send them to Africa. It was reported in the *Enquirer* that two hundred blacks wished to leave Southampton County and move to Liberia.[37] Virginia governor Floyd supported the idea of removing all of the free blacks from the state, but suggested going even further. In a letter to South Carolina governor James Hamilton, Floyd proposed a plan to emancipate the slaves of Virginia.[38] Two days after he penned this letter, Floyd wrote in his diary that he desired to gradually abolish slavery in Virginia, or at least "begin the work by prohibiting slavery on the West side of the Blue Ridge Mountains."[39]

Even editor Thomas Richie believed something needed to be done about slavery in Virginia. He wrote that "the dark and growing evil, at our doors.... The disease [slavery] is deep seated—it is at the heart's core—it is consuming, and has all along been consuming our vitals. What is to be done? Oh! my God—I don't know, but something must be done."[40]

However, by the time the Virginia legislators met in January 1832 to debate the issue of strengthening Virginia's slave codes, Governor Floyd must have changed his mind, because he never brought up his emancipation plan to the legislature. Historian Stephen B. Oates offered a theory to explain why Floyd abandoned his plan. According to Oates, Vice President John C. Calhoun, who was Floyd's hero, visited the governor on December 3, 1831, and "may have talked the governor out of any abolition moves...[and convinced Floyd] that the South could best protect its slave system from abolitionist coercion, not through emancipation, but behind the bulwark of state rights and nullification."[41] Calhoun left Richmond on December 5, and on December 6, Governor Floyd proposed in his message to the legislature "the revision of all the laws intended to preserve in due subordination the slave population of our State," making no mention of the possibility of emancipating the state's slaves.[42]

In the January and February 1832 meetings of the Virginia legislature, a debate over slavery ensued as some legislators used the discussion of slave codes as an opportunity to express their opinions on slavery. Representative James Gholson stated that slaves in Virginia are "as happy a laboring class as exists upon the habitable globe. They are well fed, as well clothed, and as well treated."[43] General William Brodnax, who participated in putting down Turner's revolt, admitted to the legislature that he believed slavery was evil, but "no emancipation of slaves should ever be tolerated."[44] Representatives from the state's western counties were outspoken against slavery and stressed the damaging effects of slave labor, and some of the representatives from these counties proposed the emancipation of Virginia's slaves.

The debate in the Virginia legislature, which lasted several weeks, was an exceptional event in Southern history. Thomas Richie summed up its significance in the Richmond *Enquirer* by stating: "And what is *more remarkable* in the History of Legislature, we now see the whole subject ripped up and discussed with open doors, and in the presence of a crowded gallery and lobby—Even the press itself hesitating to publish the Debates of the body. All these things were indeed new in our history. And nothing else could have prompted them, but the bloody massacre in the month of August."[45] Without Nat Turner's revolt, the debate over slavery in Virginia, and possibly the national debate, would not have occurred.

The Virginia legislature rejected the push from antislavery legislators to emancipate slaves and send blacks to Africa. Instead, state legislators passed a new slave code on March 15, 1832. The provisions within this new slave code were as follows: slave preachers were prohibited from preaching to any slave; free blacks could not possess a firearm; slaves were forbidden from consuming alcohol; slaves and free blacks were prohibited from assaulting a white person; any person was forbidden from distributing pamphlets advising slaves to commit insurrection; and harsh punishments would be handed down for violation of these codes.[46]

The attack of slavery by antislavery newspapers in the North in the months following Nat Turner's revolt ultimately led to the development of an organized abolitionist movement in the United States and increased the national debate over slavery in Antebellum America. On September 17, 1831, an abolitionist newspaper, the New York *Daily Sentinel*, explained that the slaves that followed Nat Turner were fighting to emancipate themselves and that the way to achieve this was to "put to death, indiscriminately, the whole race of those who held them in bondage."[47] The author of the article wondered how the state that "gave birth to the immortal Jefferson, the author of the declaration that declares that *all* men are born *free and equal*" could continue to deny blacks freedom.[48] This was one of the central arguments of abolitionists during the Antebellum Period. Slavery, in their opinion, was contrary to American principles. Abolitionists viewed freedom as an inalienable right and believed that the slaves' struggle for freedom was "the same in principle as the struggle of our fathers in '76."[49]

Some abolitionists demanded the immediate and universal end of slavery in the United States. The *Ohio State Journal* implored the people of the United States to no longer "shut their eyes to the dreadful evils of slavery" and called on them to devise a plan to remove "this curse from among us."[50] At the end of 1831, Garrison formed the New England Anti-Slavery Society, which was the first antislavery organization in the United States. By 1833, this organization had attracted so many followers from around the country that it changed its name to the American Anti-Slavery Society. Just two years after Nat Turner's revolt, a national antislavery organization was in existence. In a private letter to William Lloyd Garrison, James Forten, a free black from Philadelphia, summed up the importance of Nat Turner's revolt: "This insurrection in the south, will be the means of bringing the evils of slavery more prominently before the public."[51] Abolition was now a national movement that continued to gain more and more momentum during the Antebellum Period.

Notes

1 Herbert Apthecker, *Nat Turner's Slave Rebellion: Including the 1831 "Confessions"* (Mineola, NY: Dover Publications, Inc., 1966), 44–56.

2 "Southampton Affair," *Constitutional Whig* (Richmond, VA), September 3, 1831.

3 Mary Kemp Davis, *Nat Turner Before the Bar of Judgment: Fictional Treatments of the Southampton Slave Insurrection* (Baton Rouge: Louisiana State University Press, 1998), 42–43.

4 *Compiler* (Richmond, VA), August 24, 1831.

5 "The Insurrections," *Intelligencer* (Petersburg, VA), August 26, 1831.

6 "Extract of a letter from the Senior Editor, dated JERUSALEM, Southampton Ct. House," *Constitutional Whig* (Richmond, VA), August 29, 1831.

7 Eric Foner, ed., *Great Lives Observed: Nat Turner* (Englewood Cliffs, NJ: Prentice-Hall, Inc., 1971), 61.

8 "Southampton Affair," *Constitutional Whig*, September 3, 1831.

9 "Commotion in North Carolina," *National Intelligencer* (Washington, DC), September 19, 1831.

10 Ibid.

11 "BLOOD! BLOOD!! BLOOD!!! Another insurrection!" *The Liberator* (Boston, MA), September 24, 1831.

12 "The Banditti," *Enquirer* (Richmond, VA), September 20, 1831.

13 Montfort Stokes to James Hamilton, 18 November 1831, in *Great Lives Observed*, 61.

14 Ibid.

15 "Southampton Affair," *Constitutional Whig*, September 3, 1831.

16 Ibid.

17 Ibid.

18 Louis P. Masur, *1831: Year of Eclipse* (New York: Hill and Wang, 2001), 13–14.

19 "Miscellaneous," *Niles Register* (Baltimore, MD), September 10, 1831.

20 *The Liberator* (Boston, MA), September 17, 1831.

21 "Slave Insurrection," *Worcester* (MA) *Spy*, September 17, 1831.

22 "Southern Cruelty," *The Liberator*, October 1, 1831.

23 Ibid.

24 "Insurrection in Virginia," *Journal of Commerce* (New York, NY), September 10, 1831.

25 Ibid.

26 *Argus* (Albany, NY), September 22, 1831.

27 John Floyd to James Hamilton, 19 November 1831, in *Great Lives Observed*, 59.

28 *Mercury* (Charleston, SC), October 11, 1831.

29 Henry Mayer, *All on Fire: William Lloyd Garrison and the Abolition of Slavery* (New York: St. Martin's Griffin, 1998), 122.

30 John Floyd, diary entry, 27 September 1831, in Henry Irving Tragle, *The Southampton Slave Revolt of 1831: A Compilation of Source Material* (Amherst: The University of Massachusetts Press, 1971), 255–56.

31 Ibid.

32 Stephen B. Oates, *The Fires of Jubilee: Nat Turner's Fierce Rebellion* (New York: Harper & Row, 1975), 133.

33 "The Banditti," *Enquirer* (Richmond, VA), August 30, 1831.

34 "The Banditti," *Enquirer* (Richmond, VA), September 2, 1831.

35 "The Southampton Tragedy," *Enquirer* (Richmond, VA), September 27, 1831.

36 *Compiler* (Richmond, VA), October 11, 1831.

37 *Enquirer* (Richmond, VA), October 7, 1831.

38 John Floyd to James Hamilton, 19 November 1831, in *Great Lives Observed*, 60.

39 John Floyd, diary entry, 27 September 1831, in *The Southampton Slave Revolt of 1831*, 261.

40 "Virginia Legislature," *Enquirer* (Richmond, VA), January 7, 1832.

41 Oates, *The Fires of Jubilee*, 137.

42 John Floyd, "Fellow-Citizens of the Senate and of the House of Delegates," Governor's Message to the Legislature, December 6, 1831, in *Great Lives Observed*, 99.

43 "Speech of James Gholson," *Enquirer* (Richmond, VA), January 21, 1832.

44 "General William H. Brodnax's Assessment of Slavery and the Rebellion," *Enquirer* (Richmond, VA), January 24, 1832.

45 *Enquirer* (Richmond, VA), February 4, 1832.

46 Foner, *Great Lives Observed*, 114–16.

47 "Daring Outrage of Virginia Slavites [*sic*]," *Daily Sentinel* (New York, NY), September 17, 1831.

48 Ibid.

49 *African Sentinel and Journal of Liberty* (Albany, NY), October 1, 1831.

50 "Insurrectionary Movements," *Ohio State Journal* (Columbus), October 20, 1831.

51 James Forten to William Lloyd Garrison, 20 October 1831, in *Great Lives Observed*, 85.

4

DISUNION OR SUBMISSION?

Southern Editors and the Nullification Crisis, 1830–1833

Erika Pribanic-Smith

A wave of disunion rumblings erupted after the 1828 passage of a federal act increasing protective duties on foreign imports, which many Southerners saw as unconstitutional and unequal in its benefits. This "Tariff of Abominations" and its successor, the Tariff of 1832, caused newspapers throughout the South to debate the best response to national measures that many deemed oppressive. South Carolina alone reacted with an 1832 state convention that would adopt an Ordinance of Nullification, nullifying the tariffs of 1828 and 1832 and proclaiming that the state could decide which federal laws were enforceable in South Carolina. Congress would respond with a bill allowing the president to forcibly apply federal law, but also passed the Compromise Tariff of 1833. South Carolina then repealed its original nullification ordinance, but nullified the federal bill authorizing presidential use of force. The Nullification Crisis was over, although the issue of nullification would not be settled for another three decades. This chapter explores responses to the nullification question in South Atlantic and Gulf Coast newspapers from 1830—when nullification discussion became prevalent in the region—until the passage of South Carolina's Ordinance of Nullification in 1832.

Except for South Carolina's Nullifiers, responses differed little from state to state. While some editors called for unqualified resistance to what they perceived as gross usurpation of power by the federal government, others sought a moderate, peaceful solution. Some pushed for individual states to act alone while others called for the Southern states to work in concert. Those loyal to the US government and alarmed by revolutionary rhetoric urged patience and even submission. Often, the degree to which a newspaper supported or opposed nullification corresponded with adherence to local political parties. It also depended largely on the extent to which each state felt directly oppressed.

Because of South Carolina's unique geography, demographics, and healthy foreign trade, historians assert that its economic dependence on agriculture exceeded that of other Southern states. Furthermore, a ruling class of planters intent on preserving slavery as well as their own political and social dominance advocated disunion earlier and with more vigor than in any other Southern state.[1]

An ever-growing group threatened to dismember the Union at several key points in the decades before secession, including the 1830s Nullification Crisis.[2] Rumblings of disunion sentiment began when Congress passed protective tariffs in 1824 and 1828 to encourage domestic manufacturing. Nearly all of the Southern states protested the 1824 tariffs, which raised existing taxes to as much as 37 percent of the goods' value.[3] South Carolinians found the products prohibitively expensive; they had been hit particularly hard by a depression that began in 1819.[4] Furthermore, because they perceived the tariff as easing the depression effects for Northern manufacturers at the expense of Southern agricultural trade, protesters in South Carolina claimed sectional oppression.[5] Tariff dissenters also called the act of tariff making unconstitutional and feared that federal tyranny might extend to abolition.[6]

Protest increased in 1828, when Congress passed the highest tariff in US history.[7] This "Tariff of Abominations" raised rates to as much as 50 percent of the goods' value.[8] Arguments against the 1828 act echoed those used against the 1824 tariff, but the passion with which they were advanced had significantly increased.[9] On July 14, 1832, Congress repealed the Tariff of 1828, but a new tariff bill replaced it, establishing new duties on several items. Although many of these duties were reduced from those in the Tariff of 1828, some were increased.[10] Unsatisfied by reductions set forth in the Tariff of 1832, South Carolinians called for a convention, which assembled in November 1832 to draft the Ordinance of Nullification. The convention called the tariffs of 1828 and 1832 unconstitutional and declared them null and void within the state of South Carolina. The ordinance declared that any efforts to coerce the state by military force or interruption of commerce would result in secession.[11]

The idea to nullify the tariff emerged from an 1828 pamphlet titled the "South Carolina Exposition and Protest." Drafted in secret by then-Vice President John C. Calhoun (a Carolina native), the protest declared that the states individually had the right to declare laws they found offensive null and void.[12] Whereas the tariff and other "American System" policies had drawn a fairly uniform response from Southerners who feared a government of unlimited power and unequal benevolence, the nullification remedy sparked passionate debate. The argument was slow to spread beyond South Carolina, but by 1830, editors throughout the South had taken a position, some calling for resistance, some calling for moderation, and some urging patience and even submission.

In the early 1830s, small newspaper staffs made newsgathering difficult, so editors copied much of the news content from other papers. Only the editorial column contained original content, and placement of unsigned material there indicated these items were written by an editor.[13] During this era of partisan, personal journalism, a single strong editor generally dominated the editorial column, stamping "his principles, interests, values, and prejudices on all aspects of the newspaper," whether he wrote the content or not.[14] Therefore, ideas expressed in unsigned editorials are attributed to the newspaper's editor. Editors often extracted from exchange papers in the editorial column as well, to express agreement or disagreement with fellow editors.

South Carolina

The story of nullification in the newspapers of South Carolina involves two distinct local political parties—Free Trade/Nullification and Unionist. Although initially weaker than the Nullification Party, the Unionist Party became viable during the controversy Calhoun's "Exposition and Protest" instigated.[15] When the Nullifiers splintered over key issues and crumbled under the weight of disunionist accusations, moderate Nullification men fled to the Union Party. The bolstered Unionists dominated elections in 1830 for state and city officials, and they appeared to control the state's politics and public opinion.[16] Thus, nullification appeared unlikely at the end of 1830, but that soon changed. Free Trade partisans used their presses more effectively than the Unionists over the course of 1831, tipping public opinion in favor of a party that supported unqualified resistance, as evidenced by victories at the polls in 1831 and 1832. Free Trade control of the South Carolina government resulted in the convention that passed the Ordinance of Nullification in 1832.[17]

Unionist papers' chief argument was that nullification was treasonous and would lead to secession and civil war. Anti-nullification papers thought talk about nullifying an act of Congress while still remaining a member of the federal government was nonsense.[18] Although the editor of the *Greenville Mountaineer* clearly was against the tariff, his newspaper and the *Charleston Courier* denied that it was any worse than other acts the state had borne without rebelling.[19] Unionist editors argued that a majority of the people of South Carolina saw nullification as a greater evil than the tariff. They were not willing to endanger the Union, "which Washington, the father of this country, pronounced the great palladium of our liberties."[20] Unionist newspapers pleaded with their fellow statesmen to rise up and save their state from civil war.[21] Writers in the *Charleston Courier* proclaimed that the ballot box was their only weapon and asserted that voters would decide "whether the broad Banner of our Union, with its Stripes and its Stars, shall continue to wave over South Carolina...or whether we shall be among the first to tear asunder its folds."[22]

Nullifiers denied that their position would result in civil war. In fact, they argued that nullification would preserve the Union, calling it a middle ground between submission and secession. They asserted that the Unionists cried disunion and bloodshed to scare South Carolinians into defeating the measure.[23]

Nonetheless, Free Trade editors evoked the revolutionary spirit of 1776 as one tactic to convince readers to adopt nullification. According to an editorial in the Columbia *Southern Times and State Gazette*, no descendants of Tories could be found in the nullification ranks—only Patriots, who had inherited the "love of liberty; the inextinguishable hatred of tyranny and tyrants, the high unbending spirit of resistance to oppression in every form."[24] In addition, nullification editors declared that a number of Revolutionary War veterans supported their cause. Numerous speeches, letters, and quotations from veterans published in the nullification papers vilified the Federalist Party, glorified the founding fathers, and expressed that if Carolinians did not resist federal usurpations to preserve the Constitution, their fighting would have been in vain. The *Pendleton Messenger* gloated that Revolutionary hero Gen. Thomas Sumter was among nullification's most ardent supporters and proclaimed him still to be "vigorous in the maintenance of the principles for which, fifty years ago, he fought and bled."[25] A writer in the *Camden and Lancaster Beacon* pointed out that Sumter had not only fought on the battlefields of the Revolution, but also, as a congressman for South Carolina in 1798, he had fought "shoulder to shoulder with Jefferson" and "hurled from the seats of power the friends of *federalism and consolidation and disunion*, as the republicans of this day must do."[26]

"Republicanism" was the second pillar of the nullification cause. Writers in favor of resistance portrayed the conflict as between Federalists and "True Republicans," and they appealed to readers' reverence for the patriarchs of the Democratic-Republican Party in their arguments for nullification. They cited the Kentucky and Virginia Resolutions of 1798 and 1799—in addition to other writings by James Madison and Thomas Jefferson—as the sources of Republican principles as well as the nullification doctrine. Nullifiers particularly adhered to Jefferson's notions that the states had reserved rights, one of which was to peacefully protest federal acts deemed oppressive or unconstitutional.[27] Editorials in the *Pendleton Messenger* noted many parallels between the tariff conflict and the Alien and Sedition discord at the turn of the nineteenth century. The *Messenger* editor argued that principles never changed: supporters of the tariff espoused the same Federalism that flourished under President John Adams. The *Messenger* also contended that the language of the Unionists echoed the cries of treason, war, bloodshed, and disunion that Federalists of old uttered.[28] Free Trade newspapers asserted that the only means of combating such a foe was the same doctrine that spurred the political revolution of 1801—nullification. Free Trade editors and correspondents provided letters and manuscripts that they believed proved Jefferson both created and upheld the policy.[29]

Unionist newspapers countered that attaching the founding fathers' names to the nullification doctrine was ridiculous. A *Charleston Courier* correspondent argued that the Kentucky Resolution Jefferson wrote against the Alien and Sedition Acts did not give *one* state the power to veto the solemn acts of all others, but authorized *the several sovereign states* together to seek redress for constitutional infractions.[30] As such, the Unionists suggested gathering the Southern states in a convention to determine the will of everyone suffering from the tariff. The *Greenville Mountaineer* editor even was willing to pursue nullification if that was the consensus. He argued, "For the State to act alone is the height of folly," but to resist with the consent of all concerned would assure South Carolina the assistance of her sister states in the hour of danger.[31] The *Courier* agreed, stating, "If she *must* RESIST, let that resistance be by the CONFEDERATED SOUTH."[32]

The *Greenville Mountaineer*'s editor posited that states are not independent sovereignties because they do not have the authority to do anything and everything they please; they could be considered sovereign only in the exercise of their reserved rights, and declaring laws unconstitutional was not one of them. The editor argued that the Constitution makes it "utterly impossible for such a power to exist in the States" and noted that James Madison denied writing anything akin to South Carolina's doctrine of nullification. Instead, the Republicans of old granted each state the right to secede if it became clear that the government had adopted a fixed and settled policy "which must inevitably ruin and crush us to the earth if we continue members of this Union." He did not think that time had come.[33]

Unionist editors called for forbearance. They had no doubt Congress eventually would enact satisfactory reductions of the tariff and pleaded for the Nullifiers to wait before taking drastic measures. Each movement in Congress on the issue seemed to auger a positive future result.[34] Conversely, the Free Trade editors saw each movement in Congress as proof that the federal government intended to saddle the South with an ever-heavier burden. The Nullifiers asserted that if the state submitted quietly, the people could be oppressed forever, and the federal government could be allowed to pervert the sacred Constitution at will.[35] The Free Trade press pushed for a full repeal; nothing else would be acceptable. An editorial in the *Camden and Lancaster Beacon* declared that acceding to any compromise would be a compromise of principles and of constitutional rights, which would be a deep disgrace.[36] The *Charleston Mercury* called for the South to rise up together "as sovereign members of a violated league."[37]

Georgia

The other Southern states did not heed the *Mercury*'s call—even South Carolina's closest neighbors. Though Georgia would secede within a month of South Carolina on the eve of the Civil War, in the 1830s, the state largely disagreed with the Carolina brand of resistance, except when Georgians felt their own state's rights had been explicitly violated. Georgia clearly opposed the tariff as unconstitutional, inexpedient, and oppressive to the Southern states, but, as the *Savannah Republican* noted in 1831, hostility toward the tariff did not equate with support of nullification.[38] However, whether Georgia editors fully rejected the notion of nullification or supported the idea in a different form depended on their political

FIGURE 4.1 "Jackson and the nullifiers," 1832. This anti-nullification verse to the tune of "Yankee Doodle Dandy" was printed as a broadside (a single page or poster) and sold in New York City. Two of the verses read as follows:

Why Yankee land is at a stand,
And all in consternation;
For in the South they make a rout,
And all about Nullification.
Sing Yankee doodle doodle doo,
Yankee doodle dandy,
Our foes are few our hearts are true,
And Jackson is quite handy.
[…]

affiliation. The editors supporting one party fully opposed nullification, calling South Carolina's proceedings treasonous.[39] Newspapers on the other side, such as the *Athenian*, declared that Georgia would "not be dragged nor dragooned either into submission to the eastern and western states on the one hand, nor into a tame compliance with all the wild and precipitate movements of South Carolina on the other."[40]

Georgia's response to Native American affairs complicated the factions' positions on nullification. During the 1820s and 1830s, Georgia campaigned to remove the Cherokees, passing legislation to seize Cherokee land and distribute it among white Georgians. The Supreme Court reviewed multiple contradictions with federal treaties in *Worcester v. Georgia* in March 1832, ultimately rejecting the state's attempt to nullify federal agreements with the Cherokees. Historians assert that this led some Georgians, wishing to counter the federal decision, to view South Carolina doctrine as more attractive.[41] Others, however, worked to further distance themselves from their nullifying neighbor.[42]

Other South Atlantic Neighbors

North Carolina and Virginia overwhelmingly opposed nullification. As in other Southern states, North Carolinians and Virginians generally opposed the tariff. However, numerous editorials denied that the tariff had as detrimental an effect as the Nullifiers proclaimed. Editors asserted that the good people of North Carolina and Virginia were too devoted to the Union to follow South Carolina on its ruinous path. "The horrible idea of shedding the blood of neighbours, friends, and relatives, in a civil war, has yet found no abiding place in the mind of a North Carolinian," the *Fayetteville Observer* declared. "God forbid that a press in North Carolina should ever dare openly to promulgate such sentiments."[43] The *Newbern Spectator* noted that North Carolina was one of the first states to "step forward and assist in establishing the independence we now enjoy," and it would be the last "to aid in any manner, directly or indirectly, to destroy that independence, or

---

Caption for Figure 4.1 (Cont.)

Our country's cause, our country's laws,
We ever will defend, Sir,
And if they do not gain applause,
My song was never penned, Sir.
So sound the trumpet, beat the drum,
Play Yankee doodle dandy,
We Jackson boys will quickly come,
And be with our rifles handy.

(From the Library of Congress Rare Book and Special Collections Division, Washington, DC. <www.loc.gov/item/rbpe.11800800/> Printed Ephemera Collection; Portfolio 118, Folder 8. "Jackson and the nullifiers.")

dissolve this Union."[44] Virginia papers asserted that their statesmen would reject disunionist doctrine; the *Richmond Enquirer* averred that no one in the South would cooperate with the Nullifiers but would stand firm in the notion that the Union must be preserved.[45]

North Carolina editors pointed out which South Carolina newspapers were anti-nullification and which advanced the cause of disunion. The *Newbern Spectator* praised the *Charleston Courier* for its steadfastness in support of the Constitution and Union and copied copiously from the *Courier*'s editorial columns.[46] Conversely, the *Spectator* pointed to the *Charleston Mercury* as the chief proponent of nullification and censured its editors for their dangerous errors in judgment.[47]

Gulf Coast States

Most newspapers in the Gulf Coast states echoed the same sentiments that appeared in the Unionist South Atlantic newspapers. Though they quickly followed South Carolina's lead into the Confederacy during the winter of 1861, Mississippi and Alabama were not so hasty to adopt South Carolina nullification doctrine in the early 1830s.

Acceptance of tariff policy in Natchez, Mississippi, contributed to the rejection of nullification there. The *Natchez Weekly Democrat* (which later became the *Weekly Courier*) published multiple editorials insisting that the tariff benefited not only the state's manufacturers but also its planters. Its editor encouraged readers to objectively study the potential ruinous effects on Mississippi if the tariff were repealed and to cast their votes accordingly for representatives who would support the "American System."[48] This pro-tariff attitude could be attributed in part to chief tariff author Henry Clay's frequent stops in Natchez—viewed as a commercial center of the South in the early nineteenth century—to speak on the virtues of his economic system.[49] Kinship with neighboring Louisiana also played a role; Louisiana's sugar planters profited from import tariffs on foreign sugar.[50] Editorials in Natchez newspapers noted general support of the tariff in Louisiana and indicated that Mississippians should follow their neighbor's example.[51]

Newspapers elsewhere in Mississippi, as well as in Alabama, opposed the tariff with the same vigor as those in other Southern states, and they similarly denied that the policy warranted the chaos South Carolina was causing. Echoing the South Atlantic Unionist editors, editors in the Gulf Coast states urged forbearance. They believed that Congress would adjust the odious policy fairly; reductions in the spring of 1832 provided hope that the tariff eventually would reduce to nothing. Quoting Jefferson and Madison, editors encouraged tariff opponents to band together in peaceful protest to hasten the tariff's demise; they considered nullification foolish and unconstitutional.[52] If the tariff did appear in the future to be fixed policy, then the Vicksburg *Mississippian* advised secession as the "sacred right of throwing off the yoke of oppression," but the editor did not think that would happen.[53] An Alabama editor similarly noted that the time had not yet come to "ring the funeral knell of American grandeur, glory, and happiness!"[54]

Ten states followed South Carolina into the Confederacy in the winter, spring, and early summer of 1861. Of the eleven Confederate states, eight were US states at the time of the Nullification Crisis.[55]

Most newspapers that vehemently supported nullification in 1830–32 were in South Carolina.[56] Claiming venerated Republicans Jefferson and Madison as the originators of nullification ideology and appealing to readers' lingering sense of patriotism from the Revolutionary War, Free Trade editors in South Carolina sought to convince the public that nullification was a constitutional, peaceful, and proper remedy.

Opposing newspapers in South Carolina and other states denied that Madison and Jefferson originated nullification ideology. Unionists proclaimed that one state could not resist on its own, but that the proper course was for all afflicted states to join together. If the oppression simply could not be borne, then secession was the rightful solution, but Unionist editors could not imagine such a circumstance. For the next thirty years, with each new issue that separated the Northern and Southern states, philosophical discussions of nullification and secession would continue in Southern newspapers. The issues debated in the South during the Nullification Crisis of 1832 moved from the philosophical to reality with the secession of South Carolina in 1860.

Notes

1 Frederic Bancroft, *Calhoun and the South Carolina Nullification Movement* (Baltimore: The Johns Hopkins Press, 1925), 18–19; William J. Cooper, Jr., *Liberty and Slavery: Southern Politics to 1860* (New York: Knopf, 1983), 151–53; William W. Freehling, *Prelude to Civil War: The Nullification Controversy in South Carolina, 1816–1836* (New York: Harper & Row, 1966), 90–91.

2 See Yates Snowden and H. G. Cutler, eds., *History of South Carolina*, 5 vols. (Chicago: Lewis Pub. Co., 1920); David Duncan Wallace, *The History of South Carolina*, 4 vols. (New York: American Historical Society, 1934).

3 *Act of May 22, 1824*, chap. 136, *US Statutes at Large* 4 (1824): 25; Bancroft, *Calhoun and the South Carolina Nullification Movement*, 8; Dall W. Forsythe, *Taxation and Political Change in the Young Nation, 1781–1833* (New York: Columbia University Press, 1977), 78.

4 Freehling, *Prelude to Civil War*, 106–08.

5 Bancroft, *Calhoun and the South Carolina Nullification Movement*, 20; Forsythe, *Taxation and Political Change in the Young Nation*, 82–83.

6 Bancroft, *Calhoun and the South Carolina Nullification Movement*, 14–15; Forsythe, *Taxation and Political Change in the Young Nation*, 79–81, 85.

7 William McKinley, *The Tariff in the Days of Henry Clay, and Since: An Exhaustive Review of Our Tariff Legislation from 1812 to 1896* (New York: Henry Clay Publishing Co., 1896), 6.

8 *Act of May 19, 1828*, chap. 55, *US Statutes at Large* 4 (1828): 270.

9 Bancroft, *Calhoun and the South Carolina Nullification Movement*, 16–17, 33; John George Van Deusen, *Economic Bases of Disunion in South Carolina* (New York: Columbia University Press, 1928), 23–24, 30–33, 36–40.

10 *Act of July 14, 1832*, chap. 227, *US Statutes at Large* 4 (1832): 583.

11 *Journal of the Conventions of the People of South Carolina, Held in 1832, 1833, and 1852* (Columbia: R. W. Gibbes, 1860); *Proceedings of the Convention of South Carolina Upon the Subject of Nullification Including the Remarks of Governor Hamilton, on Taking the President's Chair: the Ordinance Nullifying the Tariff Laws and the Report Which Accompanied It: an Address to the People of the United States: an Address to the People of South Carolina* (Boston: Beals, Homer, 1832).

12 Bancroft, *Calhoun and the South Carolina Nullification Movement*, 38–50, 75–86; Forsythe, *Taxation and Political Change in the Young Nation*, 88–92; Van Deusen, *Economic Bases of Disunion in South Carolina*, 45–51. Calhoun's authorship was not revealed until after his break from the Democratic Party in 1831. Historians attribute the secrecy to Calhoun's aspirations for the presidency. (Gerald M. Capers, "A Reconsideration of John C. Calhoun's Transition from Nationalism to Nullification," *Journal of Southern History* 14 (1948): 39, 43–45.)

13 Hazel Dicken-Garcia, *Journalistic Standards in Nineteenth-Century America* (Madison: University of Wisconsin Press, 1989), 19; Carl R. Osthaus, *Partisans of the Southern Press* (Knoxville: University Press of Kentucky, 1994), 4.

14 Osthaus, *Partisans of the Southern Press*, 1–99.

15 Erika J. Pribanic-Smith, "Conflict in the South Carolina Partisan Press of 1829," *American Journalism* 30 (2013): 365–92.

16 Pribanic-Smith, "Rhetoric of Fear: South Carolina Newspapers and the State and National Politics of 1830," *Journalism History* 38 (2012): 166–77.

17 Pribanic-Smith, "South Carolina's Rhetorical Civil War: Nullification and Local Partisanship in the Press, 1831–1833," *Media History Monographs* 17, no. 2 (2014).

18 *Greenville (SC) Mountaineer*, June 4, 1831.

19 *Greenville (SC) Mountaineer*, May 14, 1831, and June 4, 1831; Berkley, "For the Courier," *Charleston Courier*, April 23, 1831; Palmetto, "To the Editor of the Courier," *Charleston Courier*, June 20, 1831.

20 *Greenville (SC) Mountaineer*, February 25, 1832, April 21, 1832, and September 22, 1832; R., "For the Mountaineer," *Greenville (SC) Mountaineer*, July 9, 1830; R., "For the Mountaineer," *Greenville (SC) Mountaineer*, July 23, 1830; "From our Correspondent," *Charleston Courier*, May 20, 1830; Moultrie, "The Times," *Charleston Courier*, June 10, 1830.

21 A Planter, *Charleston Courier*, August 8, 1831; *Greenville (SC) Mountaineer*, September 24, 1831.

22 "Union & State Rights Meeting," *Charleston Courier*, September 1, 1831.

23 *Charleston Mercury*, June 24, 1830; *Winyaw (SC) Intelligencer*, March 6, 1830; "Disunion," *Winyaw (SC) Intelligencer*, July 17, 1830; "Nullification and Its Effects," *Southern Times* (Columbia, SC), May 10, 1830; *Pendleton (SC) Messenger*, February 15, 1832, April 4, 1832, and December 5, 1832; *Columbia (SC) Telescope*, November 27, 1832.

24 *Southern Times and State Gazette* (Columbia, SC), January 22, 1831.

25 *Pendleton (SC) Messenger*, September 14, 1831.

26 *Camden and Lancaster (SC) Beacon*, September 6, 1831 (emphasis in original).

27 Jefferson and Madison drafted the Kentucky and Virginia Resolutions in response to the Alien and Sedition Acts. Passed by Federalists who felt threatened by French-sympathizing Republican opposition and French immigrants, the Alien and Sedition

Acts authorized the deportation of aliens deemed dangerous to American peace and safety, extended the duration of residence required to become an American citizen, and made it illegal to publish criticisms of certain government officials. For detailed discussions of the acts and subsequent resolutions, see Douglas Bradburn, "A Clamor in the Public Mind: Opposition to the Alien and Sedition Acts," *William & Mary Quarterly* 65 (2008): 565–600.

28 *Pendleton* (SC) *Messenger*, May 2, 1832, May 30, 1832, October 31, 1832.
29 Sidney, "Nullification," *Charleston Mercury*, March 24, 1832; *Charleston Mercury*, July 2, 1832; *Pendleton* (SC) *Messenger*, March 28, 1832, April 4, 1832, and July 11, 1832; *Camden and Lancaster* (SC) *Beacon*, April 3, 1832.
30 Rutledge, "Nullification," *Charleston Courier*, May 3, 1832.
31 *Greenville* (SC) *Mountaineer*, April 7, 1832, May 5, 1832, June 23, 1832.
32 *Charleston Courier*, April 25, 1832, June 15, 1832, July 11, 1832 (emphasis in original).
33 "The Federal Judiciary," *Greenville* (SC) *Mountaineer*, February 27, 1830; "State Sovereignty," *Greenville* (SC) *Mountaineer*, April 23, 1830; "Livingston's Speech," *Greenville* (SC) *Mountaineer*, May 28, 1830; "The Power of Nullification," *Greenville* (SC) *Mountaineer*, April 3, 1830; "This is the correct view of the whole matter," *Greenville* (SC) *Mountaineer*, July 16, 1830.
34 *Greenville* (SC) *Mountaineer*, August 4, 1832, and September 22, 1832; *Camden* (SC) *Journal*, quoted in *Charleston Courier*, August 1, 1832.
35 *Southern Times* (Columbia, SC), May 20, 1830, May 27, 1830; *Charleston Mercury*, May 29, 1830, January 22, 1831, March 7, 1832, and July 2, 1832; *Pendleton* (SC) *Messenger*, January 5, 1831, and April 4, 1832; *Camden and Lancaster* (SC) *Beacon*, February 21, 1832, April 3, 1832.
36 "The Crisis," *Camden and Lancaster* (SC) *Beacon*, February 28, 1832.
37 *Charleston Mercury*, February 19, 1830.
38 *Savannah* (GA) *Republican*, quoted in *Athenian* (Athens, GA), November 8, 1831.
39 *Federal Union* (Milledgeville, GA), July 24, 1830; *Georgia Statesman* (Milledgeville), February 27, 1830; *Macon* (GA) *Telegraph*, October 29, 1831.
40 "Gov. Troup's Letter," *Athenian* (Athens, GA), October 12, 1830. See also *Southern Banner* (Athens, GA), September 7, 1832, November 9, 1832.
41 Edwin A. Miles, "After John Marshall's Decision: Worcester v. Georgia and the Nullification Crisis," *Journal of Southern History* 39 (1973): 519–44.
42 See, for example, *Newbern* (NC) *Spectator*, April 13, 1832, and *Natchez* (MS) *Weekly Courier*, July 13, 1832.
43 *Fayetteville* (NC) *Observer*, quoted in *Newbern* (NC) *Spectator*, March 27, 1830.
44 *Newbern* (NC) *Spectator*, May 1, 1830.
45 "The Ultras," *Richmond* (VA) *Enquirer*, May 8, 1832 (quotes from the *Fredericksburg* (VA) *Arena* and *Charlottesville* (VA) *Advocate*).
46 *Newbern* (NC) *Spectator*, May 1, 1830, September 11, 1830.
47 "The Mercury Again," *Newbern* (NC) *Spectator*, July 10, 1830. An editorial in the September 25, 1830 issue of the *Newbern* (NC) *Spectator* noted that North Carolina's prominent newspapers—specifically the *Raleigh Star*, *Salisbury Journal*, *Fayetteville Observer*, and *Examiner* (likely Fayetteville)—had spoken out against the *Mercury*.
48 See *Natchez* (MS) *Weekly Democrat*, July 31, 1830, September 19, 1830.
49 Robert V. Remini, "Henry Clay and Natchez Connection," *Journal of Mississippi History* 54 (1992): 269–78.
50 David O. Whitten, "Tariff and Profit in the Antebellum Louisiana Sugar Industry," *Business History Review* 44 (1970): 226–28.

51 "The West and the South," *Natchez* (MS) *Weekly Democrat*, July 31, 1830; "Letter from a Gentleman in New Orleans," *Natchez* (MS) *Weekly Courier*, April 30, 1831; "Anti-Tariff Meeting," *Southern Clarion* (Natchez, MS), quoted in *Vicksburg* (MS) *Advocate & Register*, September 30, 1831.

52 "Nullification," *The Mississippian* (Vicksburg), January 16, 1832, and January 30, 1832; *Time's Tablet & Mississippi Gazette* (Natchez), November 10, 1830, and November 17, 1830; *Natchez* (MS) *Weekly Democrat*, December 25, 1830; *Alabama Journal* (Montgomery), March 31, 1832; "Congress—The Tariff," *Spirit of the Age* (Tuscaloosa, AL), March 14, 1832; *Mobile* (AL) *Commercial Register*, August 11, 1832.

53 "Nullification," *The Mississippian* (Vicksburg), January 16, 1832.

54 *Alabama State Intelligencer* (Tuscaloosa), quoted in *Newbern* (NC) *Spectator*, December 1, 1832.

55 Arkansas, Florida, and Texas were not yet states by 1833.

56 Newspapers outside of South Carolina alluded to newspapers in their state that upheld the doctrine, but extant issues were not available to review their contents for confirmation. The importance of such confirmation is demonstrated in editorials from the *Natchez* (MS) *Weekly Courier*, which falsely labeled the *Woodville* (MS) *Democrat* as a nullification paper and then had to retract the accusation; *Natchez* (MS) *Weekly Courier*, December 9, 1831, December 31, 1831, January 20, 1832. Though the *Courier* editor appeared to have made an honest mistake, such accusations in other newspapers may have been a means to insult an opposing newspaper.

5

ABOLITIONIST EDITORS

Pushing the Boundaries of Freedom's Forum

David W. Bulla

In the decade and a half before the American Civil War, three journalists took aim at what they considered to be the most serious problem facing the nation: slavery. William Lloyd Garrison in the *Liberator*, Frederick Douglass in *The North Star*, *Frederick Douglass' Paper*, and *Douglass' Monthly*, and Gamaliel Bailey in the *National Era* made the case for ending slavery, even though it was permitted by the US Constitution. Trying to coax the United States into eradicating slavery the way Mexico had in 1829 and Great Britain had in 1832, these three editors used their advocacy newspapers to persuade the nation to change its ways. They attracted the wrath of not only the slaveholders in the South but also conservatives in the North.

Leading up to this period, abolitionist editors in the North had seen mob violence destroy their newspapers, and in one case take the life of an editor, Elijah P. Lovejoy of Alton, Illinois. Furthermore, legislators in Southern states attempted to enact laws that would make it illegal to criticize slavery. This chapter examines how all three editors—who did not necessarily see eye to eye on how to reach the goal of slavery's extinction—pushed the parameters of press freedom forward through advocacy journalism at a time when the top legal authority—the United States Supreme Court—had yet to make any landmark decisions on the extent of such freedom.

Garrison and Douglass were accomplished orators, and Bailey's highly successful newspaper in Washington, DC, consistently defended the right to petition and fought long-standing congressional gag orders against petitions for abolition that made it illegal to even broach the subject on Capitol Hill. This chapter looks at editorials the three men wrote from 1847 to 1857, focusing on how each responded to policy issues related to the Mexican War, the Wilmot Proviso, the Compromise of 1850, the Kansas-Nebraska Act in 1854, the adoption of the Lecompton Constitution in

Kansas in 1857, and the *Dred Scott v. Sandford* US Supreme Court Decision, also in 1857. Their takes on these policy issues were well outside the mainstream of the political press of mid-century. They not only put the issue of slavery firmly in the public agenda, they also expanded the forum for freedom by suggesting revolutionary political action in the pages of their newspapers.

The War with Mexico and the Wilmot Proviso

Born a slave in Maryland in 1817 and later escaping to freedom in Massachusetts, Frederick Douglass began publishing *The North Star* in Rochester, New York, in November 1847. The orator and journalist would oversee a publication that would reach an average circulation of 3,000 a week; the newspaper would last until 1851 before being replaced by *Frederick Douglass' Paper*.[1] Douglass called the war with Mexico ("our sister republic") "disgraceful, cruel, and iniquitous" and despaired for a speedy ending to the conflict.[2] "The friends of peace," he wrote, "have nothing to hope for" from either political party, castigating the Democrats for starting it and the Whigs for supplying it.[3] Douglass called it a "slaveholding crusade," criticizing the political opposition for not questioning the ultimate purpose of President James K. Polk's war.[4]

William Lloyd Garrison also felt the war was "atrocious" and that "its real object, the extension and preservation of slavery, no intelligent man honestly doubts."[5] A key issue during the war was the defeat of the Wilmot Proviso (Rep. David Wilmot's proposed amendment to an appropriations bill), which would have made it illegal for any land captured during the Mexican War to become slave territory. In *The Liberator* (which would reach a peak circulation of 3,000), Garrison called the amendment's rejection a "temporary defeat of freedom," but said that eventually there would be an "everlasting victory over her slaveholding foes."[6]

Douglass welcomed Wilmot's amendment because he felt it put the idea of freedom foremost in the mind of Americans and that it "indicated a great principle in the national heart."[7] In response to the failure of the Wilmot Proviso, Douglass, in *The North Star*, took aim at allies of slavery in the North, saying they were "servile dough-faces." He also said those legislators who were willing to have the next presidential election contested on the basis of the arguments in the Wilmot Proviso were too few in number, too young, and not in positions of leadership in their parties. "They are the weak against the wrong," he added.[8]

In January 1847, Gamaliel Bailey began the *National Era* newspaper in Washington, DC, in order to shine a light on slavery, because many in the South felt that the institution was "exempt from investigation, discussion, opposition."[9] The paper would reach a peak circulation of 28,000. Bailey, who as an abolitionist editor in Cincinnati had faced death threats over his anti-war stance, noted that foreigners would say "Americans are tolerant upon all questions save that of slavery."[10]

Bailey emphasized that the abolition of slavery would have to come about lawfully within the boundaries of the Constitution—which Garrison said was impossible because the founding document was protective of slavery. Bailey added that he hoped to reach the mind of the South—"to disarm prejudice, correct misconception, and win respectful attention."[11]

Bailey's opposition to the war was strong. He would call any Northern congressman who opposed the Wilmot Proviso a traitor. However, Bailey praised the vigorous discussion the bill had produced. When it passed the House, the editor wrote, "It ought to be the subject of rejoicing in all the land; for if slavery be an evil, what wise man, what good man, could wish to multiply such an evil?"[12]

In May 1847, Bailey would print resolutions from Massachusetts Congressman Edward L. Keyes pinning the main cause of the war on three objectives: (1) the extension of slavery into the conquered lands; (2) of strengthening slavery's political support; and (3) the reduction of Northern political power by the slave power. Keyes called the latter "a tyranny" and the annexation of the new territory gained in the president's war as being "inconsistent with the well-being of the Union."[13]

In July 1847, Garrison printed in *The Liberator* an open letter to President Polk from Francis Jackson, president of the Massachusetts Anti-Slavery Society. In the letter, Jackson encouraged the president to emancipate his slaves because "no greater sin can be committed against God" than owning slaves. Jackson called slavery "man-stealing" and accused Polk of "kidnapping human beings." The letter writer wondered how Polk could be a Christian and accept slavery.[14]

In September 1847, Garrison tried to put the conflict into philosophical relief. He said that there could be "no genuine Union between Good and Evil," and that the Constitution, as good as it was, was conceived with one major flaw: the allowance of slavery in the American system. In his estimation, the evil side of American history had been victorious in the first decades of the nation's existence. There had been attempts to "put the monster on short allowance, but they have failed." Garrison reckoned that the Mexican War would expand slavery into the newly conquered territory and "create a boundless slave market."[15]

Garrison was not only anti-war; he would assert he was pro-Mexico. When General Winfield Scott neared Mexico City and appeared ready to capture the capital, the *Liberator* chief wrote that he had hoped the Mexicans would have "mustered enough strength" at last to give Scott "a good licking."[16] Garrison did not want the American general or his soldiers killed, but to be captured and held prisoner for a while.

The US Senate ratified the Treaty of Guadalupe Hidalgo on March 10, 1848. The terms of the treaty resulted in Mexico giving 525,000 square miles of its territory to the United States in exchange for $15 million. This represented what would today be the entire southwestern part of the continental United States. The treaty also held that the US government would take over $3.25 million in debts Mexico owed to American citizens. Garrison reported that the treaty put "no restriction on the Slave Power," nor did it appear there was "any attempt" to include "the Wilmot Proviso with the territorial acquisition." He concluded: "once more the Slave Power wields the resources and the power of the nation to accomplish its own diabolical purposes."[17]

Douglass also opposed the Treaty of Guadalupe Hidalgo. He wrote: "In our judgment, those who have all along been loudly in favor of a vigorous prosecution of the war, and heralding its bloody triumphs with apparent rapture, and in glorifying the atrocious deeds of barbarous heroism on the part of wicked men engaged in it, have no sincere love of peace, and are not now rejoicing over peace, but plunder. They have succeeded in robbing Mexico of her territory, and are rejoicing over their success under the hypocritical pretense of a regard for peace."[18]

Coverage of the Compromise of 1850

The next major issue galvanizing the abolitionist press was the Compromise of 1850. The compromise came in five different bills passed by Congress in September 1850. The compromise, the brainchild of Whig senator Henry Clay of Kentucky and Democratic senator Stephen A. Douglas of Illinois, resulted in California being admitted as a free state and Texas giving up on its claim to New Mexico, which along with Utah would be allowed to decide its slavery status under popular sovereignty. The slave trade was banned in Washington, DC, and the Fugitive Slave Law was strengthened.

Garrison found little that he liked in the Compromise, railing most intensely against the new Fugitive Slave Law. He wrote on the front page of *The Liberator* that it should be "resisted, disobeyed, and trampled under foot, at all hazards."[19] Garrison also ran the commentary of several newspapers from across the nation on the Fugitive Slave Law. The *True Wesleyan* (Boston, MA) said that every Northern legislator who voted for the law or abstained from the vote should be "arraigned and indicted at the bar of public opinion."[20] The editor of *Zion's Herald* (Boston, MA) lamented the loss of the spirit of humanity and liberty in those who passed the law.[21]

Bailey took more of an informative approach to the Compromise. He published the votes, such as the one admitting California into the Union. He noted that two-thirds of the "Southern members voted against admission" and that if California had had a tradition of slaveholding, then the Southerners would have voted for its admission.[22]

Douglass saved his venom for Kentucky's Henry Clay, one of the authors of the Compromise. Douglass called him the "crafty" Clay, the man with "soft and gentle diction" who came to the deliberations on the various laws with a spirit of compromise. He said that Clay and his colleagues' bills, "like all Southern compromises," "gives everything to Liberty, *in words*, and secures everything to Slavery, *in deeds*."[23]

Coverage of the Kansas-Nebraska Act

The Louisiana Purchase of 1803 had granted the United States vast territories in the West. The Missouri Compromise of 1820 had prevented the creation of new slave states above latitude 36°30'. What would become the huge Nebraska Territory was clearly north of the Missouri Compromise line. In January 1854, Illinois Democratic senator Stephen A. Douglas introduced a bill that divided the Nebraska Territory into two territories. Douglas interjected the concept of popular sovereignty by proposing that the settlers of the new Kansas and Nebraska territories should decide democratically whether to come into the Union as free or slave.

Bailey saw Douglas's law as one that would "convert into Slavery Territory a vast tract of land which for thirty-three years has been consecrated by American Law to Freedom and Free Labor."[24] If the freemen of the North accepted this bill, it would mean their damnation, Bailey argued. Douglass blazed away against the Kansas-Nebraska Act, castigating "the audacious villainy of the slave power, and the contemptible pusillanimity of the North."[25]

After both houses passed the Kansas-Nebraska Act and only a few days before President Franklin Pierce signed it in May 1854, Garrison wrote: "The deed is done, and the Slave Power is again victorious." He went on to write that the legislation countered "the laws of God and the rights of universal man."[26] For Garrison, the Kansas-Nebraska Act was a turning point. Now he began to publish stories with the words "disunion" and "agitation" in the headlines. It was now a time of action, not inaction. The week after his strong response to the passing of Kansas-Nebraska, Garrison noted in an article titled "Disunion at the North" that abolitionist Wendell Phillips was now openly talking about changing the government—that the Union would be saved only if the abolitionists were successful. Six years before Abraham Lincoln's consequential election, Garrison was beginning to see that civil war was coming.[27]

Lecompton and Dred Scott

The next major policy issue was the proposed Lecompton Constitution in Kansas in 1857. It took a proslavery stance, in large part because the territorial legislature had a majority of slave advocates, even though most settlers in Kansas were pro-free labor. The proposed Lecompton Constitution (named for the town that served as a temporary capital of Kansas) stipulated that the property of slave owners would be protected once Kansas became a state and introduced a referendum on allowing more slaves to enter the territory upon the commencement of statehood.

Bailey jumped on Lecompton, covering the Grasshopper Falls mass meeting of free state supporters in his September 1857 editions. In that conference, the speakers denounced the territorial legislature. In an article that originated in the *Chicago Tribune*, the Grasshopper Falls attendees denounced "a portion of the people of Missouri" who invaded Kansas in 1855 and "established the oligarchy which has since claimed to exercise the functions of government amongst us."[28]

In August 1859, Douglass chastised Kansas for having gone through so many constitutions as a young state. He noted that the black man was the theme that led to all these constitutions. The Negro, he wrote, "balks every exertion, defeats every plan, and baffles all their wisdom. He is the rock of offence, the stone of stumbling, and the severest test of all their political skill." He went on to say there was little difference between the two main parties on Lecompton. He commented that the Democrats had declared themselves the enemies of the black man, and the Republicans "have not declared themselves our friends."[29]

The last policy issue examined here was the US Supreme Court case *Dred Scott v. Sandford*, which had wound through the courts for years and was eventually decided in March 1857. Scott, a black slave who had been taken by his owners from a slave state to a free state and free territories, wanted his freedom and sued for it. In a 7–2 decision, the court, led by Chief Justice Roger B. Taney, held that "a negro, whose ancestors were imported into [the US], and sold as slaves," whether he was now free or a slave, was not legally an American citizen and accordingly could not sue in federal court. In other words, the court ruled that Scott could not sue for his freedom. Moreover, the federal government had no power to regulate slavery in the territories acquired after the original creation of the United States.

Douglass said the "devilish decision" obviously came from "the Slaveholding wing of the Supreme Court." Now Congress could no longer "prohibit slavery anywhere." He said the "National Conscience will be put to sleep by such an open, glaring, and scandalous tissue of lies as that decision."[30] Douglass said the Dred Scott case was a wakeup call, and it was "another proof that God does not mean that we shall go to sleep, and forget that we are a slaveholding nation."[31]

In Washington, Bailey had been intimately involved in the suit, helping pay Scott's legal expenses. Bailey saw nothing but ominous tones from the decision and told his readers it was time for action—that they now must turn to the ballot box for a remedy. The *National Era* editor declared it was time for the voters of the North to no longer be "political slaves of the slaveholders." An abolitionist revolution was now at hand, he held.[32]

In the *Liberator*, Garrison called the court's decision "infamous and tyrannical." He praised the two dissenting judges, stating that they alone had protested against the "usurpations and encroachments of slavery."[33]

Throughout the period examined here, the three abolitionist journalists maintained their steady coverage of slavery, slaveholders, and the federal and state policies that served the interest of the institution. Indeed, Bailey, Douglass, and Garrison were not merely criticizing an economic system protected by the Constitution; by 1857, they were contemplating an alternative America, one that would be free of slavery perpetually. Douglass believed that this would not happen without violence. The Maryland native and former slave had been kidnapped by political opponents and knew full well the extent to which those opponents would go to silence him and other abolitionist journalists. Douglass repeatedly told his readers that violence was inevitable in this struggle and that abolitionists would have to use "the proper means, fight with the right weapons."[34]

Garrison had lost a libel suit when he insulted a proslavery man and spent time in jail when he lived in Baltimore. He knew violence would be at the core of whatever ultimately decided the fate of slavery.

In the mid-1850s, disunion became a word that Garrison and the abolitionist editors began to use with some frequency, although they did not all agree on how it should happen. For example, in March 1857, the *Liberator* began to print articles about petition calls for Massachusetts to secede from the Union in response to Lecompton, Dred Scott, and Kansas-Nebraska. In response to the Dred Scott verdict, Garrison wrote: "Give us Disunion with liberty and a good conscience, rather than Union with slavery and moral degradation. What! Shall we shake hands with those who buy, sell, torture, and horribly imbrute their fellow-creatures, and trade in human flesh! God forbid!"[35]

The more moderate Bailey, whose audience included non-slaveholders in the South, was not as extreme in his position on the solution to ending slavery. Bailey's main goal was dialogue. Yet the Washington editor, too, expected mob action against his newspaper.[36] Bailey would not live to see John Brown, the election of Lincoln, the coming of the Civil War, emancipation, or the Thirteenth Amendment. He would die in June 1859 at the age of 51 on a journey to Europe, where he hoped to get some rest from his journalistic battle against slavery.

The war of words that Bailey, Douglass, and Garrison waged from 1847 until 1857 proved to be active, informal defenses of freedom of the press, even as some of their opponents argued that they had overstepped their boundaries and that the abolitionists' degree of government criticism was unacceptable. By injecting the conversation about ending slavery into nineteenth-century political discourse in the United States, they actively expanded press freedom; that is, these journalists reinforced the protections for publishing guaranteed in the First Amendment through their deeds. These three abolitionists bravely upheld freedom's forum and provided a place in public to discuss a noxious topic in a revolutionary way. In fact, abolitionist journalism was not just nudging freedom's forum; it was pushing its gate fully open. Garrison even believed that he should give the other side an opportunity to air its point of view. He ran a column titled "Refuge of Oppression" where he would publish articles by proslavery newspapers, usually from the South.

Perhaps Douglass, the former slave and master rhetorician, expressed the need for a free forum best in his lecture about freedom of speech, printed in Garrison's *Liberator* in December 1860: "To suppress free speech is a double wrong. It violates the right of the hearer as well as those of the speaker. It is just as criminal to rob a man of his right to speak and hear as it would be to rob him of his money."[37]

Notes

1   Carter R. Bryan, *Negro Journalism in America before Emancipation,* Journalism Monographs, no. 12 (Lexington, KY: Association for Education in Journalism, 1969), 22.

2 "The War with Mexico," *The North Star* (Rochester, NY), January 21, 1848, in Philip S. Foner, *The Life and Writings of Frederick Douglass: Early Years, 1817–1849* (New York: International Publishers, 1950), 292.

3 Ibid., 293.

4 Ibid., 293, 295.

5 "Letter on the Mexican–American War," William Lloyd Garrison to Richard Daniel Webb, July 1, 1847, *Teaching American History*, accessed September 28, 2017, http://teachingamericanhistory.org/library/document/letter-on-the-mexican-american-war/.

6 "Slavery Again Triumphant," *The Liberator* (Boston, MA), March 12, 1847, 42.

7 Frederick Douglass, quoted in Wu Jin-Ping, *Frederick Douglass and the Black Liberation Movement: The North Star of American Blacks* (New York: Garland Publishing, 2000), 49.

8 "The North and the Presidency," *The North Star* (Rochester, NY), March 17, 1848, in Foner, *The Life and Writings of Frederick Douglass*, 299.

9 Ford Risley, *Abolition and the Press: The Moral Struggle against Slavery* (Evanston, IL: Northwestern University Press, 2008), 103.

10 Ibid., 111.

11 "Introductory," *The National Era* (Washington, DC), January 7, 1847, 2.

12 "The Wilmot Proviso," *The National Era* (Washington, DC), February 25, 1847, 2.

13 "The Mexican War—Massachusetts," *The National Era* (Washington, DC), May 13, 1847, 2.

14 "To James K. Polk, President of the United States," *The Liberator* (Boston, MA), July 2, 1847, 106.

15 "The United States and Mexico," *The Liberator* (Boston, MA), October 15, 1847, 166.

16 "The New Conquest of Mexico," *The Liberator* (Boston, MA), September 17, 1847, 150.

17 "Ratification of the Treaty," *The Liberator* (Boston, MA), March 17, 1848, 42.

18 "Peace! Peace! Peace," *The North Star* (Rochester, NY), March 17, 1848, in Foner, *The Life and Writings of Frederick Douglass*, 300.

19 "The Fugitive Slave Bill," *The Liberator* (Boston, MA), September 27, 1850, 154.

20 "The Law for Catching Men," *The Liberator* (Boston, MA), September 27, 1850, 155.

21 "The Slave Controversy—Perversion of Public Sentiment," *The Liberator* (Boston, MA), September 27, 1850, 155.

22 "The California Bill—the Votes," *The National Era* (Washington, DC), September 26, 1850, 2.

23 "Henry Clay and Slavery," *The North Star* (Rochester, NY), February 8, 1850, in Foner, *Frederick Douglass: Selected Speeches and Writings* (Chicago, IL: Lawrence Hill Books, 1999), 153–54.

24 "The Nebraska Bill—Agitation in Prospect," *The National Era* (Washington, DC), January 12, 1854, 2.

25 "The End of Compromises with Slavery—Now and Forever," *Frederick Douglass' Paper*, May 26, 1854.

26 "The Nebraska Bill Passed—Another Triumph of the Slave Power," *The Liberator* (Boston, MA), May 26, 1854, 82.

27 "Remarks of Mr. Garrison," *The Liberator* (Boston, MA), March 12, 1858, 43.

28 "From Kansas," *The National Era* (Washington, DC), September 24, 1857, 1.

29 Douglass, "The Kansas Constitutional Convention," in Foner, *The Life and Writings of Frederick Douglass*, 451.

30 Douglass, "The Dred Scott Decision," in Foner, *The Life and Writings of Frederick Douglass*, 410–11.

31 Ibid., 412.

32 "The Supreme Court and Slavery—The Duty before Us," *The National Era* (Washington, DC), March 12, 1857, 2.

33 "The Decision of the Supreme Court," *The Liberator* (Boston, MA), March 13, 1857, 42.

34 Douglass, "The Doom of the Black Power," in Foner, *The Life and Writings of Frederick Douglass*, 362.

35 "Remarks of Mr. Garrison," *The Liberator* (Boston, MA), March 12, 1858, 43.

36 Risley, *Abolition and the Press*, 107.

37 "Frederick Douglass at Music Hall," *The Liberator* (Boston, MA), December 14, 1860, 199.

6

WHEN THE PEN GIVES WAY TO THE SWORD

Editorial Violence in the Nineteenth Century

Abigail G. Mullen

In 1839, Frederick Marryat sardonically observed in his *Diary in America*, "The newspaper press is the most mischievous, in consequence of its daily circulation, the violence of political animosity, and the want of respectability in a large proportion of the editors."[1] In this statement, Marryat points to three important features of American newspaper culture: strong ties to the political community, the rigors of a newspaper job (though most newspapers did not in fact have daily circulation), and the question of respectability amongst editors. As well as pointing to unique aspects of newspaper culture in the nineteenth century, these three features represent major pressure points for newspaper editors.[2] The way they responded to job pressures demonstrates their commitment to their political and journalistic ideals.

Many editors lasted less than five years in the job. Very few editors made a career at one newspaper. Even men who thrived as editors tended to move frequently, as in the case of the Locke brothers, who moved from the *Hancock Jeffersonian* (Findlay, OH) to the *Tiffin (OH) Tribune* to the *Toledo Daily Blade*, or Julius Orrin Converse, who purportedly edited twelve different newspapers throughout his life. Many also died in the business, some by violent means, but most just of illness or old age. All of them faced similar pressures throughout their tenure as editors.

Political Pressures

Newspapers were political organs, and as such their editors were political beasts. Journalism historian Patricia Dooley reports that between 1800 and 1830, approximately 50 percent of newspapers in the United States were affiliated with a political party in some way, whether officially or unofficially.[3] However, the research conducted for this chapter found that partisanship seems to have increased during the years after 1830. Only fourteen of a survey of eighty-three newspapers published between 1836 and 1860—16 percent—did *not* have a clear partisan affiliation, and of those fourteen, two were abolitionist, which is in itself something of a partisan affiliation. Though most editors claimed independence from a particular party, the masthead of *Brownlow's Knoxville Whig* summed up the political position of most newspapers: "Independent in All Things, Neutral in Nothing."[4]

Many editors had explicit personal connections to politics, before, during, or after their tenure at the newspaper. One of the most famous examples is Clement Vallandigham, who edited the *Dayton* (OH) *Daily Empire* before becoming the leader of the radical Copperhead faction of the Democratic Party. Parson William Brownlow had to hand over the reins of *Brownlow's Knoxville Whig* to his son when he was elected governor of the state of Tennessee in 1865. James H. Birch, of the *Western Monitor* (Fayette, MO), actually held a seat in the Missouri Senate while remaining the editor of his paper. After a political shift to Whig, Birch received an appointment as register of lands in 1843, which he could not manage along with the paper—so he left the paper.[5] Many other editors, having demonstrated their commitment to their party through their newspapers, received similar political appointments or were elected to political office. Editors labored under the pressures of their own political ambitions, but political pressure sometimes came from external sources as well.

Political opposition could have drastic effects: editors were forced out of their jobs, papers were forced to close, and a number of editors were actually killed. Sometimes an editor's own party opposed him politically. In 1837, Democratic leaders in Nashville removed Samuel Laughlin from the editorship of the Democratic *Nashville Union* for incompetence.[6]

Some papers faced political opposition writ large. The *Democratic Banner* of Bowling Green, Missouri, found itself without funding after many people in the county started voting Whig in 1846. The *Banner* was able to scrape by until 1852, when it had to close its doors. In Ohio, when the government threatened to shut down Democratic newspapers, the *Holmes County Farmer* replied, "For every democratic paper that is suppressed by the tyrants the people will suppress an abolition paper."[7] The newspaper promised to fight political fire with fire.

Violence against newspapers and their editors was often politically motivated. Cassius Clay, editor of the emancipationist newspaper *The True American* (Fayette, KY), expected violence and so he mounted two cannons in his newspaper office.[8] Similarly, in 1856, a raid known as the "Sack of Lawrence" resulted in the destruction of the office of the *Kansas Herald of Freedom*, an abolitionist newspaper, while its editor, G. W. Brown, was in prison in another town.[9] Though abolitionism was a touchstone for the worst violence, abolitionist papers were not the only victims. The *Bangor* (ME) *Democrat* was destroyed in 1861 by a mob who did not appreciate the *Democrat*'s political views. In 1862, the editor of the *Dayton* (OH) *Daily Empire*, a Copperhead newspaper, was murdered by a member of the Republican Party.[10]

Newspapers' close proximity to each other both polarized political opinions and allowed for easy physical contact between rival editors. Nearly every town that had a newspaper had two—one for each political party. Larger towns, such as Nashville, had even more papers, with multiple papers representing a spectrum of political ideas. When the *Sunbury American* was started in 1840, it was the third Democratic paper in a Pennsylvania county of only approximately 20,000.[11] This type of physical and political proximity led to a great deal of sniping between newspapers.

In cases where a town had multiple newspapers, it was not always just the opposing political parties that feuded, but sometimes papers that shared political ideals. Since newspapers with the same basic politics were fighting for the same readers, the pressure to distinguish one's newspaper was much more intense. In 1859, a dispute between two such editors ended in extreme violence. The *Nashville Union*'s editor, George Poindexter, got into a fight in the streets with Allen A. Hall of the Nashville *Daily News*, in an extension of a print war between the two editors. After the confrontation in the street, Poindexter headed back to his office. Later that evening, Hall found Poindexter in his office and shot him with a gun loaded with buckshot. The fight in which Poindexter was killed was over slavery—but the question was not whether one was proslavery and the other antislavery, but rather whether Poindexter's proslavery views were radical enough. Both Poindexter and Hall were Democrats.

In the case of a dispute between Henry Rives Pollard and Robert Kelley, again, the sides were not proslavery versus antislavery, but rather proslavery versus radical proslavery. Papers, and thus editorships, were so closely linked to politics that editors had to be willing to passionately defend their views—even to their own party members—in order to get the subsidies from their party that allowed them to stay afloat.[12] As editor Weston F. Birch remarked, "'To be beaten upon principles can be endured,' but to compromise was too much for 'a man of feeling to risk.'"[13]

While actual violence was rare, editors could be extraordinarily insulting in print. For instance, the *Kansas Weekly Herald* (Leavenworth) described the editor of the *Squatter Sovereign* (Atchison, KS) as "the low, silly, garrulous numbskull of the *Squatter Sovereign*, yclept Kelley—the contemptible, whining, blind puppy of Atchison, that answers to the name of 'Bob.'"[14] James Birch and Nathaniel Patten, editors in Howard County, Missouri, engaged in a print war that lasted three years until Patten finally gave in and packed his bags.[15]

Physical Pressures

A second pressure point for editors was simply the hardship of the job. The newspaper trade posed incredible financial difficulties, as well as problems with machinery, supplies, and communications. Out of financial necessity, nearly every newspaper published regular pleas for subscribers to pay their fees.[16]

Natural disasters such as fire or flood were always concerns, and many news-paper offices were damaged or even destroyed at some point. One newspaper lost an estimated $25,000 when its office and presses burned to the ground.[17]

Some newspapers did not even have offices at first: the *Freemen's Champion* (Prairie City, KS) started in a tent, and the *Kansas Weekly Herald* (Leavenworth) issued its first papers under an elm tree—the newspaper actually predated the establishment of the town.[18] Newspapers in the Western territories struggled to keep a steady supply of paper: Arizona and Utah papers often had to wait weeks for their paper shipments, and sometimes the shipments never arrived.[19] Editors had to be able to withstand extreme adversity just to keep the presses running.

Newspaper editors responded to physical pressures in different ways. Often, they packed up and moved to another paper. In his valedictory editorial, Clark H. Green, of the *Glasgow* (MO) *Weekly Times*, described his reason for suspending the paper this way: "a general depression in business, the impossibility of making collections, and the fact that but little is doing of which money can be realized."[20] Green never returned to the paper. Others found their health destroyed by the arduous labor and had to retire. Still others practiced multiple professions at once, as in the case of George Grenville and George Wyllys Benedict of the *Burlington* (VT) *Free Press*, who practiced law while operating the paper.

Social Pressures

In order to maintain their readers, editors had to walk a fine line between maintaining their own integrity and pleasing their often-demanding constituency. One editor summed up the ethical dilemma this way: "From the hour he started his paper to the present time he has been solicited to lie upon every subject, and can't remember ever having told a wholesome truth, without diminishing his sub-scription list, or making an enemy."[21]

These concerns about morality and ethics feed into a larger concern: respect-ability. The *Nashville Union and American* described the fallen George Poindexter as "gentle as a child," a description which seems inappropriate considering that he got into a street brawl with a rival editor. But it also stated that "Life itself he was ready at any moment to lay down rather than submit to a stigma upon his honor."[22] While not all editors were fractious or abrasive, questions of honor and respect-ability were always at the forefront of their minds. Whatever their background, editors were often viewed as lower class, not gentlemen enough to be challenged to a duel by a gentleman, but fit to be caned.[23] Some editors were indeed working class—Samuel P. Ivins, for example, was the foreman on the *Knoxville Standard*'s press before becoming editor of the *Athens* (TN) *Post* in 1848.[24] But others were actually from respectable upper-class families and were trained in law or other "respectable" professions.[25] Nevertheless, with their status in question, editors were sensitive about their public image. Historian Ryan Chamberlain argues that editors felt tremendous community pressure to uphold a code of honor, which led to disputes and duels.[26]

The time-honored practice of dueling helped many editors settle questions of respectability amongst themselves. Though the origins of their quarrels may have been political, the way these editors chose to resolve their disagreements indicates that the issue was personal honor, not party politics. Many challenges were issued that never actually came to a duel, as in the case of the Atchison *Squatter Sovereign*'s Robert S. Kelley and the *Kansas Weekly Herald*'s Henry Rives Pollard.[27] In other cases, the duels were fought without harm to either party. A duel between Edward E. Cross and Sylvester Mowry resulted in no injury, but Cross declared himself defeated and relinquished control of the *Weekly Arizonian* (Tubac) to Mowry.[28] But some editors were injured or even killed in duels.[29]

Of course, not every editor settled his disputes in a gentlemanly fashion. In 1861, a fight between R. C. Satterlee of the *Leavenworth* (KS) *Herald* and Daniel Read Anthony, younger brother of Susan B. Anthony and editor of the Leavenworth *Conservative*, roughly resembled a wild Western shootout in the middle of town. Satterlee was killed.[30] Anthony avoided a murder conviction because Satterlee shot first.

The three pressure points—political, physical, and social—did not occur singly. Every editor had to deal with all of them every day, and greater pressure in one area exerted pressure in other areas as well. The combination of these pressures can help explain why there was such high turnover in the job. But the fact that many editors moved on to other papers indicates that they did not see the odds as insurmountable on the whole—merely that their ideals could be better realized elsewhere. They were committed to their political and social goals enough to put themselves under the same pressures over and over again. In fact, the mobility of these editors as they sought for the perfect editorial job allowed them to create and shape American culture across a wide area.

Notes

Author's note: The research for this chapter was done under the auspices of Viral Texts, a project of the NULab for Texts, Maps, and Networks at Northeastern University. The author is especially grateful for the assistance and encouragement of project directors Ryan Cordell, Elizabeth Maddock Dillon, and David Smith.

1 Frederick Marryat, *Diary in America, Series Two*, Project Gutenberg (2007), www.gutenberg.org/files/23138/23138-h/23138-h.htm.

2 Nearly all the papers referenced in this article are part of the Library of Congress's *Chronicling America* data set of digitized newspapers beginning in 1836. The data set has significant holes: major publishing centers such as Boston and Philadelphia are completely absent, and New York is woefully incomplete. For that reason, examples come mostly from newspapers not in major cities, but rather in places where the creation and upkeep of a newspaper was no small feat, where supplies such as paper and spare parts could not be taken for granted.

3 Patricia L. Dooley, *Taking Their Political Place: Journalists and the Making of an Occupation* (Westport, CT: Greenwood Press, 1997), 82–83. Dooley found that from 1870–1900, only 35 percent of newspapers declared a particular affiliation. (Dooley's sample size

for papers from 1800–1830 was one hundred and twenty newspapers; the 1870–1900 sample was eighty-six papers.)

4 See the prospectus for *Brownlow's Knoxville Whig and Rebel Ventilator*, September 14, 1864.

5 Jerry E. Wilson, "James H. Birch," *Dictionary of Missouri Biography*, ed. Lawrence O. Christenson (Columbia: University of Missouri Press, 1999), 74.

6 Paul H. Bergeron and Jeanette Keith, *Tennesseans and Their History* (Knoxville: University of Tennessee Press, 1999), 96.

7 *Holmes County Farmer* (Millersburg, OH), August 18, 1864.

8 Gene Murray, "The Lion's Roar: Cassius Clay's *The True American*," in *The Civil War and the Press*, ed. David B. Sachsman, S. Kittrell Rushing, and Debra Reddin Van Tuyll (Piscataway, NJ: Transaction Publishers, 2000), 112.

9 Pearl T. Ponce, *Kansas's War: The Civil War in Documents* (Athens: Ohio University Press, 2011), 14.

10 *Dayton* (OH) *Daily Empire*, November 13, 1862.

11 "Total Population by County, 1840 United States Census," *National Historical Geographic Information System: Version 2.0* (Minneapolis, MN: University of Minnesota, 2011).

12 This is not a universal statement. There were a few papers that managed to maintain political neutrality, or independence, such as the *Cincinnati* (OH) *Daily Press*, which barely ever made its political views known. But most 'neutral' papers ended up taking a political stance within a few years of their beginning. The *Democratic Banner* (Bowling Green, MO) itself started out politically neutral as the *Radical*, but within two years, under pressure from the community, editor James D. Henderson started supporting the Democratic Party.

13 William Henry Lyon, *The Pioneer Editor in Missouri, 1808–1860* (Columbia, MO: University of Missouri Press, 1965), 64. Birch was editor of the Fayette *Monitor*.

14 *Kansas Weekly Herald* (Leavenworth), June 1, 1855.

15 Lyon, *The Pioneer Editor in Missouri*, 66.

16 For example: *Clearfield* (PA) *Republican*, December 25, 1854; the *Examiner* (Louisville, KY), February 11, 1840, which begged four hundred subscribers to pay their fees; and the *Columbia Democrat* (Bloomsburg, PA), February 10, 1838.

17 *Evening Star* (Washington, DC), February 12, 1869. Other examples include the *Palladium* (Richmond, IN), recorded in the *Perrysburg* (OH) *Journal*, September 2, 1854; the *Alta California* (San Francisco), recorded in the *Mountain Sentinel* (Ebensburg, PA), August 14, 1851; the *Express* (Petersburg, VA), recorded in the *Cleveland* (OH) *Daily Leader*, June 14, 1866.

18 *Kansas Weekly Herald* (Leavenworth), September 15, 1854; notes for *Freemen's Champion*, Library of Congress, *Chronicling America*, accessed April 4, 2018, https://chroniclingamerica.loc.gov/lccn/sn95063180/.

19 This lack of paper caused the demise of the Salt Lake City paper *The Mountaineer*, whose last issue announces a suspension of publication due to lack of paper.

20 *Glasgow* (MO) *Weekly Times*, August 22, 1861.

21 *St. Cloud* (MN) *Democrat*, March 12, 1863.

22 *Nashville* (TN) *Union and American*, November 22, 1859.

23 Joanne B. Freeman, *Affairs of Honor: National Politics in the New Republic* (New Haven, CT: Yale University Press, 2001), 172.

24 *East Tennessee: Historical and Biographical* (Chattanooga, TN: A.D. Smith and Co., 1893), 162.

25 For example: Julius Orrin Converse of the *Jeffersonian Democrat* (Chardon, OH); Lucian J. Eastin of the *Kansas Weekly Herald* (Leavenworth); and Edward Bobo Murray of the *Anderson* (SC) *Intelligencer*. Joseph Patterson Smith, *History of the Republican Party in Ohio* (Lewis Publishing Company, 1898), 408; Walter Williams, *A History of Northwest Missouri* (Lewis Publishing Company, 1915), 750; *The Record of Sigma Alpha Epsilon* (Sigma Alpha Epsilon, 1894), 228.

26 Ryan Chamberlain, *Pistols, Politics and the Press: Dueling in 19th Century American Journalism* (Jefferson, NC: McFarland, 2009), 72.

27 The *Kansas Weekly Herald* (Leavenworth) reported that the dispute had been settled through the intervention of the principals' friends, June 29, 1855.

28 Several articles from the *Weekly Arizonian* (Tubac) of July 14, 1859, address the duel and its fallout.

29 See Freeman, *Affairs of Honor*, 172.

30 See the *Independent* (Oskaloosa, KS), June 19, 1861.

7

AN EDITORIAL HOUSE DIVIDED

The Texas Press Response to the Compromise of 1850

Mary M. Cronin

The Compromise of 1850 averted civil war for a decade, but the nine-month long congressional debate roiled Southerners, and Texans in particular, and unleashed yet again vigorous and very public discussions about the soul and direction of the American nation. The United States was embroiled in conflict, as decades of festering tensions revolving around slavery, states' rights, and the South's growing estrangement from the Union came to the fore in the aftermath of the war with Mexico.

The discourse that ensued mirrored earlier arguments about the expansion of slavery into newly acquired Western territory, a concern that had implications for the balance of power between Northern and Southern states. That previous, highly charged rhetoric ultimately produced the Missouri Compromise in 1820, which attempted to maintain the balance of power between the North and the South in Congress by admitting Missouri as a slave state and Maine as a free-soil state, and banning slavery in the Louisiana Purchase territory north of the 36°30' latitude line.[1]

The acquisition of new Western land following the war with Mexico again threatened to upset the balance of power between free and slave states. It also set up a border dispute between the US government and Texas, a dispute that Congress attempted to resolve as part of the compromise legislation. Texans had claimed since 1836 that their state's western border was the Rio Grande, a claim which, if acknowledged by the federal government, would have placed a significant portion of New Mexico territory, including the city of Santa Fe, under Texas's control.[2] The Treaty of Guadalupe Hidalgo, signed on February 2, 1848, ended the war with Mexico and provided the United States with a vast amount of new territory, but both the Zachary Taylor and Millard Fillmore administrations had steadfastly refused to acknowledge Texans' boundary claims—claims that the Texans themselves had been unable to enforce.[3] By 1850, tensions were so high that some fire-eating politicians and editors were calling for war with the US government as the only means of establishing Texas's boundaries.[4]

Texas's Whig and Democrat editors avidly followed the congressional debate to forge a compromise and attempted to lead the discourse within their home state regarding the various compromise proposals. These editorial discussions reveal both deep political schisms and common ground between the state's Democratic and Whig journalists. Equally important, the debate also allowed Texas's editors, fire-eaters, and moderates alike, to establish their positions on Union membership, secession, states' rights, and the possibility of civil war—issues that would continue throughout the 1850s, despite the passage of the compromise.

This chapter provides an overview of the compromise and outlines partisan approaches on a national level, looking at samples of editorials from Texas newspapers in particular. This study of Texas's peculiar interests at the time reveals the sectionalism that had affected newspapers throughout the United States. Specifically, it reveals that the Taylor administration's political missteps and support for policies unfavorable to Southerners allowed fire-eaters the opportunity to dominate political discourse during the summer of 1850. However, a combination of rising moderate voices in Washington, DC, and throughout the South, as well as the realization that new President Millard Fillmore would not back down on threats of force to protect New Mexico territory from Texans, ultimately allowed Texas's moderate editors to shift the narrative toward compromise by stressing the benefits of remaining with the United States.[5]

The editorial and political discourse also demonstrates that in the decade before the war, there was no unanimous call for disunion by Texas's editors. On the whole, Texas's editors largely supported remaining within the Union. However, they expressed their perceptions that Texans were giving more than they were receiving from their nation, their fears that abolitionists increasingly dominated Congress, and their concerns that the federal government was disrespectful of Southern honor and the Southern way of life.

The Missouri Compromise of 1820 had not been centered solely on the issue of slavery. Larger economic, social, and political concerns had also entered into the debate. In a period marked by growing national prosperity, the 1820 Compromise legislation had focused on "security, property, and power."[6] Over the next three decades, sectional concerns about those issues continued to grow. During the Mexican-American War, rising abolitionist sentiment in the North directly confronted Southern demands for the expansion of slavery into the land that would be ceded by the Mexican government. Many Southerners were convinced that abolitionists were growing in power in Washington, DC, and that they could eventually put an end to Southern society as Southerners knew it.[7]

The push by Northern politicians to keep slavery out of the West increased national tensions as Southerners felt increasingly threatened, isolated, and disrespected by what they saw as Northern machinations aimed at reducing

Southern political power.[8] With the Mexican War over and millions of acres available for settlement, the issues of slavery, secession, and states' rights once again came to a head. As historian John C. Waugh has stated, "The South had come to believe by 1849 that its political identity and destiny were riding on the struggle for more such [slave] states. Political equilibrium had come to mean equilibrium for slavery."[9]

Texans were as deeply concerned about the eventual shape of any compromise legislation in 1850 as were other Southerners—and perhaps more so. Southerners largely settled Texas, thus the state's citizens were as supportive of the "peculiar institution," as the rest of the South, even though only 30 percent of all Texas families owned slaves.[10] Proof of Texans' devotion to its economic system could be seen at the ballot box. The wealthy planter class, who owned 72 percent of the state's real property, dominated politics.[11] But, the potential expansion of slavery in the West was just one aspect of the compromise talks that concerned Texans. An equally pressing, and intertwined, issue was the final determination of Texas's borders.

The Compromise of 1850

Both national prosperity and regional discontent had increased during the 1840s. As Joel H. Sibley has stated, "Partisan conflict was intense, elections were bitterly fought everywhere, and the hostility between members of the opposing parties carried over beyond the elections into subsequent relations in Congress and elsewhere."[12] President James K. Polk's expansionist dream of obtaining California for the US was realized following the war with Mexico. The 1848 Treaty of Guadalupe Hidalgo ceded Mexico's northern territories to the United States, but the land transfer led to federal handwringing over Texas's boundaries, as well as to strident arguments about whether or not slavery would be permitted in the newly acquired territory.[13] The treaty did not address Texans' claims that the state's boundaries included a large swath of New Mexico territory that was located east of the Rio Grande.[14]

Congressional infighting further complicated the situation. Members of the House of Representatives voted sixty-three times before selecting a Speaker in December 1849. With ten Free Soil members (who opposed the introduction of slavery in new Western territories), neither Whigs nor Democrats held a clear majority.[15] With the House divided, new President Zachary Taylor lobbied Congress to admit California as a free state and encouraged New Mexicans to begin the process of applying for statehood.[16] Taylor hoped that the Texas–New Mexico boundary crisis would be resolved by the nation's courts if New Mexico was admitted as a state. The response from Southerners was both fierce and fearful, as some threatened secession. The nation's sectional equilibrium was in jeopardy. Although Taylor was a slaveholder, he put the Union first. Stating that he would brook no talk of disunion, the president said of the nation, "…I shall stand by it and maintain it in its integrity."[17]

As historian David Potter states, Southerners faced "a crucial choice: they must somehow stabilize their position in the Union, with safeguards to preserve the

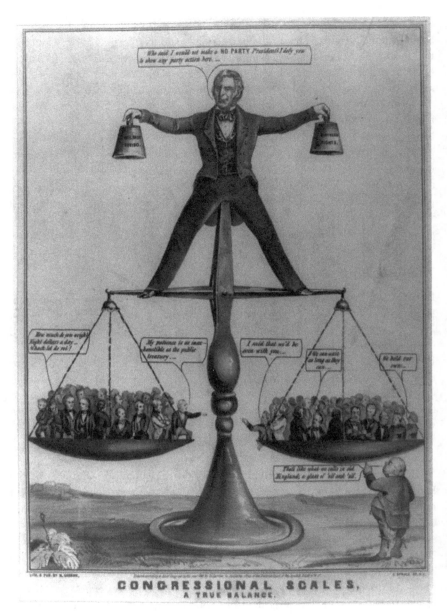

FIGURE 7.1 "Congressional scales, a true balance," lithograph published by Currier & Ives, 1850. This satirical cartoon shows President Zachary Taylor striving to balance opposing viewpoints on slavery between the North and South and depicts the struggle that occurred in Congress to meet an agreement on key issues relating to the expansion of slavery into Western territory. Eventually, a series of five bills—known as the Compromise of 1850—were passed that sought to settle these issues. In the cartoon, Taylor stands on

security of the slave system, or they must secede before their minority position made them impotent....They now wanted, therefore, not just a settlement of the territorial question, but a broad sectional adjustment."[18] The president, therefore, faced a national crisis. Not surprisingly, subsequent debates by members of Congress, the general public, and the press, to forge a compromise were filled with highly charged, emotional rhetoric. The addition of more free-soil states and territories would upset the delicate political and social balance that had been established with the Missouri Compromise of 1820. The discovery of gold in California, coupled with Californians' petition to become a free state, further heightened political tensions.

As he did in 1820, Henry Clay, the man known as the "Great Compromiser," once again attempted to forge consensus. In January 1850, Clay and Daniel Webster developed a series of five resolutions aimed at keeping the nation intact.[19] Members of Congress spent much of 1850—302 exhausting days—proposing and debating Clay's plan, as well as a series of other proposals, including those developed by senators Thomas Hart Benton, John Bell, and James Pearce. With the help of Stephen A. Douglas, enough votes were obtained to pass the legislation.[20]

The eventual compromise was comprised of a series of five bills. Collectively, the legislation allowed for the question of the expansion of slavery into Western territories to be decided by popular sovereignty by the citizens of those new Western states and territories. California was admitted as a free state; territorial governments were established for New Mexico and Utah; the slave trade was abolished in the District of Columbia; Texas's boundary claims were settled in exchange for $10 million for Texans to pay off their debts; and, in a concession to Southerners, a strict fugitive slave act was established that required Northern citizens to assist in the apprehension of runaway slaves. By September 17, 1850, all of the bills had been passed and the crisis of 1850 was publicly proclaimed to have been resolved.[21]

Texas, New Mexico, and the Boundary Crisis

The political and sectional wrangling over the issue of expanding slavery into the land acquired from Mexico began long before the war had even concluded. Pennsylvania Congressman David Wilmot had fired the first salvo into the

---

Caption for Figure 7.1 (Cont.)

a set of scales, holding a weight in each hand. The left weight reads "Wilmot Proviso" (a proposed amendment to an appropriations bill that would have banned slavery from land captured during the Mexican War), while the right weight reads "Southern Rights." On the equally balanced scales, several members of Congress are grouped. On the left scale, Senator Henry Clay, a primary architect and supporter of the compromise resolutions, is visible. On the right, senators Lewis Cass and John Calhoun are depicted. (From the Library of Congress Prints and Photographs Division Washington, DC. <www.loc.gov/item/90716208/> Reproduction Number: LC-USZ62-8230 (b&w film copy neg.). Currier & Ives, "Congressional scales, a true balance.")

national debate regarding slavery in the American West on August 8, 1846, a mere four months into the war with Mexico. The freshman congressman attached a one-sentence amendment to President James Polk's request for $2 million to persuade the Mexican government to part with their northern territories. Seeking to keep any new Western lands free for white settlers to thrive, Wilmot's "proviso" attempted to block slavery from any land purchased or seized from Mexico. Wilmot's amendment to Polk's funding request, which was given a stamp of approval by many Northern Democrats, was passed by the House of Representatives, but failed in the Senate. Although the congressional attempt to ban slavery in any newly acquired Western territories died, the fact that Congress had attempted to take legislative action on slavery led to an outcry by Southern politicians, members of the press, and the public. A growing chorus of fire-eaters across the South began demanding secession.[22]

While regional factions of Whigs and Democrats wrestled to construct compromise legislation, Texans supported expanding slavery throughout the West and hoped for a quick and positive resolution of the issue. But the slavery issue was intertwined with the state's border issue. Texans viewed the signing of the Treaty of Guadalupe Hidalgo as a victory—initially. The treaty stated that the Rio Grande would serve as the boundary between the US and Mexico, an agreement that Texans believed would allow them to extend their influence, and slavery, into the eastern section of New Mexico—land that Texans had claimed since 1836. Santa Fe residents rejected the Texans' claims, however, as did the commander of federal troops in Santa Fe.[23]

Texas's citizens angrily denounced what they perceived as treacherous machinations from the federal government, US troops, and New Mexico's citizens. If Texans' long-standing claim to the land was not honored, slavery could not be expanded, and the specter of New Mexico being admitted to the Union as a free state or territory meant that the South's slave states would be encircled by their free-soil counterparts.[24] As compromise talks began, Texas's editors realized that new President Taylor was no friend of the Lone Star State.[25]

Partisanship and the Antebellum Texas Press

Editorial debate about the various compromise proposals was loud and vigorous throughout Texas during 1850. Few shied away from registering their opinions on the ongoing debate, and many proposed solutions to the crisis. However, the state's editors were far from unified. Some editors were ardent Unionists who pinned their hopes on the federal government for an amicable compromise, while others argued for Texas to secede, or even go to war in order to preserve the institution of slavery and protect what Texans perceived were their rightful borders.

The state's editors took democracy seriously. They regularly reprinted politicians' speeches and put forth factual information when it could be obtained, telling readers they should make up their own minds on issues. But, such claims didn't stop editors from also trying to lead public opinion via their own editorial viewpoints on the political, social, and economic concerns of the day.

At the time, eight Texas publications supported the Democratic Party, four editors proclaimed their support for the Whig party, four publishers declared themselves and their journals politically neutral, and two journals (according to historian Marilyn McAdams Sibley) "defied classification because of their equivocal stands."[26]

Texas's Democratic editors did not present a united front during the months of debate. Some considered themselves John C. Calhoun Democrats who put states' rights and the protection of the institution of slavery at the forefront of their concerns. Others were Andrew Jackson Democrats who sought first and foremost to protect the Union.[27]

Texas's Whig editors were outnumbered by their Democratic colleagues, but they had influence, serving as the voice of the influential business class in Texas's growing cities. Although they were less ultra-Southern in their positions on sectional issues, they were conscious that their fellow Texans were not tolerant of anything that smacked of abolition.[28] For example, the prospectus put forth by Robert H. Howard, the editor of the Galveston *Semi-Weekly Journal*, reflected the reality that Howard was a member of a minority political party in a largely Democratic-leaning state. Howard said in March 1850 that he recognized Democratic publications outnumbered Whig ones in Texas, but that nonetheless, he would support "the leading doctrines of the Whig party." Carefully explaining his political leanings on sectional issues, he stated: "…we will at the same time hold southern institutions and southern interests paramount to any mere party tenets," while also promising that the newspaper would be a "zealous advocate" of Texas's dignity and honor.[29]

Howard's careful prospectus clearly took into account the views of other outspoken editors whose political allegiances were with the Democratic Party. For example, Robert W. Loughery, a Calhoun Democrat and the co-editor of the Marshall-based *Texas Republican*, while not a fire-eater, was very public in his views that anyone who did not support slavery should not remain in a slave state.[30] Such statements led some Whig editors to switch political allegiances in the late 1840s as tensions heated up between slavery and free-soil supporters. For example, Austin-based editors John S. Ford and Michael Cronican discontinued their Whig newspaper, the *National Register*, which had lobbied against Texas's annexation by the United States, and established a new, Democratic-supporting journal, the *Texas Democrat*.

Many of the state's editors remained suspicious of their Whig-leaning colleague's intentions. Editor A. W. Canfield of the San Augustine-based *Red-Lander* publicly stated that "no true friend of Texas" could be a member of the Whig Party.[31] Not surprisingly, then, Texas's successful Whig editors often urged citizens to vote for a particular candidate or issue, rather than for the party.

Coverage of the Compromise in Texas Newspapers

Speeches and motions by members of Congress regarding potential compromise legislation crowded out much of the other news in Texas's newspapers during the first nine months of 1850. Editors put forth their opinions on all aspects of the compromise bills, but the boundary dispute spurred the most pugnacious responses by the state's politicians, journalists, and citizens, none of whom wanted to lose acreage to Free Soilers.[32]

Texas's editors agreed on what they saw as the key issues. States' rights, federal recognition of Texas's boundary claims, the value of Union membership, and concerns about congressional and presidential disrespect of Southern honor were vigorously discussed within the pages of the Texas press. The rhetoric became so heated that the fire-eating editor of the Austin-based *Texas State Gazette*, William H. Cushney, evoked the memory of Texans' fight for independence at San Jacinto, to openly encourage war with federal troops stationed in Santa Fe, rather than cede claims to part of New Mexico territory as part of the compromise.[33]

Other editors in Texas, including Francis Moore, Jr. of the *Democratic Telegraph and Texas Register* in Houston and Charles DeMorse of the Clarksville *Northern Standard*, attempted to be voices of reason and moderation for much of the debate. Both men saw great benefits for their state's development if Texas remained in the Union, and both men hoped that members of Congress would work out an "advantageous compromise."[34]

Texas's politically moderate editors were heartened by Massachusetts Senator Daniel Webster's speech to the US Senate on March 7. A reprint of the speech took up half of the space in the March 22 issue of the Galveston *Semi-Weekly Journal*.[35] Although Webster denounced Southern talk of secession, the senator's plea for compromise proved something of a salve to many Southerners. Editor Robert H. Howard, in a carefully worded editorial told readers that he had excluded most of the week's news in favor of publishing Webster's full speech. "It is a document that merits a most careful perusal, and to Texans in particular, from the deep interest of their state in the politics of the day, it is doubly interesting," Howard stated.[36]

Even the fiery Editor Cushney of the *Texas State Gazette*, who was not yet at the point of calling for war or secession, praised Webster's speech, pronouncing it "manly, frank, fearless, and truthful." He saw the speech as politically restorative and told readers so: "If a better feeling be not restored between the North and South by this speech, if a better spirit is not awakened in Congress by it, we shall almost despair of such a consummation."[37]

Strong views on the Union and nationhood further complicated the editorial debate. Texans viewed the Union in emotional terms. The nation "was a thing of the heart" to Texans who respected the blood that was shed during the Revolutionary War. The state's citizens recognized that the US also served an important role as a protector, as well as a source of much-needed resources.[38]

Thus, moderate editors like Francis Moore remained watchful of events in Congress, yet publicly worried that Henry Clay's compromise proposal was failing Southern interests.[39]

President Taylor's desire to admit New Mexico as a free state gave Texas's fire-eaters an opportunity to express their outrage while making loud demands for military action.[40] Cushney of the *Texas State Gazette* called on citizens to resort to a "force of arms" to regain their honor and preserve their boundary claim.[41] To Cushney, the state's interests came first. He supported his state-centric viewpoint by telling readers they had given up much to join the Union and had received little in return, including no protection from Indian attacks. The outspoken editor said no compromise was acceptable if it surrendered "blood-bought soil" to Free Soilers.[42]

The editors of the *Democratic Telegraph and Texas Register* and the *Semi-Weekly Star* agreed. W. H. Ewing of the *Semi-Weekly Star* called for force to be used against New Mexicans and US soldiers stationed there to prove "that Texas is not to be trifled with."[43] The *Telegraph*'s editor denounced the president's proposal as an attack on states' rights: "Such a flagrant act of usurpation and tyranny on the part of the General Government has never before [been] known."[44]

Other editors were less certain about going to war, but they agreed that Texans' honor had been besmirched by the "fanatics" in the nation's capital. Editor P. Cordova of the Austin *South-Western American* supported his state's boundary, but told readers he hoped that the threats of force would be enough to make Congress and the president understand the justness of the Texans' cause.[45] Editor Benjamin Neal of the *Nueces Valley* agreed that Texans had been treated with "indignity" by the president and Congress, whom he called despots, but, while calling for action, Neal tempered his response by making clear that the "action" involved a meeting of citizens to determine a unified response to the latest federal outrage against Texans' honor.[46]

The moderate Charles DeMorse maintained that the time was not yet ripe for military action. He encouraged Texans to give Congress more time to work out a compromise, although he admitted that the president and many members of Congress appeared to harbor "a most unchristian hatred" toward Texas."[47] The Whig-leaning editors of the Victoria *Texian Advocate* and the Galveston *Semi-Weekly Journal* agreed with DeMorse, holding out hope that the president would adopt a conciliatory policy toward Texas.[48]

Cooling Temperatures and the Resurgence of Moderate Voices

James Pearce unveiled a compromise plan with terms more favorable toward Southerners on August 5. At the same time, the realization that the new Millard Fillmore administration was not about to back down regarding the status of New Mexico lowered the tempers of Texans, which, in turn, brought a level of moderation back to the public narrative on the compromise legislation. The editor of the Houston *Democratic Telegraph and Texas Register* admitted that the time for talk of violence was over.[49]

The editor of the *Semi-Weekly Star* resigned himself to the news that Congress would most likely approve the bills that formed the compromise, but he warned that the legislation was unfair to Texas, "inconsistent with historical truth," and contradictory. The future tranquility of the state was uncertain, he said, even with a compromise in place.[50] Although the *Star*'s editor argued against any compromise, stating that signing such a deal would "forge the chains of oppression for posterity," he admitted that the Pearce plan offered Texans needed money to pay off the state's debt. "We must weigh dollars against honor, expediency against principles, the pocket patriotism of citizens is about to be called to requisition, and we may find only the spurious coin of Union conservatives," he cautioned.[51]

The news that President Fillmore had signed off on all of the compromise legislation was slow to reach Texans, so the debate continued, but the focus was shifting toward acceptance—if, in some quarters, grudgingly—that a compromise was on the horizon. For example, a letter published in the September 26, 1850, issue of the *Texian Advocate* who identified himself by the moniker Old Reality, praised the congressional passage of the compromise legislation, stating that the $10 million given by Congress for payment of Texas's debts was an acceptable trade-off for Texans' willingness to fix its northern and western boundaries.

Despite months of war-like rhetoric by fire-eaters, the letter's author admitted Texans had never exercised jurisdiction over Santa Fe. And, he reminded readers that Texas "entered the Union because she loved the Union." Texans, the writer opined, received a worthy and honorable deal. He portrayed New Mexico's eastern territory as "poor, barren, timberless…waste" that was unfit for whites. The letter's author, steeped in the era's prevailing racial views, called the territory's population "miserable," noting it consisted of "disappointed demagoguing [*sic*] Americans," Indians, and Mexicans. The author noted with appreciation that the compromise legislation left Texas an "American" state. "Our state will grow up *strictly American* in feeling. Think of that, fellow citizens! Not only so, but *Southern* in feeling also," he gushed.[52]

Fire-eaters were not cowed into silence, however. The September 21, 1850, issue of the *Texas State Gazette* contained carefully selected excerpts of articles from newspapers throughout the state that gave the erroneous impression that Texas's editors were united in their unwillingness to give up Santa Fe in exchange for $10 million. One such excerpt, from the *Nueces Valley* stated: "Seas of blood would have to be waded through before any such project as the use of force to compel Texas to surrender her claims and yield up the Santa Fe territory, however feasible it may appear, could be consummated, even by the powerful nation of the United States. Fierce and angry would be the contest, before Texas will submit to an invasion of her rights." Despite the article's grandiloquence, the editor said secession, rather than war, was preferable to "dishonorable submission to an unjust encroachment…."

Another excerpt, from the Galveston *News*, revealed Editor Willard Richardson's concerns that a "collision" might occur after the president had denied Texans their rights to possess eastern New Mexico. "Can it be seriously declared that a President of the United States will have a recourse to arms, to prevent a sovereign State of this Union, from enforcing its laws within its own limits?" Such an act, Richardson stated, was tyranny and he called on the state legislature to preserve Texans' honor.

The Matagorda *Tribune* and the *Star State Patriot* agreed. The editor of the *Tribune* called the compromise legislation "that idol of the Abolitionists" and called on Texans to stop procrastinating and assert "her just claims...which none but a despotic and unjust power would dare deny, or attempt to withhold." The *Star State Patriot* pronounced Texas "most grossly insulted, most outrageously treated by the Executive in attempting to wrest from her the territory to which she has an undoubted right. If she should tamely submit, and crouch spaniel like at the feet of her oppressor, what guarantee has she that her rights will be vindicated and her wrongs redressed?" He added that when constitutional rights are respected, the state of Texas would remain faithful to the Union, but warned, "touch her liberty and you arouse the rattlesnake."[53]

A Compromise Achieved and a Looming Battle

Moderate voices ultimately prevailed in the short term and averted war. The Pearce bill, with its promise of $10 million for Texans to pay their old debts in exchange for settling their boundary claim, ultimately proved acceptable to most Texans. The editors of *The Northern Standard*, the *Texian Advocate*, the *Democratic Telegraph and Register*, and the *Civilian* stayed the course in promoting acceptance of the compromise, stating it was essential to Texas's peace, prosperity, and the preservation of the Union. The *Texian Advocate*'s editors called the compromise "fair and honorable" in their sales pitch to readers, before adding that almost $3 million would be left over after the state's debts had been paid. That money could then be used to improve the state's schools and infrastructure.[54] The two San Antonio newspapers also supported the compromise, as did three publications once opposed to the legislation.[55] Editor DeMorse confidently announced on October 12 in the Clarksville *Northern Standard* that the Red River area would "roll up a cool 500 majority" of votes in favor of approving the compromise legislation.[56]

A number of Texas editors remained outspoken in their rejection of the Compromise of 1850. Both Austin newspapers, the Galveston *News*, the *Texas Monument*, the Washington *Lone Star*, the Matagorda *Tribune*, and the *Marshall Republican* all remained opposed to Texans' acceptance of the compromise, arguing that slaveholders' rights would be curbed and that Texans had been disrespected. The editor of the La Grange-based *Texas Monument* printed the compromise legislation in full in the October 2, 1850, issue of his newspaper, then asked that readers not allow partisan considerations to color their personal views on whether or not to support the congressional legislation. "Every true Texian should stand by his State, without asking who is for, or who is against her," the editor opined. Although he acknowledged that the final legislation passed by Congress was more favorable to Texas than any of the previous compromise plans, upon full and rational consideration, the editor encouraged citizens to vote against the measure, arguing that Congress had not treated Texans with "courtesy and justice" in resolving the border issue.[57]

The editor of the *Texas State Gazette* also remained opposed to the end. Two articles in the October 26 issue denounced compromise. One of the articles, a letter purportedly written by a man who had fought at San Jacinto, evoked the memory of that battle in calling for rejection of a compromise that the author said denied Texans their rights and was penned by "a few vagabond fanatics of the North." Texans, the author said, "should proudly hurl back in the teeth of the President the mean and cowardly threats made against her and brand with scorn the disgraceful and foul proposition made by our Congressmen. Texas did not so gallantly break the lance with the Mexican Despot, to disgrace herself so soon in the sisterhood of which she is now one of the proudest and brightest members."[58]

Despite the editorial outcry, Texas voters approved the legislation by a margin of three to one. The legislature then approved an act of acceptance, which Governor Bell signed on November 25, 1850. Although many Texans felt that the immediate crisis had been resolved, in reality the Compromise of 1850 merely slowed the eventual march toward civil war. The issues raised from January through September would continue to resonate throughout the rest of the decade. Growing Southern discontent and resistance to free-soil supporters made the specter of secession loom.

Notes

1 John C. Waugh, *On The Brink of Civil War: The Compromise of 1850 and How it Changed the Course of American History* (Wilmington, DE: SR Books, 2003), 14; Senate, 16th Cong., 1st sess., *Annals of Congress*, 367–426.

2 Mark J. Stegmaier, *Texas, New Mexico, and The Compromise of 1850: Boundary Dispute and Sectional Crisis* (Kent, OH: Kent State University Press, 1996), 8–20; Holman Hamilton, *Prologue to Conflict: The Crisis and Compromise of 1850* (Lexington: University Press of Kentucky, 1964), 17–18, 20.

3 Scott A. Silverstone, *Divided Union: The Politics of War in the Early American Republic* (Ithaca, NY: Cornell University Press, 2004), 184; K. Jack Bauer, *Zachary Taylor: Soldier, Planter, Statesman of the Old Southwest* (Baton Rouge: Louisiana State University Press, 1985), 292–93.

4 Bauer, *Zachary Taylor*, 309; Randolph B. Campbell, *Gone to Texas: The History of the Lone Star State* (New York: Oxford University Press, 2003), 234.

5 On Fillmore and his views on compromise, the issue of slavery, and his enmity toward the Whig faction led by William H. Seward, see Hamilton, *Prologue to Conflict*, 107–09; Robert J. Rayback, *Millard Fillmore: Biography of a President* (Newtown, CT: American Political Biography Press, 1992), 219–23.

6   Robert Pierce Forbes, *The Missouri Compromise and Its Aftermath: Slavery and the Meaning of America* (University of North Carolina Press, 2009), 5.

7   Silverstone, *Divided Union*, 188–89; David M. Potter, *The Impending Crisis, 1848–1861* (New York: Harper Torch Books, 1976), 93.

8   Hamilton, *Prologue to Conflict*, 48; Potter, *The Impending Crisis*, 93.

9   Waugh, *On The Brink of Civil War*, 14.

10 Campbell, *An Empire for Slavery: The Peculiar Institution in Texas, 1821–1865* (Baton Rouge: Louisiana State University Press, 1989), 209–10; T. R. Fehrenbach, *Lone Star: A History of Texas and Texans* (New York: Da Capo Press, 2000), 279.

11 Campbell, *An Empire for Slavery*, 209–10; William Earl Weeks, *Building the Continental Empire: American Expansion from the Revolution to the Civil War* (Chicago: Ivan R. Dee, Inc., 1996), 52, 87–93; Steven E. Woodward, *Manifest Destinies: America's Westward Expansion and the Road to the Civil War* (New York: Vintage Books, 2011), 111.

12 Joel H. Sibley, *The Partisan Imperative: The Dynamics of American Politics Before the Civil War* (New York: Oxford University Press, 1985), 40.

13 Richard Griswold del Castillo, *The Treaty of Guadalupe Hidalgo: A Legacy of Conflict* (Norman: University of Oklahoma Press, 1992), 3, 13.

14 Griswold Del Castillo, *The Treaty of Guadalupe Hidalgo*, 8–14.

15 Potter, *The Impending Crisis*, 90.

16 Bauer, *Zachary Taylor*, 298.

17 Quoted in Potter, *The Impending Crisis*, 91.

18 Ibid., 94.

19 Bauer, *Zachary Taylor*, 301.

20 Waugh, *On the Brink of Civil War*, 178–79, 182–84; Roger A. Griffin, "Compromise of 1850," *Texas State Historical Association*, accessed July 13, 2017, www.tshaonline.org/handbook/online/articles/nbc02.

21 Hamilton, *Prologue to Conflict*, 156–65.

22 Campbell, *Gone to Texas*, 233; Waugh, *On the Brink of Civil War*, 19–30.

23 Wallace, *Charles DeMorse: Pioneer Statesman and Father of Texas Journalism* (Kessinger Publishing, LLC, 2010), 107; Executive Documents, 29th Cong., 2nd sess., *Congressional Globe* no. 19, 6.

24 Stegmaier, "Zachary Taylor versus the South," *Civil War History* 33, no. 3 (1987): 219–20.

25 Ibid., 228; William J. Cooper, Jr., *The South and the Politics of Slavery, 1828–1856* (Baton Rouge: Louisiana State University Press, 1978), 276–77.

26 Marilyn McAdams Sibley, *Lone Stars and State Gazettes: Texas Newspapers before the Civil War* (Texas A&M University Press, 2000), 240.

27 Ibid., 260; Cooper, *The South and the Politics of Slavery*, 253.

28 Campbell, *An Empire for Slavery*, 214, 218.

29 "Prospectus of the Journal," *Semi-Weekly Journal*, March 22, 1850.

30 Sibley, *Lone Stars and State Gazettes*, 260; Campbell, *An Empire for Slavery*, 223; James Marten, *Texas Divided: Loyalty and Dissent in the Lone Star State, 1856–1874* (Lexington: University Press of Kentucky, 1990), 12–13.

31 *Red-Lander*, November 13, 1845.

32 "The Sale of Our Territory," *Texas State Gazette*, June 29, 1850. The article was a reprint from the Galveston *Journal*.

33 *Texas State Gazette*, December 1, 1849, and December 5, 1849.

34 "Affairs at Washington," *Democratic Telegraph and Texas Register*, May 9, 1850; Wallace, *Charles DeMorse*, 71–73, 109–10.

35 "Speech of Mr. Webster, of Massachusetts," *Semi-Weekly News*, March 22, 1850.

36 *Semi-Weekly Journal*, March 22, 1850.
37 "Mr. Webster's Speech," *Texas State Gazette*, March 30, 1850.
38 Walter L. Buenger, *Secession and the Union in Texas* (Austin: University of Texas Press, 1984), 42–43.
39 *Democratic Telegraph and Texas Register*, March 14, 1850.
40 Bauer, *Zachary Taylor*, 293–95; "Proclamation," *Democratic Telegraph and Texas Register*, July 4, 1850.
41 "The Issue," *Texas State Gazette*, June 8, 1850.
42 "The Compromise!" *Texas State Gazette*, June 1, 1850.
43 *Semi-Weekly Star*, July 11, 1850.
44 "Government Usurpation," *Democratic Telegraph and Texas Register*, June 20, 1850.
45 "Gov. Bell's Message," *South-Western American*, August 16, 1850.
46 "The Governor's Proclamation," *The Nueces Valley*, July 13, 1850.
47 "Excitement at Austin," *The Northern Standard*, July 13, 1850.
48 "The Texas Boundary Question," *Texian Advocate*, August 16, 1850.
49 "The Boundary Question," *Democratic Telegraph and Texas Register*, August 21, 1850.
50 "The President's Message," *Semi-Weekly Star*, September 9, 1850.
51 Ibid.
52 "The Ten Million—Peace," *Texian Advocate*, September 26, 1850.
53 "Spirit of the Texas Press on the Santa Fe Question," *Texas State Gazette*, September 21, 1850.
54 *Texian Advocate*, September 26, 1850.
55 Stegmaier, *Texas, New Mexico, and the Comprise of 1850*, 303, 308.
56 "Something for the Sovereigns," *The Northern Standard*, October 12, 1850.
57 "The Texas Boundary," *The Texas Monument*, October 2, 1850. The *Texas Monument* was produced by a committee who claimed the publication was "devoted to no sect in religion, or party in politics" ("Prospectus of the Texas Monument," *Texas Monument*, October 2, 1850.).
58 "The Boundaries" and "Letter from a Soldier of San Jacinto," *Texas State Gazette*, October 26, 1850.

8

"THE GOOD OLD CAUSE"

The Fugitive Slave Law and Revolutionary Rhetoric in *The Boston Daily Commonwealth*

Nicole C. Livengood

The *Boston Daily Commonwealth* was founded in January 1851 to serve as the organ of the anti-slavery Free Soil Party, and to mobilize support for Charles Sumner, the Free Soil Party candidate for senator (who was chosen in April 1851 by the Massachusetts state legislature, after more than twenty ballots). With the goal of unifying and energizing its readers to redeem the Commonwealth's shame of having to return escaped slaves, as required by the Fugitive Slave Law of 1850, the newspaper consistently harnessed the post-Revolutionary generation's use of Revolutionary tropes and symbols as it responded to the geographical and political fragmentation that preceded the Civil War.

The *Commonwealth's* prospectus forecast its ideological investments and its rhetorical strategies. The editors announced that the paper would not be a "bond-servant of any cause, or party, except that of Freedom, Truth, and Humanity. The POLE STAR toward which it will ever point will be The RIGHT; but the right of ALL."[1]

The words "bond-servant" and "POLE STAR" asserted the *Commonwealth's* investment in the abolition of slavery, communicated its kinship to the Revolutionary generation and its privileging of higher law, and subtly chastised the merchants and Whigs (especially native son Daniel Webster) who had prioritized money and politics over principles by supporting the Fugitive Slave Law of 1850. The *Commonwealth's* emphasis on "Freedom, Truth, and Humanity" was also an early indication of its strategic use of the memory of "those who crossed the ocean on the Mayflower for the rights of conscience and fought for freedom at Bunker Hill."[2] Moreover, it constructed a local and regional "grand master narrative" that placed its audience "within the broader story of human progress."[3]

The *Commonwealth's* rhetorical strategies can be delineated by focusing on its response to the capture and legal trials of three escaped slaves. Each of these slaves—Shadrach Minkins, Thomas Sims, and Anthony Burns—were early Boston test cases of the Fugitive Slave Law, and the *Commonwealth's* treatment of these cases and their reverberations illustrates the way that the *Commonwealth* used the "legacy…of the exemplary founders"[4] to alternately affirm and shame its readers. The rare inclusion of three engravings in their retrospective of the events surrounding Thomas Sims's return to slavery powerfully demonstrates the way the *Commonwealth* used the nation's glorified past to generate indignation and move a contemporary generation of "desponding patriots" to action.[5] Finally, its treatment of the Sims case, and later, its coverage of Anthony Burns's return to slavery, demonstrate an evolution in its philosophy as it became increasingly militant. As such, the *Commonwealth* reflected the spirit of the times and the coming war, which it greeted with optimism, predicting that the "American people *will* become ashamed" of slavery and "the wrong will be righted."[6]

In an early issue of the *Commonwealth*, the editors asserted that "resistance to tyranny is the good old cause" that Boston's "liberty was trained in." It continued, "she was brought up to rebellion for the right, and has schooled us all in that learning."[7] Boston's abolitionists put the lessons of the past into practice soon after the Fugitive Slave Law took effect. Samuel Gridley Howe, Minister Theodore Parker, and other members of the Vigilance Committee heeded the voice of higher law and helped fugitive slaves William and Ellen Craft escape slave catchers.[8]

Shortly thereafter, the dramatic and public rescue of Shadrach Minkins affirmed the right of rebellion. In what Theodore Parker referred to as "the most noble deed done in Boston since the Boston tea-party in 1773,"[9] a group of men— many of them free blacks—mobbed the courthouse and rescued Minkins (who would find his way to Montreal, Canada, where, the *Commonwealth* reported, he opened a barbershop).[10] One month after Minkins's rescue, the *Commonwealth* implicitly celebrated the event when it quoted the Latin Massachusetts motto and "Scripture of rebellion in 1776," which in English translated to "By the sword we seek peace, but peace only under liberty." In an early preview of its later militancy, it indirectly justified Minkins's rescue by reading it through the past: "Liberty had to fight; there was no alternative; and she did fight, and by victory, made rebellion in favor of abstract right against unjust law, the most respectable thing in history."[11]

To President Millard Fillmore and Secretary of State Daniel Webster the rescue of Minkins was treason, "an outbreak against the Constitution and legal authority of the government."[12] The president's proclamation against the rescue was printed in full by the *Commonwealth*, which argued it was "nothing less than a proposition to put the whole country in a stage of siege, or under martial law."[13]

The *Commonwealth* editors believed that Massachusetts's formerly revered Senator Daniel Webster had, with his support of the Fugitive Slave Law, exchanged "cherished sentiments of freedom" for the "glittering baubles of high official honor."[14] The *Commonwealth* viewed the proclamation as an alarming echo of colonial Boston's oppression under British law.[15] It equated the president and

Webster with King George III and Lord North, who after the Boston Tea Party enacted a series of harsh tax laws as retribution.[16]

What Fillmore and Webster viewed as treason, the *Commonwealth* viewed as a triumph of principle befitting the sons of Puritans and Revolutionaries. It most clearly articulated this point of view when it printed Richard Henry Dana's closing arguments on behalf of Minkins's attorney, Charles Davis, who was accused of participating in the rescue. Dana addressed the wider implications of the Minkins rescue by drawing parallels between it, the Boston Tea Party, and the War of Independence: "if the act of 1850 [the Fugitive Slave Law] had been imposed upon us, a subject people, by a monarchy...as a province, by a mother country, without our participation in the act, we should have rebelled as one man." He proclaimed it a "mortifying contradiction" that Britain had more liberties than America.[17]

The *Commonwealth* frequently pointed to the "mortifying contradiction[s]" that saturated their reality as Bostonians and Americans. But nothing was more mortifying than the aftermath of Thomas Sims's arrest on April 3, 1851, and nothing in the *Commonwealth*'s history so galvanized it or inspired it. Sims, a "bright looking young man,"[18] with a wife and children, had escaped Georgia on a ship. He arrived in Boston after enduring harsh conditions and mistreatment by the crew, who had discovered him. After only one month as a free man in Boston, he was captured by the agents of his alleged owner, James Potter, and arrested on the order of US Commissioner George T. Curtis.[19]

What followed Sims's arrest was "one of the most disgraceful scenes ever witnessed in this city,"[20] and the *Commonwealth* responded with urgency and indignation. The *Commonwealth* announced the travesty in an extended headline, reprinted from the previous day's extra in bold, fully capitalized font: "The Arrest of Sims! Great Excitement Around the Court House! The Building in Chains! The Entire Police and Watch Department Brought Into Requisition!"[21] Concerned about the possibility of a rescue attempt, the authorities placed the courthouse under guard, a move that caused Gamaliel Bailey, editor of the antislavery *National Era*, to reflect that "Boston was a garrisoned town while the poor boy was in custody."[22] Fearing a mob, the Boston municipal government also closed Faneuil Hall, a popular meeting place, to abolitionists.

The curtailment of free speech, the garrisoning of the courthouse, the fact that Boston's merchants took slave owner James Potter's agents "in the hollow of her hand and treated them...tenderly and deferentially"[23] and, of course, Sims's return to slavery with an escort of Massachusetts soldiers, "desecrate[d] the soil of Massachusetts."[24] The *Commonwealth* demonstrated its anger through the rare publication of engravings that accompanied its three-issue retrospective of the Sims affair. Generally, antislavery publications used images to generate additional "sympathy for the slave."[25] By contrast, it seems that the *Commonwealth*'s illustrations were meant to generate political ire. In two of the three engravings, the engraver modeled his images on Paul Revere's engraving of the Boston Massacre, which historians agree is one of the most powerful pieces of war propaganda of all time.[26]

The engraver understood the immense resonance of Revere's Boston Massacre engraving. As with Revere's original image, the soldiers in "Sims Leaving the Boston Court House" appear well ordered. Rather than appearing "sneaky,"[27] however, they appear jovial, willing and eager to serve the cause of slavery "beneath the shadow of Faneuil Hall and Bunker Hill."[28] Unlike Revere's engraving, which features one unthreatening, disinterested dog at the foreground, there are two dogs in "Sims Leaving the Boston Court House." These dogs represent the canine creatures so well known for hunting runaway slaves. More significantly, they symbolize the slave catchers, known by the scornful epitaph "bloodhounds," who profited from the capture and return of slaves on Boston's sacred ground. In short, they are the exact opposite of the "loyalty and fidelity" of Revere's dog.[29] They are in this way particularly resonant, as they indict Boston's judges, police force, and citizens for enforcing the Fugitive Slave Law. In the end, it is they, rather than the slave catchers, who are the real bloodhounds.

The bloodhounds at the foreground of "Sims Leaving the Boston Court House" are significant for another reason as well. The dogs' rabid attention to one spot on the ground may be meant to suggest that they smell blood already shed, reminding the reader of the Paul Revere engraving and recalling the blood of African American and Natick Indian Crispus Attucks, believed to be the Boston Massacre's first victim.[30] By the 1850s, Attucks had been reclaimed from obscurity and fashioned into the symbol of "African American patriotism, military service, sacrifice, and citizenship"—and, most important for the *Commonwealth*, "the first martyr of the American Revolution."[31]

In "Sims Passing Through State Street," the *Commonwealth* positioned Sims as the first martyr of the Fugitive Slave Law, explicitly articulating what the first engraving only implied. The *Commonwealth*'s haunting caption equated Sims and Attucks: "Over the ground where Crispus Attucks, a colored citizen of Boston, fell, the first victim of a massacre, by British Troops." In calling Attucks a citizen, the *Commonwealth* claims Sims a citizen of the United States as well, and thus makes a case for the political rights of African Americans. The unspoken story is this: Thomas Sims, an African American citizen of Boston, fell, the first victim of the Fugitive Slave Law, not by British Troops, but by American troops, sponsored and approved by a government founded on the principle that "All Men are Created Equal." Upon Sims's return to slavery, the *Commonwealth* announced, "The Victim has Been Sacrificed"[32]—Boston's first martyr to a law that sent eleven out of thirteen fugitives back to slavery in the first three months of existence.[33]

The engravings also highlight the Fillmore administration's subversion of individual freedoms, local sovereignty, and states' rights. In Revere's engraving, Attucks is highly visible, *the* embodiment of the massacre's destruction. By contrast, Sims is invisible in both of the engravings modeled after those of Revere. He is subsumed by the soldiers in "Sims Passing Through State Street."[34] In addition to being an accurate representation of the wall of bodies the soldiers built around him, it is symbolic, for the obliteration of his body represents the obliteration of his and

Bostonians' rights. Further, the latter engraving's focus on the soldiers rather than the crowd accurately depicts the conditions of Sims's removal. He was secreted from the Courthouse "between the Setting of the Moon and the Rising of the Sun."[35] Although some abolitionists, including Theodore Parker, witnessed his march toward Long Wharf, there were not enough people to mount a true protest or rescue. The *Commonwealth* reported that "the only demonstrations made by the spectators as the procession passed were frequent cries of 'Shame' and questions of '*Where is Liberty?*'"[36] The final engraving, which bears no resemblance to Revere's, depicts a small group of onlookers gazing at the *Acorn* and slumped in defeat.[37]

The 1853 inauguration of President Franklin Pierce did not alter the Fillmore administration's patterns. In the spring of 1854, Massachusetts's—and America's—subversion of rights and increasing national fragmentation were even more evident: the Kansas-Nebraska Act threatened to open slavery's doors to the Western part of the continent; Charles Sumner's attempts to speak on the Senate floor were denied; and former slave Anthony Burns was captured in Massachusetts, resulting in nearly two hundred men surrounding the courthouse. As the *Commonwealth* despaired, "for the third time Boston witnesses the disgraceful spectacle of a man… accused of nothing but obedience to the upward yearnings of a spirit that bade him reach the same inalienable right[s]…which we enjoy."[38] The *Commonwealth* again invoked the specter of Crispus Attucks: Burns remained a prisoner, "hedged in" and "within a pebble's throw of that spot, where nine and seventy years ago the red life-blood of a man of his own color spirited into the sunlight in defence of New England liberty."[39]

The arrest of Anthony Burns provided *Commonwealth* readers a chance to redeem themselves from the ignominy of the Sims affair. To many, the failure to rescue Sims had signified a lack of manhood and a betrayal of the principles of 1776, so much so that the *Commonwealth* predicted that neither Bunker Hill nor the Revolution would have occurred if they "had been postponed" until 1851.[40] Its question regarding Burns—"Shall Boston Steal Another Man?"—therefore, held both Boston's government and its citizens culpable for Sims's return to slavery.[41] Five thousand people answered the *Commonwealth*'s question by first gathering at Faneuil Hall and then rioting at the courthouse, where the public had been barred from observing the legal proceedings against Burns. Excitement and indignation fueled the people's response, but Vigilance Committee members were unable to agree on a coherent rescue plan.[42] Under cover of the mob, a small group battered down the courthouse door. Fights ensued, shots were fired, and one man was killed.[43] But Burns remained a prisoner in Boston's "*own* courthouse…guarded by *their own* officers."[44]

According to James Trent, biographer of abolitionist Samuel Gridley Howe, the almost simultaneous arrest of Burns and the passage of the Kansas-Nebraska Act turned more New Englanders to antislavery than any other cause.[45] The Burns case, intoned the *Commonwealth*, had "se[t] aside the jury…pu[t] our towns and cities under martial law, transform[ed] men into bloodhounds, mad[e] freedom a crime and mercy a treason."[46] It was tangible proof of the long arm of slavery, evidence of a future in which the Kansas-Nebraska Act would, in the words of the newly formed Republican Party, "reduce the North…to the mere province of a few slaveholding Oligarchs of the south."[47] It was, moreover, part of a larger pattern of moral lassitude and unmanliness, which Minister Theodore Parker mercilessly catalogued in a sermon that the *Commonwealth* printed. Parker excoriated "non-resistant" Boston, charging his audience to "REMEMBER" the many ways its slow reaction to the Fugitive Slave Law had betrayed the "spirit of our fathers." "Today," he asserted, "we have our pay for that conduct."[48] The *Commonwealth* likewise condemned Bostonians for once again failing to live up to their predecessors, who had removed the yoke of oppression. By contrast, "we are but serfs: pliant supple menials of the slaveholders." Nonetheless, the *Commonwealth* felt that the actions taken in the Burns case indicated progress, and "honored" the men who in the "name of outraged liberty" gathered at Faneuil Hall and used violent means in an "ill-timed" effort to rescue Burns.[49]

In its first issue, the *Commonwealth* had encouraged readers not to despair over the "temporary reverses" that characterized its time. "All history," it observed, "shows that great principles establish themselves not by an uninterrupted progress, but by alternations of triumph and defeat."[50] It reaffirmed this perspective many times in its short publication history, and as it evolved, it became more militant and forward thinking, perceiving the immense local and national setbacks of the early 1850s as a sign of future victory and vindication. This perspective is best exemplified by its June 2, 1854, publication of Charles Sumner's "Final Protest Against Slavery in Kansas and Nebraska," which was printed on the same day Burns was marched from the courthouse toward slavery. In it, Sumner reflected on the ways that the Kansas-Nebraska Act was "at once the worst and the best bill on which Congress ever acted." The "Present victory of slavery" was, of course, the worst aspect of the bill, but it "prepares the way" to a "future, when…the slave power will be broken." Sumner envisioned a time when "slavery will then be driven from its usurped foothold" by "the great *Northern Hammer*" which "will descend to smite this wrong."[51]

In her memoir, *Reminiscences*, Julia Ward Howe—Samuel Gridley Howe's wife, *Commonwealth* co-editor, and future author of the "Battle Hymn of the Republic"—claimed that the *Commonwealth* "did a good service in the battle of opinion which unexpectedly proved a prelude to the most important event in our history as a nation."[52] With this, she not only attributed to it immense persuasive power, but also presented the *Commonwealth* as a crucial instrument in effecting the Civil War and emancipation. It is difficult to determine the actual impact the *Commonwealth* had on public opinion and the arc of history. It seems likely, however, that the *Commonwealth* offered its readers more than scholars' usually cursory discussions have suggested. Through its use of the tropes and symbols of the American Revolution, it established a powerful genealogy of leadership and a legacy of rebellion that inspired "desponding patriots" to take heart because "Liberty is never in extreme peril so long as its friends are true."[53]

FIGURE 8.1 "Anthony Burns," printed by R. M. Edwards in Boston, 1855. In this woodcut illustration, a portrait of fugitive slave Anthony Burns drawn from an early daguerreotype is ringed by drawings, including: his being sold as a youth at an auction, his arrest on May 24, 1854 in Boston, Burns in prison, his "address" to the court, and his removal from Boston by federal marshals and troops. Burns's arrest and trial in the spring of 1854 prompted protests and riots in Boston. (From the Library of Congress Prints and Photographs Division Washington, DC. <www.loc.gov/item/2003689280/> Reproduction Number: LC-USZ62-90750 (b&w film copy neg.). John Andrews, engraver, "Anthony Burns / drawn by Barry from a daguereotype [*sic*] by Whipple & Black.")

The *Commonwealth* also captured the local progression of a move toward what Albert J. von Frank recognizes as a "broader revolutionary impulse"[54] that six years later resulted in Civil War. Samuel Gridley Howe asserted in a letter to educator Horace Mann, "Poor Burns has been the cause of a great revolution. You have no idea of the change of feeling here."[55] Although the *Commonwealth* cited financial difficulties when it announced its final issue on September 21, 1854, it is possible that those involved realized that philosophy was not enough; words could not accomplish what action could. *Commonwealth* reader and Boston Vigilance Committee member Thomas Wentworth Higginson certainly endorsed that idea. Shortly after the Burns case, he announced, "*a revolution* has begun!" Drawing a distinction between kinder, gentler reform and revolution, he encouraged his audience to make their political decisions with the goal to "bring nearer the crisis which will either save or sunder this nation."[56]

Higginson's advice crystalized a philosophy toward which the *Commonwealth* and a number of its editors, readers, and writers had been moving. Shortly after the passage of the Kansas-Nebraska Act resulted in bloody warfare and the suppression of Free Soil politics on the states' borders, Higginson, Samuel Gridley Howe, and Theodore Parker expanded their energies from local resistance (each had been involved in some if not all of the above slave cases) to national rebellion. They lent their weight to "violence-laden politics" by providing rifles to the antislavery settlers.[57] As three of the "Secret Six" who helped fund John Brown's raid on Harper's Ferry, they took another step toward the abolition of slavery, which on July 4, 1854, the *Commonwealth* had linked with what it foresaw as the coming Civil War. Commemorating "this day, seventy-eight years ago," it had imagined that, one day, the "bloody Revolutionary struggle, will become a practical, living reality to this whole people."[58]

Notes

1 "New Daily Paper," *Boston Daily Commonwealth*, January 1, 1851, 2. (All references to the *Commonwealth* come from the Boston Public Library's microfilms of the paper.)
2 "Shall the Cause be Surrendered?" *Boston Daily Commonwealth*, June 17, 1852, 2.
3 Gordon S. Barker, *Fugitive Slaves·and the Unfinished American Revolution: Eight Cases, 1848–1856* (Jefferson, NC: McFarland, 2013), 3.
4 Ibid.
5 "New Daily Paper," *Boston Daily Commonwealth*, January 1, 1851, 2.
6 "Fourth of July," *Boston Daily Commonwealth*, July 4, 1853, 2.
7 "Resistance to Unjust Laws," *Boston Daily Commonwealth*, March 18, 1851, 1.
8 Barker, *Fugitive Slaves and the Unfinished American Revolution*, 31–32.
9 Theodore Parker, quoted in Barker, *Fugitive Slaves and the Unfinished American Revolution*, 44.
10 "Shadrach," *Boston Daily Commonwealth*, April 3, 1851, 2.
11 "Resistance to Unjust Laws," *Boston Daily Commonwealth*, March 18, 1851, 1.
12 "President's Message," *Boston Daily Commonwealth*, February 22, 1851, 1.

13 "Calling Out the Militia," *Boston Daily Commonwealth*, February 26, 1851, 1.

14 "Mr. Webster and His Reception," *Boston Daily Commonwealth*, July 29, 1852, 2.

15 "Resistance to Unjust Laws," *Boston Daily Commonwealth*, March 18, 1851, 1.

16 David Lee Russell, *The American Revolution in the Southern Colonies* (Jefferson, NC: McFarland, 2000), 45.

17 "Dana's Closing Arguments," *Boston Daily Commonwealth*, February 27, 1851, 2.

18 "Kidnapping of Sims," *Boston Daily Commonwealth*, April 25, 1851, 1.

19 Barker, *Fugitive Slaves and the Unfinished American Revolution*, 54–56.

20 "Arrest of Sims," *Boston Daily Commonwealth*, April 5, 1851, 1.

21 Ibid.

22 Gamaliel Bailey, quoted in Mitch Kachun, "From Forgotten Founder to Indispensable Icon: Crispus Attucks, Black Citizenship, and Collective Memory, 1770–1865," *Journal of the Early Republic* 29, no. 2 (2009): 272.

23 "Kidnapping of Sims," *Boston Daily Commonwealth*, April 26, 1851, 1.

24 "Anniversary of the Kidnapping of Sims," *Boston Daily Commonwealth*, April 12, 1852, 2.

25 Bernard F. Reilly, "The Art of the Antislavery Movement," in Donald M. Jacobs, ed., *Courage and Conscience: Black and White Abolitionists in Boston* (Bloomington: Indiana University Press, 1993), 63.

26 Nancy Osgood, "Josiah Wolcott: Artist and Associationist," *Old-Time New England* (Spring/Summer 1998): 9. The identity of the engraver is unknown, although the signature of "Wolcott" on one of the engravings suggests that jack-of-all arts, abolitionist, and friend of escaped slave Henry "Box" Brown is the most likely candidate.

27 Scott Casper, "The Boston Massacre: Paul Revere's Most Famous Engraving Transformed a Minor Colonial Melee into a Revolutionary Cause Celebre," *American History* 43, no. 3 (August 2008): 22.

28 "Justice in Chains," *Boston Daily Commonwealth*, April 9, 1851, 2.

29 "Paul Revere's Engraving of the Boston Massacre, 1770," *Gilder Lehrman Center*, accessed June 26, 2014.

30 Stephen Kantrowitz, "A Place for 'Colored Patriots': Crispus Attucks among the Abolitionists," *Massachusetts Historical Review* 11 (2009): 110.

31 Mitch Kachun, "From Forgotten Founder to Indispensable Icon: Crispus Attucks, Black Citizenship, and Collective Memory, 1770–1865," *Journal of the Early Republic* 29, no. 2 (2009): 249, 279.

32 "Victim Has Been Sacrificed," *Boston Daily Commonwealth*, April 14, 1851, 4.

33 James Oliver Horton and Lois E. Horton, "A Federal Assault: African Americans and the Impact of the Fugitive Slave Law of 1850," in Paul Finkelman, ed., *Slavery and the Law* (Lanham, MD: Rowman and Littlefield, 2002), 149.

34 "Sims Passing Thorough State Street," *Boston Daily Commonwealth*, April 21, 1851, 2.

35 "Sims Leaving the Boston Courthouse," *Boston Daily Commonwealth*, April 25, 1851, 1.

36 "Victim Has Been Sacrificed," *Boston Daily Commonwealth*, April 14, 1851, 4.

37 "Sims Shipped for Savannah!" *Boston Daily Commonwealth*, April 28, 1851, 1.

38 "Faneuil Hall Tonight," *Boston Daily Commonwealth*, May 26, 1854, 2.

39 "From the Evening Edition," *Boston Daily Commonwealth*, May 27, 1854, 2.

40 "Anniversary of the Kidnapping of Sims," *Boston Daily Commonwealth*, April 12, 1852, 2.

41 "Another Man Seized in Boston," *Boston Daily Commonwealth*, May 26, 1854, 2.

42 Albert J. von Frank, *The Trials of Anthony Burns: Freedom and Slavery in Emerson's Boston* (Cambridge, MA: Harvard University Press, 1998), 12.

43 Ibid., 66–68.

44   "Sovereignty of the People!" *Boston Daily Commonwealth*, May 26, 1854, 2.

45   James W. Trent, *The Manliest Man: Samuel G. Howe and the Contours of Nineteenth-Century American Reform* (Amherst: University of Massachusetts Press, 2012), 199.

46   "Progress of Despotism," *Boston Daily Commonwealth*, June 19, 1854, 2.

47   Michigan Republican State Platform, quoted in Paul Calore, *The Causes of the Civil War: The Political, Cultural, Economic and Territorial Disputes Between North and South* (Jefferson, NC: McFarland, 2008), 221.

48   Theodore Parker, "A Lesson For the Day," *Boston Daily Commonwealth*, May 29, 1854, 2.

49   "Fugitive Case," *Boston Daily Commonwealth*, June 3, 1854, 2.

50   "New Daily Paper," *Boston Daily Commonwealth*, January 1, 1851, 2.

51   Charles Sumner, "Final Protest Against Slavery in Kansas and Nebraska," *Boston Daily Commonwealth*, June 2, 1854, 2.

52   Julia Ward Howe, *Reminiscences, 1819–1899* (Boston: Houghton, Mifflin, 1900), 253.

53   "New Daily Paper," *Boston Daily Commonwealth*, January 1, 1851, 2.

54   von Frank, *The Trials of Anthony Burns*, xiii.

55   Samuel Gridley Howe, quoted in Trent, *The Manliest Man*, 201.

56   Thomas Wentworth Higginson, quoted in von Frank, *The Trials of Anthony Burns*, 261.

57   James Brewer Stewart, *Holy Warriors: The Abolitionists and American Slavery*, rev. ed. (New York: Hill and Wang, 1997), 168.

58   "All Men are Created Equal," *Boston Daily Commonwealth*, July 4, 1852, 2.

9

FRANKLIN PIERCE AND THE FAILURE OF COMPROMISE

Newspaper Coverage of the Compromise Candidate, the "Nebraska Act," and the Midterm Elections of 1854

Katrina J. Quinn

Stepping into the White House at a time of escalating sectional strife, Franklin Pierce had been hailed as a compromise candidate. Nominated on the 49th ballot of the 1852 Democratic National Convention, it was hoped he could appease the quickly diverging factions of the Democratic Party and mitigate the nation's simmering sectional discord. Like Pierce's candidacy, the Kansas–Nebraska Act he signed in 1854 was conceived as a compromise; and like Pierce, the legislation failed miserably, with the paradoxical effect of cementing sectional divisions and driving the country toward civil war.

Pursuing the notion of "failed compromise," this chapter examines press coverage of Franklin Pierce following his unlikely nomination through the Kansas–Nebraska debate and midterm elections of 1854. It begins by establishing the edifice of "compromise" by assessing newspaper coverage of the Pierce campaign and the 1852 election. The chapter then gauges coverage of the Kansas–Nebraska debate in both the North and the South, demonstrating that newspapers in both regions found the act to be controversial and dissatisfactory, their attack on the legislation often accompanied by verbal assaults on the president. The press's representation of Pierce in both Democratic and oppositional newspapers quickly evolved from that of a minister of compromise to an agent of division, with Democratic losses in the 1854 midterm elections laid squarely at his feet.

This study considers reportage and editorial content, native or reprinted, from June 1, 1852, the first day of the 1852 Democratic National Convention, through December 31, 1854, following the midterm elections. It excludes letters to the editor and transcripts of congressional debate or speeches, which do not necessarily reflect the opinions of the host newspapers.

From the Granite State to the 1852 Nomination

Part of the Pierce appeal was his relative obscurity. He was born November 23, 1804, the sixth of eight children, in a cabin now at the bottom of a lake in Hillsborough, New Hampshire. His father, Benjamin Pierce, was a soldier in the Revolutionary War, present at the Battle of Bunker Hill, and twice served as governor of New Hampshire. Trained as a lawyer, Franklin Pierce was undistinguished in his tenure in the US House of Representatives and Senate, and then as US Attorney for New Hampshire. He later volunteered to serve in the Mexican-American War, from 1846 to 1848, achieving the rank of brigadier general, a position which brought him modest renown, particularly in his home state. Pierce was welcomed back to New Hampshire as a hero and resumed his law practice, which he continued until called as an improbable candidate for president in June 1852.

The 1852 Democratic Convention in Baltimore convened under the shadow of controversy. The Compromise of 1850 had divided the party into factions, largely but not solely along geographic lines. The delegates presented no fewer than fourteen candidates, including prominent and promising prospects such as former Secretary of State James Buchanan of Pennsylvania, Senator Lewis Cass of Michigan, and the former Secretary of War William L. Marcy of New York. Pierce was first offered without fanfare by Virginia's delegation following the thirty-fourth ballot, but the nomination was finally adopted after an impassioned speech by North Carolina's James C. Dobbin prior to the forty-ninth ballot. Dobbin urged compromise, perhaps more than the merits of the candidate, bringing the protracted deliberations to a close.[1]

Pierce was a compromise—but why Pierce? His obscurity, his lack of a substantive political record made him the perfect candidate. Although the initiative to ordain Pierce seemed to spring organically from the tired Democratic delegates, Pierce had been speaking "compromise" even before the convention, and his ameliorative or noncommittal rhetoric was now aggressively reprinted in Democratic newspapers in both the North and the South. Most prevalent in the coverage was a statement Pierce made in support of the Compromise of 1850 and the Fugitive Slave Act, a move sure to strengthen allegiance of Southern Democrats at the convention. Another widely reprinted statement was a toast Pierce delivered at a Democratic gathering in New Hampshire, articulating a vision for the party and the nation. "No North, no South, no East, no West under the Constitution," he announced, "but a sacred maintenance of the common bond, and true devotion to the common brotherhood."[2]

Quickly on the heels of the nomination, the public relations wheels began to turn. An official portrait was commissioned, with Pierce in military garb, striking a manly pose astride a horse.[3] Two book-length biographies were published, one by Pierce's college pal, the renowned Nathaniel Hawthorne.[4] And many fawning and magniloquent biographies were printed in Democratic newspapers, painting the small-town New Hampshire lawyer as an archetypal hero, using grandiose language that established Pierce as a candidate of destiny, sure to draw the disparate corners of the Union together. The *Semi-Weekly North Carolina Standard* of April 3 is an example, publishing an extended narrative biography of Pierce, saluting "New Hampshire's favorite son—wise in council and pure in patriotism. The country is possessed of his title to renown in the record of his gallant deeds upon the tented field, his boundless attachment to the Constitution and the Union, and a life-long devotion to the principles of the Democratic party."[5] Like all hero stories, these narratives launched a discourse of heroism, intellect, and humility, connecting Pierce to the very fiber of the nation through his personal résumé and family heritage. They drew their hero as bigger than life—certainly bigger than Franklin Pierce.

Complementing their flowery assessment of the candidate, Democratic newspapers in the North also printed articles enthusiastically praising the convention and the wisdom of the Pierce nomination as a compromise move. After many ballots, according to the Washington *Daily Union*, "the assembled democracy, as if with one consent, turned its eye to General Pierce as the only man who could at this time unite in his support every democratic element and interest of the whole Union."[6] The *Hartford Times* offered praise for the compromise: "Gen. Pierce was taken up as a compromise candidate, on whom the friends of the different candidates could unite in harmony and good feeling. ...As a candidate he is unexceptionable, and we shall look for his triumphant election."[7] Called "INCORRUPTIBLE" and "true-hearted," Pierce heralded an age of national reconciliation according to the Albany *Atlas*. "We doubt not [the selection] will evoke the same spirit throughout the country," the *Atlas* proclaimed, "and...we predict the most auspicious results for the party and the permanent interests of the republic." From the Troy *Budget*, "This nomination will unite all conflicting interests, and secure the unanimous and cordial support of the democratic party from Maine to California." Similarly, the *Boston Transcript* quixotically predicted that the Pierce compromise could even elicit support from those outside the party, including Whigs. The paper and others concurred with the *Boston Post*, which predicted that citizens "from Canada line to California...will hail Franklin Pierce as the compromise and harmony candidate...."[8] The "compromise" chorus among Northern Democratic newspapers was loud and strong.

As anticipated, the Pierce nomination also brought widespread support from Southern, proslavery members of the party, and Southern Democratic newspapers joined the enthusiastic "compromise" refrain of their Northern counterparts. The *Richmond Enquirer* stated, "the nomination is a thoroughly national one.... Gen. Pierce, it is honestly believed, will unite the Democractic [sic] party of the country more fully than any other candidate."[9] North Carolina's *Wilmington Journal* reported that the Onslow County Democrats issued a formal resolution of support, noting Pierce to be a "statesman of the right stamp," whose accomplishments and personal characteristics "have endeared him to us and the people of the whole country."[10] Described as "distinguished" and possessing "unobtrusive worth and talent," the Fayetteville *North Carolinian* compared Pierce to George Washington, for "office has sought him and not he office."[11]

But not everyone was delighted with the outcome of the convention, as coverage in opposition newspapers demonstrates. However, even while these newspapers launched pointed criticism of the candidate, they, too, invested in the symbolic rendering of Pierce as a compromise.[12] In a June 12 editorial, *The Cecil Whig*, of Elkton, Maryland, for example, wrote a scathing condemnation of Pierce, and counted vice-presidential nominee William King of Alabama as part of the inglorious compromise. "It is useless for the locofocos to attempt to conceal the fact, that the nominations have been made solely with a view to reconcile the seceding factions of their party—the northern Abolitionists and the southern Disunionists— by giving each a candidate," the newspaper wrote. The editorial used imagery that suggested the impossibility of compromise: "In one common effort for party power, extremes are brought together—the oil and water are combined—Herod and Pilate are made friends."[13] An editorial from the *Hartford Courant* reprinted in Washington's *Republic* indicted the Democratic delegates for a hand well played.

FIGURE 9.1 "Ornithology," lithograph published by John Childs, 1852. This political cartoon depicts Whig presidential candidate Winfield Scott as a turkey (left) and Democratic candidate Franklin Pierce (right) as a gamecock on either side of the "Mason & Dixson's Line." Scott yells, "Get out of the way fellow! I want the whole of the road!" Pierce replies, "Cock a doodle do----oo! Don't you wish you may get it! But you can't get over this line," insinuating that Scott would not win the South because of his antislavery views. (From the Library of Congress Prints and Photographs Division Washington, DC. <www.loc.gov/pictures/item/2008661551/ > Reproduction Number: LC-USZ62-5784 (b&w film copy neg.). John Childs, "Ornithology.")

"The friends of Gen. Pierce have managed their game most adroitly," the newspaper declared. "By successively killing off the prominent candidates, and thus bringing the party to the very verge of destruction, they have induced them, when in a moment of desperation, to seize upon their nominee as the last means of re-uniting their divided ranks."[14] Montepelier, Vermont's *Green-Mountain Freedman* called Pierce "one of the most utterly servile of northern doughfaces...."[15] In addition to targeted denunciation and name-calling, a slew of newspapers ran unflattering biographies of the nominee, contradicting those in the Democratic press at every turn. New Hampshire's *Claremont Eagle* noted that Pierce was not such a grand hero, for his exploits on the battlefield included fainting and falling from his horse.[16]

At the ballot box, the Pierce compromise proved successful, with the Pierce–King ticket receiving 254 electoral votes to forty-two for the Whigs' own compromise ticket, featuring Gen. Winfield Scott of New Jersey and William A. Graham of North Carolina. Pierce may have entered the White House optimistically, the beneficiary of a compromise strategy, but turmoil over the issue of slavery and states' rights continued to grow.

Putting Compromise to the Test: The Nebraska Act of 1854

By the end of 1853, the debate over what to do about Nebraska Territory was simmering. Legislative action was necessary to determine the fate of millions of acres in the center of the continent, lands of inestimable value for the completion of a transcontinental railroad. But determining a political configuration for the vast area exacerbated unresolved issues of states' rights, particularly the issue of slavery. Pierce's political efforts to build compromise amidst growing sectional conflict reached their climax as Congress debated what contemporary newspapers almost universally referred to as the "Nebraska Act," a measure which would have allowed residents to vote on slavery and would have overturned the provision of the Missouri Compromise of 1820 prohibiting slavery above latitude 36°30'.

The Kansas-Nebraska bill lifted the shroud of compromise. The abolitionist press in the North immediately condemned it as a proslavery move with unpredictable, disastrous consequences. Philadelphia's *The Friend* predicted that should the legislation pass into law, "it will be the severest, the most paralyzing blow that has been inflicted upon the cause of freedom."[17] Recalling a strengthening sense of brotherhood between the North and South in the wake of the compromise acts of 1850, The *New York Observer and Chronicle* predicted the Nebraska Act "can work only mischief in the North and in the South. It will tear open wounds which are but partially healed. Should such a measure be adopted it will be regarded at the North as a breach of faith...."[18] Publications such as New York's *The Independent* interpreted the act as authorizing a reincarnated slave trade, asking, "What villainy of the kidnappers who prowl along the coast of Africa and whom we hang for their piracy, is baser than this, or more to be abhorred by any honorable mind?"[19]

Washington's abolitionist newspaper, the *National Era*, urged its readers to "break asunder the bands of Party, rise in their full might, and, trampling upon all compromises, all time-serving expedients, all tricksters and traitors, rally as one man, in defence [*sic*] of Freedom, Free Labor, Free Institutions...."[20] In "Who are to Blame?" *The Independent* called on lawmakers to take a stand. "You told us, gentlemen, that you hated slavery, but desired peace; that you were opposed to the extension of slavery, but feared the dissolution of the Union. Now is the time to prove your words." The same story predicted, "A crisis is impending more fearful than any this country has ever witnessed."[21] Similarly, *The Christian Inquirer* foretold, "A struggle is impending between the principle of Liberty and that of Slavery, an awful struggle, the *end* of which we well know, but the intermediate steps of which are veiled in darkness."[22]

In the North, President Pierce became the target of the abolitionist press, as well as Whig-sympathizing Northern newspapers, which impugned him as a co-conspirator with Senator Stephen A. Douglas in promoting the interests of the South—a move, they predicted, which would have enduring consequences. *The Independent* predicted that the legislation would be remembered as a "GREAT CRIME by which the administration of Franklin Pierce is to be made memorable in history."[23] Ohio's *Fremont Journal* called the Democrats the "Pierce and Douglas Slaveocratic party."[24]

The issue of failed compromise was highlighted in the Southern press as well. Take, for example, an article in Tennessee's *Athens Post*, which depicted the Nebraska Act as a bipartite failure:

> The measure was Janus-faced from the start—designed by Douglass [*sic*] and the administration to be a Southern measure at the South, and a Northern measure at the North. But some how or other it has failed in both sections.... Whatever may be thought of the principle involved in the question, contempt for the men who have thus attempted to deceive and betray both sections, is unbounded, and will manifest itself unmistakeably [*sic*] whenever an opportunity occurs.[25]

The *Charleston Mercury* represented the opinion of many by calling out the president as a traitor, who had illustrated his support of Free Soilers and abolitionists by signing the legislation. Coverage conveyed a sense of imperialism on the part of the North, more reprehensible because of the early verbal support of the administration for compromise—now, in the eyes of the Southern newspapers, an obvious lie. North Carolina's *Wilmington Herald* called the act "a death blow for the Administration in the North and North-west."[26] The *Mercury* referred to the act as one of "colonization," representing an effort on the part of "true anti-slavery men."[27]

Among the most vocal Southern opponents of Pierce, Shreveport's weekly Whig newspaper, *The South-Western*, characterized the administration as one of demagoguism and ineptitude. The "ship of state" is on the "very verge" of destruction, the newspaper wrote, "brought by the mismanagement of the democratic party and Pierce administration."[28] The newspaper quoted a speaker at a Democratic meeting in Cincinnati, who said, "President Pierce has been kneeling to the free-soilers of the north, and then to the disunionists at the south, and he is not worthy of the support of either portion of the Union. His administration has been a miserable, skulking, sneaking, cowardly one—neither Mr. Pierce nor judge Douglas deserve popular respect." A number of newspapers noted that midterm Democratic candidates were distancing themselves from the unpopular administration in their campaign speeches. "Funny times," the *South-Western* mused, "when at 'regular democratic' meetings, called to advance the cause of administration candidates, the administration itself has to be denounced by the speakers. President Pierce has a singular sort of hold upon his party, and no hold whatever on the people, and the latter are by no means loth to show it. Take away the 'loaves and fishes' and there will not be left even a solitary Piercite to tread the deserted banquet hall."[29]

Coming to the defense of the president was a sympathetic correspondent for the *New York Journal of Commerce*, acknowledging, like a voice in the wilderness, "The violent abuse which President Pierce, in common with all his predecessors, has received from a portion of the press, has been a matter of surprise to foreigners not acquainted with the unscrupulous character of partisan warfare; but the truth is, that he and his household are universally popular amongst the residents of the metropolis, and all who come within the circle of the White House bear away kindly remembrances of its occupants."[30] Also supportive of the administration, the *North Carolinian* argued that Pierce, Douglas, and Senator Lewis Cass "are entitled to our admiration because they supported the Nebraska bill as a measure of justice to the south—principle governed them in the matter."[31] The Washington *Daily Union*, the administration's organ, pointed out how frivolous were some of the complaints against the president.[32] The *Daily Union* also downplayed the assault of the president by several New York newspapers, noting the larger successes of the administration in the areas of legislation, foreign policy, and fiscal accountability. The newspaper praised Pierce for his "uncompromising and fearless stand" in favor of the Nebraska bill, a test of his leadership. "The champions of the democracy in Congress, the men who have always been true in the midst of troubles, are now, as at the beginning, the champions of Franklin Pierce."[33]

Political Fallout and the Midterm Elections of 1854

Despite the hopeful proclamations of the *Union*, however, newspapers reported increasing partisan strife and what some interpreted to be the end of the Whig and Democratic parties. In "The Condition of the two Old Parties," Mississippi's *Yazoo Democrat* said, "It is no longer a debatable question whether the whig party continues to exist as a distinct and united national organization."[34] The *Louisville*

Journal opined, "Unquestionable the whig party is, for the present at least, severed at the divided line between the North and the South, and we do not know that a party, cut in two in the middle, can properly be considered a live party." And more, if the party were to attempt to reunite, the Southern wing would "probably feel that…they would also be uniting with a vast number of abolition fanatics…."[35] The editor of the *Journal* saw a similar disunion among Democrats, and concluded that "the democratic party is as effectually severed as the whig party," predicting "that upon the proposition to repeal the Nebraska act all the Nebraska-bill democrats will unite with the whigs and the abolitionists."[36] Tennessee's *Athens Post* reported July 7, 1854, on the disunity of the Whig party, after a vote to repeal the Nebraska Act was held up as "the test given of orthodoxy in the Whig party…." US Senator James C. Jones, a Whig and former governor of Tennessee, proclaimed, according to the story, "if to trample on the rights of any section or State was necessary to be a Whig, he had to say to his former friends and associates, good bye. He would rather die than follow them in such rule of action."[37]

Coming on the heels of the Nebraska Act, newspapers reported that Democratic losses in the summer and fall midterm elections of 1854 reflected the failure of the Pierce administration to sustain a Northern Democratic coalition. The *Burlington Free Press* leveled criticism directly at the administration and Pierce. "The Democratic Ticket is a ticket to *endorse the Administration*. Who are going to vote it?"[38] The *Windsor Journal* noted that Iowa voted in Whigs at every level, and Vermont's upcoming election promised to "'bombard' the administration. She has the distinction (such as it is), of being the native State of Douglas, and also of being *next door neighbor* to the native State of the President."[39]

Although the humiliating and widespread defeat of Democratic, pro-Nebraska Act candidates in the North elicited some sympathy from Southern Democratic newspapers, the Southern press was not restrained in its derisive assessments of the president. The Democratic *Independent Press* of Abbeville, South Carolina, noted the rising tide of sentiment against the president, paradoxically uniting disparate factions in opposition:

> *The truth is, persons at the North who are deadly hostile on all other points save opposition to Pierce, are uniting to vote him down.* And even in the South, where he seems to have gained several elections, he owed it to the character of his northern opposition, rather than to any popularity of his own. … *There never was more general dissatisfaction with a man than with Pierce—not even with Van Buren in* 1840. And it is not so much on account of his principles—for he is generally admitted to be a correct man—as on account of the damnable clique of imbeciles and corrupts with which he unwisely though unwittingly, surrounded himself in the beginning.[40]

Other Southern newspapers provided coverage of the perceived dissolution of party affiliation among Northern political factions in the recent elections. "Why is it that the passage of the Nebraska Act has made the Administration odious in the North and North-west?" the *North Carolinian* asked. "Why is it that whigs, free-soilers and abolitionists have united in opposition to the democratic candidates? The reason is obvious. The present Executive has shown himself eminently capable of administering the Government upon broad and national principles. He has redeemed the pledge, under which he was inaugurated, to know no North, no South, no East, no West, but to administer equal and impartial justice to all sections alike." By attempting to maintain a field of compromise, "he has made himself more obnoxious than ever to the anti-slavery element in the North."[41] The move was interpreted to be a harbinger of sectional tumult: "By this measure of justice the Administration has incurred anew the odium of all those destructive and anti-national elements at the north which have long threatened the existence of our Republic. The tide of fanaticism now rolls triumphantly over its adherents at the north."[42]

Throughout the summer of 1854, reportage shows an accelerating rhetorical breakdown of the *compromise* refrain and a reification of sectional alliances. Establishing a paradigm of otherness, newspapers presented information in dialectic structure that highlighted sectional affiliations. The *Fayetteville Observer* reported, "We cannot shut our eyes to the fact that the whole North is in a state of excitement unparalleled in the history of the country." To the contrary, "The South is cool and prepared for any emergency that may arise. ... We are opposed to all rash and extreme measures, but...we are resolved to do what we can to maintain the honor and the rights of our section."[43] The *Weekly North Carolina Standard* reported, "The indications now are that the people of North Carolina will be more united for the future on the question of slavery than they have ever been at any former period."[44]

The rhetoric of disunion in some newspapers reached such a pitch that it became the subject of editorials in peer publications. The *North Carolina Standard* criticized the Whig-sympathizing *Raleigh Star* and the *Fayetteville Observer* for stoking the fires of sectionalism.[45] The editor of the *Yazoo Democrat* also chastised his peers: "If the Union could be dissolved by an editorial or a speech in Congress it would not last many days. Indeed it would have ceased to exist long ago." The article suggested that the rhetoric of the press may have been out of step with the feelings of the public: "There is moral treason in this habit of continually depreciating the value of the Union, and it ought to be frowned upon by every American citizen who feels proud of his country and its freedom."[46] The Democratic *Richmond Enquirer* claimed the discord was merely "mischievous machinations of the fanatics," while the "large mass of the Democracy at the North are willing and ready to stand by the constitutional rights of the South and the preservation of the Union." The paper blamed the Northern Whigs for stirring up fruitless agitation. "The Chicago Tribune attempts to excite a mob-spirit against 'the renegade' Judge Douglas," the *Enquirer* stated, but it concluded that the senator should have no fear of public opposition in Chicago since he had previously confronted abolitionists with "manly blows" and "heroic boldness."[47]

Rising levels of discord led Illinois's *Ottawa Free Trader* to ask, "Who are the Agitators?" It wrote, "The whole country is sick of, and cries out against the present useless agitation on the slavery question. The Nebraska bill is charged as the cause of it, and yet all admit that it will never make or unmake a slave. Then all the leading whig and free soil papers admit, and the whole democratic press contend that the repeal of that bill is impracticable, and to insist on it would only prolong the useless and dangerous agitation."[48] The *Washington Daily Union* agreed that the Nebraska issue was being used as a means of hyperbolic attack on the Democrats by their political opponents: "The avidity with which all the forces which are combined against the administration seize upon every appearance of division in the democratic party upon the Nebraska question, and magnify it, and fan it into a flame, indicate the point upon which they have fixed all their hopes of victory." The controversy took an ugly turn in heated speeches and public demonstrations. The *Daily Union* lamented, "The system of denunciation, of hanging in effigy, of mobbing and gagging, so often adopted heretofore by the opponents of the great democratic measures which have controlled and shaped our government, is again put in operation, with all its temporary terrors."[49]

The Failure of Compromise and the Coming of War

Newspaper coverage of the Franklin Pierce nomination, the Kansas-Nebraska debate, and the midterm elections of 1854 suggests that the notion of compromise adopted in 1852 was an illusion. The *Washington Daily Union* inadvertently articulated the issue by stating that Pierce was "the second choice of *every* State, and therefore the first choice of *all* the States"[50]—in other words, he was *nobody's* choice. In fact, the delegates at the 1852 convention adopted not a candidate so much as the notion of compromise, a construct into which the obscure Pierce could be molded. And when compromise proved to be a failure in 1853 and 1854, so did the president.

Newspapers participated in this debate by giving voice to the disparate political elements swirling around the issues in an increasingly complex web of affiliation and dissension. Reportage served as a discursive battlefield for political opponents in all corners of the country. In their editorializing and subjective coverage of events, newspapers deployed a rhetoric that suggests a sensationalist mode, both in their initial praise and condemnation of compromise-candidate Pierce, and in their reactions to the Nebraska legislation and the midterm elections. These pieces were markedly different from companion columns of congressional testimony, and much closer in tone to nearby reportage on crimes and fires. The legislation in particular struck an emotional chord that inspired not only the printed text but gave rhetorical structure and meaning to both sides of the slavery issue.

While the slavery issue had long festered in the halls of public discourse, Pierce had entered the debate relatively unknown—a blank rhetorical slate upon which the Democratic press inscribed a hero's story. But because

Pierce-the-compromise-candidate was largely a construct, the narrative was easy to deconstruct and reconstruct. Thus, the story was rewritten, with Pierce portrayed as a hapless general at the helm of a disordered party. So despite its failures to quell the forces of disunion, the Pierce administration and the Kansas-Nebraska debate had the unanticipated effect of prompting a coalition of Northern Democrats, abolitionists, and soon, Republicans, essentially setting the playing field for the impending conflict.

The reportage from this period suggests that the press—and perhaps the public as well—was philosophically divided and aware of the impossibility of compromise. Editorial content that openly engaged the issue of disunion, crystallized ideological factions, and questioned presidential authority weakened the Union and foreshadowed a nation on its path toward war.[51]

Notes

1   *Proceedings of the Democratic National Convention, Held at Baltimore, June 1–5, 1852, for the Nomination of Candidates for President and Vice President of the United States* (Washington: R. Armstrong, 1852), 57–58, https://archive.org/details/proceedingsofde00demo.
2   "New Hampshire," *Semi-Weekly North-Carolina Standard* (Raleigh), April 3, 1852.
3   Painted by Waterman Lilly Ormsby (1834–1908). Library of Congress, Prints and Photographs division, digital ID cph.3a08969. For more on the idiom of masculinity in the 1852 election, see Amy S. Greenberg, "The Politics of Martial Manhood. Or, why falling off a horse was worse than falling off the wagon in 1852," *Common-Place: The Interactive Journal of Early American Life* 9.1 (October 2008).
4   See Nathaniel Hawthorne, *Life of Franklin Pierce* (Boston: Ticknor, Reed, and Fields, 1852). The other volume was by David W. Bartlett (better known for a biography of candidate Abraham Lincoln in 1860). See Bartlett, *The Life of Gen. Franklin Pierce of New-Hampshire, The Democratic Candidate for President of the United States* (Auburn: Derby & Miller, 1852).
5   "New Hampshire," *Semi-Weekly North-Carolina Standard* (Raleigh), April 3, 1852. Also printed in the *Weekly North-Carolina Standard* (Raleigh), April 7, 1852.
6   "The Nominations!" *Washington Daily Union* (Washington, DC), June 6, 1852.
7   All these quotes are from "Voice of the Democratic Press," *Washington Daily Union* (Washington, DC), June 8, 1852.
8   "Two More Whig Flags Struck!" *Richmond* (VA) *Enquirer,* June 15, 1852. The story is also reprinted in the *North Carolina Standard* and other newspapers.
9   "The Nominations," *Richmond* (VA) *Enquirer,* June 8, 1852.
10   "First Gun!—Ratification Meeting in Onslow," *Wilmington* (NC) *Journal,* June 11, 1852.
11   "The Nominations of the Baltimore Convention," *The North-Carolinian* (Fayetteville), June 12, 1852.
12   Mott estimates that the partisan press was about equally divided in its opinion on Pierce, with 48 percent in support and 52 percent opposed. See Frank Luther Mott, "Newspapers in Presidential Campaigns," *Public Opinion Quarterly* 8, no. 3 (Autumn 1944), 348–67.
13   "The Locofoco Candidates," *The Cecil Whig* (Elkton, MD), June 12, 1852. King (1786–1853) died of tuberculosis six weeks after the election.

14 "Franklin Pierce and the Freesoilers," *The Republic* (Washington, DC), June 14, 1852.

15 "The Hunker Democracy—its Convention and Candidates," *Green-Mountain Freeman* (Montpelier, VT), June 17, 1852. A Hunker was a derogatory term denoting members of the conservative wing of the Democratic Party, especially in New York State, 1845–1848, a faction which tended to minimize the slavery issue. The *Freeman* was edited by Daniel P. Thompson, a well-known author and attorney, and early member of the Liberty Party in the 1840s.

16 See the *Claremont Eagle*, June 10, 1852, quoted in "Gen. Pierce's Military Career," *The Spirit of Democracy* (Woodsfield, OH), June 30, 1852.

17 *The Friend; a Religious and Literary Journal* (Philadelphia, PA), 27, no. 20 (January 28, 1854): 160.

18 "The Nebraska Bill," *New York Observer and Chronicle* (New York), 32, no. 5 (February 2, 1854): 38.

19 "THE INTENTION TO COMMIT A CRIME," *The Independent … Devoted to the Consideration of Politics, Social and Economic Tendencies, History, Literature, and the Arts* (New York, NY), 6, no. 270 (February 2, 1854): 36.

20 "The Nebraska Bill—The Argument," *National Era* (Washington, DC), 8, no. 371 (February 2, 1854): 22.

21 "WHO ARE TO BLAME?" *The Independent … Devoted to the Consideration of Politics, Social and Economic Tendencies, History, Literature, and the Arts* (New York), 6, no. 271 (February 9, 1854): 44.

22 "The Nebraska Bill and Its Future Consequences," *Christian Inquirer* (New York, NY), 8, no. 23 (March 11, 1854): 2.

23 "THE INTENTION TO COMMIT A CRIME," *The Independent … Devoted to the Consideration of Politics, Social and Economic Tendencies, History, Literature, and the Arts* (New York, NY), 6, no. 270 (February 2, 1854): 36.

24 *Fremont Journal* (Fremont, Sandusky County [Ohio]), August 25, 1854.

25 "Elections," *The Athens* (TN) *Post*, November 17, 1854.

26 Quoted in *The North Carolinian* (Fayetteville), November 18, 1854.

27 Quoted in "Northern Democrats and Slavery," *The South-Western* (Shreveport, LA), November 29, 1854.

28 *The South-Western* (Shreveport, LA), November 29, 1854.

29 *The South-Western* (Shreveport, LA), November 29, 1854.

30 "The Administration," *Washington Daily Union* (Washington, DC), June 30, 1854. The story also appeared in other newspapers, including the *Nashville* (TN) *Union and American*, July 6, 1854.

31 *The North-Carolinian* (Fayetteville), November 18, 1854.

32 The *Washington Union* or *Washington Daily Union* was an "administration organ," according to Emery, founded in 1845 under the Polk administration, edited from 1853 to 1856 by Alfred O. P. Nicholson, a journalist and democratic politician and lawyer from Tennessee. See Fred A. Emery, "Washington Newspapers," *Records of the Columbia Historical Society, Washington, D.C.* 37/38 (1937): 41–72. See also Carroll Van West, "Alfred Osborne Pope Nicholson," *Tennessee Encyclopedia of History and Culture*, online ed. (University of Tennessee Press, 2018).

33 "The New York Democracy," *Washington Daily Union* (Washington, DC), September 10, 1854.

34 "The Condition of the two Old Parties," *The Yazoo Democrat* (Yazoo City, MS), July 19, 1854.

35 Quoted in "The Condition of the two Old Parties," The Yazoo Democrat (Yazoo City, MS), July 19 1854.

36 Ibid.

37 *The Athens* (TN) *Post*, July 7, 1854.

38 "Owning Up," *Burlington* (VT) *Free Press*, September 1, 1854.

39 "All Hail Iowa?" *Windsor* (VT) *Journal*, quoted in *Burlington* (VT) *Free Press*, September 1, 1854.

40 Quoted in *The Athens* (TN) *Post*, November 17, 1854. Emphasis in original.

41 "The Late Elections," *The North-Carolinian* (Fayetteville), October 21, 1854.

42 Ibid.

43 "Northern Fanaticism," *Weekly North Carolina Standard* (Raleigh), September 13, 1854.

44 *Weekly North Carolina Standard* (Raleigh), October 11, 1854.

45 "Northern Fanaticism," *Weekly North Carolina Standard* (Raleigh), September 13, 1854.

46 "Dissolving the Union," *The Yazoo Democrat* (Yazoo City, MS), July 19, 1854.

47 "The National Democracy," *Richmond* (VA) *Enquirer*, September 1, 1854.

48 "Who are the Agitators?" *The Ottawa* (IL) *Free Trader*, November 4, 1854.

49 "The Democratic Party and the Nebraska Question," *Washington Daily Union* (Washington, DC), September 26, 1854. (Reprinted from *The People's Advocate* (Salem, MA), September 23, 1854.)

50 "The Man for the Times!" *Washington Daily Union* (Washington, DC), June 8, 1852. Emphasis in original.

51 For more on Bleeding Kansas and the consequences for the Pierce administration, see Michael J. C. Taylor, "Governing the Devil in Hell: 'Bleeding Kansas' and the Destruction of the Franklin Pierce Presidency (1854–56)," *White House Studies* vol. 1, no. 2 (2001), 185–95. Taylor argues that the president was an unwitting victim of the situation and not personally culpable—indeed, that "no one leader could have bound the nation together at this juncture…" (193).

10

ABOLITIONISM, THE KANSAS-NEBRASKA ACT, AND THE END OF COMPROMISE

Dianne M. Bragg

If newspaper articles at the time are any indication, passage of the Kansas-Nebraska Act in 1854 fueled the expansion of antislavery politics more than almost any other Antebellum event. Stephen Douglas's backroom deals to help secure Southern support for the act quickly ignited Northern opposition. The decision to allow the Kansas and Nebraska territories to determine their own position on slavery, under the veil of excluding federal oversight, proved to be a fateful one. The Kansas-Nebraska Act dismissed the Missouri Compromise of 1820 (which prohibited slavery above the 36°30' latitude) and affirmed slaveowners' rights to demand the return of fugitive slaves.

The sacking of Lawrence, Kansas, John Brown's infamous actions, and other horrors of "Bleeding Kansas" in the mid- to late-1850s were front-page stories across the nation, focusing the public's attention on the growing political crisis. Newspaper coverage during this time gauges to some extent the effect of the slavery debate on the nation's political and legislative landscape. With the outbreak of violence in Kansas, newspapers began to more frequently question not if, but when, the two sides in the slavery debate would find it impossible to cohabitate not only in Kansas, but in the halls of national government as well.

In their coverage of abolitionists, newspapers North and South initially often referred to them as "fanatics" and sometimes as "ultras," a term usually reserved for Southern extremists.[1] An indication of this mindset was how closely newspapers watched for the election of abolitionists to political positions. In July 1851, as the United States was on the eve of its seventy-fifth Independence Day celebration, the *Raleigh* (NC) *Register* printed an article praising the Massachusetts Democratic State Central Committee for taking a stand of "non-interference with the rights of the South." The *Register* had a specific reason for publishing this article—the Democratic Central Committee had taken action in response to a group of Massachusetts Free Soilers and Democratic legislators who had formed a coalition that resulted in abolitionist Charles Sumner's election to the US Senate.[2] The *Register* article encouraged Northern Whigs and Democrats to break their alliances with the abolitionist movement and its supporters. It was in the South's interest to move the political debate away from sectional issues, of which slavery was chief. Establishing a connection between abolitionists and sectionalism would help the South in its efforts to portray abolitionists as agitators willing to sacrifice the Union for their cause.

Closely aligned with concerns over abolitionists were fears of Northern threats to repeal the Fugitive Slave Law, which expanded a clause in Article IV of the Constitution requiring that escaped slaves be returned to their owners.[3] Northerners were incensed by the law's passage and Southerners had serious doubts that it would ever be enforced. The *Daily Alabama Journal* was not alone when it said the South would resist any "unconstitutional aggression—the repeal or disregard of the fugitive slave bill."[4] The *Mississippian and State Gazette* accused abolitionists of supporting these and other Northern "agitations" toward the South.[5]

As the 1850s progressed, prominent abolitionist leaders had become so well known that some newspaper articles began using their last names when referring to abolitionists in general, knowing that their readers would understand the reference. The *Pensacola* (FL) *Gazette* referred to Boston abolitionists as "the Garrisons and Phillipses, and other miserable fanatics," and joined other newspapers in publishing critical articles aimed directly at prominent abolitionists and their actions.[6] New York's *Weekly Herald*, owned by James Gordon Bennett, criticized prominent New York abolitionist Gerrit Smith.[7] The paper referred to abolitionism as a "clique" and claimed that even black voters in the North were not supporting the abolitionist party.[8] In response to abolitionist calls for freedom, Southern newspapers frequently carried articles that questioned whether slaves were interested in the abolitionist cause and emancipation, or if they might actually prefer their current position in the South.

By 1854, the abolitionist movement had forged inroads into the political mainstream with the debate over the organization of the Nebraska and Kansas territories. The Kansas-Nebraska Act, which President Franklin Pierce signed into law on May 30, 1854, organized these territories on the basis of popular sovereignty, allowing them to make their own decisions about slavery. The act essentially repealed the Missouri Compromise of 1820, which, among other things, prohibited slavery north of the 36°30' parallel (excluding Missouri), and upheld the Fugitive Slave Law from the Compromise of 1850. Southern newspapers reported serious concerns over allowing the citizens of Kansas and Nebraska to hold elections to decide their positions on slavery. Northern newspapers reported that Northerners were incensed over the act, and abolitionists wasted no time in channeling that anger.

Abolitionists argued that it was unconstitutional to pass an act that would repeal the Missouri Compromise of 1820. To bolster their position, abolitionists began a campaign to uphold the Missouri Compromise and joined with members of the Free Soil Party, which was opposed to slavery and its Western expansion. Missourians, not surprisingly, quickly became involved in the debate over the compromise that bore its name. The *Missouri Courier* published an article that cited support for the Kansas-Nebraska Act and criticized abolitionists, Free Soilers, and Whigs for fighting against the legislation. The *Courier* took the opportunity to take Southern Whig papers to task for avoiding coverage of "slave agitation" and failing to comment on the political debate because it had slavery as its focus. The *Courier* said the Whig press hoped to "impress the Abolition hordes with the belief that they can trample the Constitution and ride rough-shod over the South, who will not resist their encroachments for fear of agitating the subject of slavery."[9]

The Kansas-Nebraska Act created a political stew that would keep Kansas in a state of confusion and violence for years to come. One especially long-lasting political result that can be attributed to the Kansas-Nebraska Act was the creation of the Republican Party, which was based on an antislavery platform that appealed to Free Soilers and some former Whigs. Newspaper articles indicate that political chaos reigned during this period as the Whigs fell apart and the new Republican Party was being born.

The *Augusta* (GA) *Chronicle*, along with other Southern newspapers, reported on a "Missouri Pro-Slavery Convention" held in St. Louis on July 18, 1855. The convention platform noted that Missourians who owned slaves feared that Kansas might be overrun with abolitionists who were hoping to influence the decision on slavery. Their concern was that abolitionist settlers might spill over into Missouri, the article said, and create problems for slaveholders, especially those who lived on the border. The *Chronicle* cited reports of "monied associations" planning to colonize Kansas for the purpose of keeping out slavery. Missouri slaveholders made it clear that if abolitionist settlers emigrated from the Northern states to Kansas, they were prepared to fight for their rights. The concern focused on the economic impact on Missouri slaveholders if the abolitionists gained a strong foothold in Kansas. "While disclaiming any intention to interfere with actual settlers," the *Chronicle* maintained, "they will protect themselves and property, as the eighteen border counties of Missouri contain 50,000 slaves, which will be valueless if Kansas becomes the abode of abolition fanatics."[10]

In Boston, a city that was a hotbed of abolitionism, the *Boston Daily Atlas* played down the seriousness of the abolitionists' threats and turned its editorial pen on Missourians for overreacting and threatening violence.[11] Other newspapers asserted that it was the abolitionists who were working to force their will upon Kansas. The *Pittsfield* (MA) *Sun* reprinted a *Chicago Times* article containing an excerpt from the *Peoria* (IL) *Press* that expressed outrage over the threat of violence in Kansas, but laid most of the blame on abolitionists. The *Peoria Press* wrote that it did not "defend acts of 'lawless violence' alleged to have been perpetrated in that territory," but asked, "who is responsible for this state of things?"[12]

Newspapers everywhere reported that Kansas had become a battleground between the two opposing factions. According to some newspaper reports, Nebraska would be spared because its government had quickly passed legislation to ensure that only "bona fide settlers and citizens of Nebraska" would be allowed to vote on whether the state would have slavery.[13] Kansas's government remained in flux, though, with ongoing arguments over who was actually in charge. Nebraska's agricultural landscape, with crops such as wheat, did not require the amount of manpower that Southern crops, such as cotton, needed. This essentially made the slavery argument there a moot point. In Kansas, although the agriculture may have been similar to Nebraska, the state's proximity to Missouri ensured that it would become a battleground.

Newspapers reported that the lack of any firm political leadership led to the hostile climate that ultimately resulted in the attack on Lawrence, Kansas. President Pierce offered federal support to calm the Kansas territory, but it was too little and too late. Pierce's failure in Kansas opened him up for criticism from both free-state and proslavery supporters. As Missouri slaveholders had feared, Northern immigrants had been successful in making Lawrence a Free-State stronghold in Kansas. In May 1856, a territorial grand jury issued indictments against three Free-State leaders in Lawrence, along with two Lawrence newspapers and a hotel that was thought to be a Free-State fortress. All of the charges were linked to Free-State efforts to organize and train military forces. On May 21, a lack of cooperation from the town's citizens thwarted a federal marshal's attempt to make arrests in Lawrence. Further complicating matters, the marshal had recruited proslavery volunteers from Missouri to assist him, a fact that did not help him in his effort to arrest Free-State men.[14] Georgia's *Daily Morning News* came at the story from a proslavery perspective and denounced the citizens of Lawrence for not cooperating with federal authorities. "The conduct of the abolitionists," the paper declared, "has been such to arouse the law and order men to some definite action. The people of Lawrence have resolutely refused to give up those against whom there are warrants out for crimes of any sort." It was impossible to get testimony from any proslavery men in Lawrence, the article added, because "it was dangerous for any pro-slavery man to remain there over night."[15] Ultimately, the marshal's proslavery volunteers were dismissed, but they soon reorganized and headed back to Lawrence on the afternoon of May 21 under the guidance of Sheriff Samuel Jones, who had also been thwarted in a previous attempt to make arrests in Lawrence. Jones, who had been shot a few days after that failure, promised that this time he and the proslavery men with him would not go quietly.[16]

Several newspapers offered an abolitionist and Free-State perspective of events. The *Daily Cleveland* (OH) *Herald* reported that John Speer, the editor and proprietor of the *Kansas Tribune*, and another man from Lawrence had visited the *Herald*'s offices. The two men had fled Lawrence to avoid arrest for "treason" and "resisting the officers." They said the "bogus" sheriff and marshal had arrest warrants for all the "leading Free-State men in the Territory." As a result, the people of Lawrence were now at the mercy of an "enormous posse of ruffians who are now encamped around the city," and the people had no leadership. "The ruffians have come from Missouri," the *Tribune* editor told the paper, "and they *are not* and *have not been* inhabitants of the Territory!"[17] A Massachusetts newspaper, the *Lowell Daily Citizen and News*, stated that a "dark cloud hangs over the noble settlers in this territory." United States territorial officials had engaged "southern emigrants" to form a militia, the *News* declared, and were "endeavoring to provoke Lawrence to some act that might serve as a pretext for the destruction of that place and the shooting of its inhabitants. We must be prepared any day to hear of a horrible massacre of the inhabitants of that city, unless the iniquity of our government is overruled by Providence."[18]

In Washington, DC, the *Daily National Intelligencer* warned that the "reign of terror" was complete in the territory, and reported on a failed attack directed toward a "Mr. Brown," editor of the *Herald of Freedom*, a Free-State paper in Lawrence. Brown had written that "a mob had entered his hotel at Kansas city and dragged off a man supposed to be himself, but having discovered their mistake returned and demanded him."[19] Early newspaper reports on events in Lawrence often contained numerous inaccuracies. On May 26, the *New York Herald* printed the following sensational headline: "Important from Kansas. The Town of Lawrence Destroyed. Intense Excitement in the Territory. Are We to Have Civil War?" The article came from the *Herald*'s Lecompton, Kansas, correspondent and included statements on the proslavery forces and their actions in Kansas. "The Kansas war is again in full blast," the *Herald* declared: "the excitement is at its height—the pro slavery forces are coming in from every quarter."[20] The *Daily Scioto Gazette* (OH) published an article, purported to be from an eyewitness in Lawrence, that detailed several murders, along with horse stealing and pillaging.[21] The *Bangor* (ME) *Daily Whig & Courier* also reported that several Lawrence citizens were murdered.[22] Eventually, however, the truth, which was sensational enough on its own, began to come through.

Later newspaper reports put together the following, more accurate, account. As planned, the Missouri men had reentered Lawrence with the sheriff. They proceeded to loot the town, reportedly stealing valuables and whiskey, and then fired cannons at the hotel. Finding the cannons not to be very effective, they burned the hotel, along with at least one other building, and destroyed the printing presses of Lawrence's two newspapers, the *Herald of Freedom* and *Kansas Free State*. Despite early newspaper accounts to the contrary, no one was murdered at this time. The *Daily National Intelligencer* (Washington, DC) carried an article that was critical of the attack and accurately reported that one member of the "posse" who attacked the town was killed, but that the death occurred as a result of an accident. The *Intelligencer* article gave far more detail about what had happened, including a chronology of the attack.[23] Some newspapers wisely admonished the public to take care what they believed and to pay attention to the political motives behind some news stories. "Most of the horrible stories about Kansas constantly manufactured by the Black Republicans," the *Pittsfield* (MA) *Sun* asserted, "are understood by the public generally as being gross impositions upon the community, and designed for political effect." Stories about murder and outrages in Kansas, the paper added, should be viewed with caution. "Very little confidence is to be placed in any thing the Kansas agitators say in regard to that Territory."[24]

Overall, Northern newspapers expressed anger about what had happened in Lawrence, even if no Free-State men had been murdered. Some said that the events in Kansas were a blow against freedom and a victory for Southerners. There was anger that the federal government had played a role in the Lawrence attack and that federal officials had joined forces with the Missouri "ruffians." "To please and appease the Carolinians," the *Farmer's* (NH) *Cabinet* wrote, "Lawrence is sacked and burnt by U.S. officials, or those acting under them, and her free presses sunk in the depths of the Missouri, simply because they were free!"[25]

Evidence seemed to be mounting that Pierce's administration was far too supportive of the Missouri "ruffians." Without question, there was a political backlash over the federal government's handling of affairs in Kansas as newspapers on both sides blamed Pierce for the escalation of violence. The *Bangor* (ME) *Daily Whig & Courier* said the administration had deliberately lied about events there in the hope that most Free-State people would "believe it was done by the fault of the Emigrant Aid Society."[26] In the South, the *Macon* (GA) *Weekly Telegraph* printed a long article detailing the US military's futile and insufficient efforts to quell the violence in Kansas. The article stated that it did not fully blame the president or his administration for the situation, but thought that federal officials had not carried out their duties. However, the *Telegraph* did express indignation that the US army was initially sent to Kansas under Col. Edwin V. Sumner's command, and the paper accused Sumner of protecting those truly responsible for any bloodshed in Kansas. "Peace will never be restored until the Abolitionists are 'paid in their own coin'—one *good* whipping will settle the whole controversy, and it will never be done unless the citizens *are let alone*," according to the *Telegraph*. "When nothing will do a man, or set of men but a fight, there is no recourse but to accommodate them."[27]

Amidst all this turmoil, some newspapers noted another incident in Kansas, but the coverage was not as prominent as might be expected. On the evening of May 24, 1856, just days after the sacking of Lawrence, abolitionist John Brown, along with some cohorts, raided several cabins belonging to proslavery settlers in the Osawatomie area along the Pottawatomie Creek. On that fateful night and into the early morning hours of May 25, Brown's gang committed five atrocious murders. In the first incident, they dragged James Doyle (referred to as John Doyle in several newspapers) and his two eldest sons out of their cabin, leaving Doyle's wife and his youngest son inside. Once outside, the gang shot Doyle and killed his two sons with broadswords. They then hacked and mutilated the bodies, all within earshot of Doyle's wife and youngest son.[28] The gang then moved on and attacked the cabin of Allen Wilkinson, a member of the territorial legislature. The gang hacked Wilkinson to death while his wife pleaded for her husband's life. Brown and his men continued on their murderous rampage to the cabin of James Harris, where they killed William Sherman, a guest, and mutilated his body. Brown's gang seemed very specific about their targets, and some newspapers reported that the victims were going to testify against Brown in relation to charges against him for various abolitionist activities in Kansas.

Although the attacks were horrific, newspaper coverage of them was almost lost in the midst of other news from Kansas. Of those that did carry the story, many were filled with factual errors. Most of the articles mistakenly reported that eight men were killed, rather than five. The *New York Herald* carried a few short articles and reported that "eight pro-slavery men were killed by a company of abolitionists on Pottawatomie creek." One of the *Herald*'s articles did refer to the murders as a massacre.[29] On June 3, the *Herald* reported on the murders briefly as part of a long article with the headline, "Exciting News from Kansas." This report made some mention of the graphic nature of the crimes and the fact that horses belonging to the victims were stolen, but the information was buried far down in the article. Other newspapers, both Northern and Southern, carried articles similar in content, size, and placement.[30] The *Daily National Intelligencer* (Washington, DC) mentioned the murders in only one sentence in an article that said help was needed for proslavery settlers in the territory in light of the Pottawatomie creek murders.[31] The *North American and United States Gazette* (Philadelphia, PA) carried a small article, taken from the *Westport* (MO) *Border Times*, with the headline "War! War! Eight Pro-slavery Men Murdered by the Abolitionists in Franklin County, K. T.—Let Slip the Dogs of War!" The article and its headline were placed several paragraphs beneath a bold headline, "The Kansas Troubles."[32]

A few newspapers paid greater attention to the incident, with the *Semi-Weekly Mississippian* carrying two articles on June 13 that relayed the brutal details and sought revenge for the deaths. "For every Southern man thus butchered," the article said, "a decade of these poltroons should bite the dust."[33] The *Charleston* (SC) *Mercury* published an article criticizing the murderers for being cowardly for committing their crimes at night. "The Abolitionists are again under arms," the *Mercury* said. "So soon as the militia dispersed, they began to murder in the dark. There is no open fight in them." The *Mercury* warned that such incidents would soon lead to war.[34] On June 8, the *New York Herald* published a lengthy article with graphic descriptions of what had occurred. "We are in the midst of a terrible state of things," the article said. "A warfare which would disgrace the brutal Indian is now being waged by the free State party in Kansas." The article's author noted that it would be hard for people to believe the "horrors which I have to communicate."[35]

Some newspapers carried reports of arrests that were made in the case. The *Boston* (MA) *Daily Advertiser* ran a small piece noting that the men responsible for the murder of the proslavery men had been arrested and that their leader had been identified as John Brown.[36] The *Charleston* (SC) *Mercury* carried a few lines on July 4 noting that all of the men except three had been released, "as nothing could be found against them."[37] One of the men who remained in custody was John Brown, Jr., although he was not present during the murders.[38] Why didn't these murders receive as much newspaper attention as the attack on Lawrence or other such incidents in Kansas? False or inaccurate information may be partly to blame. "The reports are indeed conflicting and contradictory as to specific facts," the *Daily Picayune* of New Orleans stated.[39] It could be that confusion over what really happened left editors and readers unsure as to the truth and the story never garnered the attention it deserved. Some newspapers and their readers probably believed the incident was exaggerated for the purpose of exciting more anger against the Free-State men. And it is possible that, just as the *New York Herald* writer said, the incident was so horrific that it was simply too hard to believe.

Newspaper coverage shows that the entire country was transfixed by events in Kansas as it had, indeed, become a battleground. The rush of emigrants from the North to settle the territory and influence its future created chaos. Armed altercations, or just the threat of them, between Free-State and proslavery men became regular occurrences. It was in this environment that newspapers began to use terms like "Bloody" and "Civil War" when writing the news from Kansas. Although the *New York Herald* is often given credit for the gruesome title "Bloody Kansas," it was not alone in using such terminology. The *Ripley* (OH) *Bee* carried the following headline on October 13, 1855: "Exciting News from Kansas—The Cut Throats Commencing Their Bloody Work—Southern Hordes Pouring In— Gov. Shannon Taking Sides with the Ruffians." The *Herald* and other newspapers used headlines such as "The Bloody Code of Kansas," "Bloody Work in Kansas Close at Hand," "Bloody News in Kansas," and "Another Bloody Chapter in the History of Kansas."[40] Newspapers often referred to the strife in Kansas as a "civil war," a description used repeatedly in headlines and articles. "Civil War with all its horrors, now rages in Kansas Territory," the *Macon* (GA) *Weekly Telegraph* declared in June 1856.[41]

Events in Kansas ensured that slavery would be front and center in the 1856 presidential election. The *Macon Telegraph* gave an indication of just how crucial the slavery issue would be for the South. The paper was extremely critical of Millard Fillmore, who would be the American Party candidate, and asserted that he was the least deserving of Southerners' support. "According to his own admissions he drank in a prejudice to slavery with his mother's milk," the *Telegraph* stated. "The whole history of his life proves his bitter hostility to the South and her institutions."[42]

Although it first appeared the South had won victories with the Fugitive Slave Act and the Kansas-Nebraska Act, newspapers show that the aftermath proved otherwise. Antislavery forces used Northern anger over the Kansas-Nebraska Act and the subsequent violence to their advantage. In cases such as the sacking of

Lawrence, Free-State settlers in Kansas received favorable press, while newspapers often portrayed the Missouri "ruffians" as violent aggressors. Inexplicably, abolitionist John Brown's murderous Osawatomie attack did not garner enough accurate attention in either the Northern or the Southern press to aid the proslavery settlers' cause. Abolitionists managed to get the press attention they wanted, and the South could clearly read the impact.

Northern newspapers often portrayed the violence in Kansas as the fault of Southern slaveholders. Even though these papers sometimes considered abolitionist settlers to be troublemakers, they put the brunt of the blame for the violence on slaveholders. Thus, readers in the North saw Southerners as "ruffians," while abolitionists like John Brown received a pass.

On June 11, 1856, the *Charleston* (SC) *Mercury* reprinted a letter to the editor from D. G. Fleming that had appeared in the *Carolina Times*. The letter was sent from Leavenworth City, Kansas, and described the heightened tension in the region, particularly in Lawrence. "It is hard to say what all this will lead to," Fleming wrote, "I think that one party or the other will have to be driven from the Territory; they cannot live together."[43] As Northern and Southern readers gathered their information from their newspapers' pages, the nation's division over slavery clearly appeared to have deepened to such a degree that a political compromise or resolution seemed a distant and unlikely hope.

Notes

1 Merton Dillon, "The Failure of the American Abolitionists," *Journal of Southern History* 25, no. 2 (May 1959): 159–77; Jane H. Pease and William H. Pease, "Confrontation and Abolition in the 1850s," *Journal of American History*, 58, no. 4 (March 1972): 923–37.

2 Frederick J. Blue, *The Free Soilers; Third Party Politics, 1848–54* (Urbana: University of Illinois Press, 1973); Eric Foner, *Free Soil, Free Labor, Free Men: the Ideology of the Republican Party before the Civil War* (New York: Oxford University Press, 1995).

3 The text of the Constitution's Fugitive Slave Law clause appears in Article IV, section 2, clause 3, of the US Constitution and reads: "No Person held to Service or Labour in one State, under the Laws thereof, escaping into another, shall, in Consequence of any Law or Regulation therein, be discharged from such Service or Labour, but shall be delivered up on Claim of the Party to whom such Service or Labour may be due." This clause would later be superseded by the Thirteenth Amendment. (www.archives.gov/exhibits/charters/constitution_transcript.html)

4 "The Latest Phase," *Daily Alabama Journal* (Montgomery), October 23, 1850.

5 "The Wilmot Proviso again Threatened—Agitation for the Repeal of the Fugitive Slave Law still Increasing—another Warning to the South," *Mississippian and State Gazette* (Jackson), August 22, 1851.

6 "Another Fugitive Slave Case in Boston," *Pensacola* (FL) *Gazette*, April 19, 1851.

7 Laurence M. Vance, "A Radical Nineteenth-Century Libertarian," *Independent Review* 13, no. 3 (Winter 2009): 431–39.

8 "Gerrit Smith's Manifesto to the Wooly Heads," *Weekly Herald* (New York, NY), December 28, 1850.

9   "The Great Question," *Missouri Courier* (Hannibal), March 9, 1854.
10 "Missouri Pre-Slavery Convention," *Augusta* (GA) *Chronicle*, July 22, 1855.
11 *Boston* (MA) *Daily Atlas*, April 25, 1855.
12 "From the Chicago Times. Parties in Kansas. —Massachusetts and Missouri Emigrants," *Pittsfield* (MA) *Sun*, September 27, 1855.
13 Ibid.
14 David M. Potter, "Two Wars in Kansas," in *The Impending Crisis, 1848–1861* (New York: Harper Torch Books, 1976), 207–09.
15 "Kansas Affairs," *Daily Morning News* (Savannah, GA), May 23, 1856.
16 Potter, "Two Wars in Kansas."
17 "Just from Kansas," *Daily Cleveland* (OH) *Herald*, May 23, 1856.
18 "Kansas," *Lowell* (MA) *Daily Citizen and News*, May 23, 1856.
19 *National Intelligencer*, May 22, 1856.
20 "Important from Kansas," *New York Herald*, May 26, 1856.
21 "The Mob at Lawrence. Particulars by an Eye Witness," *Daily Scioto Gazette* (Chillicothe, OH), May 31, 1856.
22 *Bangor* (ME) *Daily Whig & Courier*, June 4, 1856.
23 "The Outrages in Kansas," *Daily National Intelligencer* (Washington, DC), May 31, 1856.
24 "The Lies about Kansas," *Pittsfield* (MA) *Sun*, June 19, 1856.
25 "To What Are We Coming?" *Farmer's* (NH) *Cabinet*, June 5, 1856.
26 *Bangor* (ME) *Daily Whig & Courier*, May 23, 1856.
27 "The U.S. Army in Kansas," *Macon* (GA) *Weekly Telegraph*, September 23, 1856.
28 The young son who survived, John Charles Doyle, moved to Tennessee with his mother, Mahala Childress Doyle. In June 1861, after Tennessee seceded, Doyle joined the Confederate army and served until 1865.
29 "Kansas Affairs. Eight Pro-Slavery Men Killed—All Quiet at Lawrence, Lecompton and Franklin," *New York Herald*, June 4, 1856; "From Kansas," *New York Herald*, June 5, 1856; "Kansas City, Mo.," *New York Herald*, June 8, 1856.
30 "Exciting News from Kansas," *New York Herald*, June 3, 1856; "Intelligence from Kansas," *Boston* (MA) *Investigator*, June 11, 1856; "Exciting Rumors from Kansas," *Charleston* (SC) *Courier, Tri-Weekly*, June 10, 1856; "From *The Carolina Times*. From Kansas," *Charleston* (SC) *Mercury*, June 11, 1856.
31 "Missourians Preparing for a Foray," *Daily National Intelligencer* (Washington, DC), June 9, 1856.
32 "The Kansas Troubles. From the Westport Border Times. *War! War! Eight Pro-slavery Men Murdered by the Abolitionists in Franklin County, K. T. —Let Slip the Dogs of War!*" *North American and United States Gazette* (Philadelphia, PA), June 6, 1856.
33 "Still Later from Kansas. —Eight Men Brutally Murdered by Abolitionists," *Semi-Weekly Mississippian* (Jackson), June 13, 1856.
34 "Exciting News from Kansas," *Charleston Mercury*, June 16, 1856.
35 "The Outrage and Murder Committed by the Free State Men upon the Pro-Slavery Inhabitants," *New York Herald*, June 8, 1856.
36 "From Kanzas [sic]," *Boston* (MA) *Daily Advertiser*, June 6, 1856.
37 "Letter from Kansas," *Charleston* (SC) *Mercury*, July 4, 1856.
38 C. Vann Woodward, "John Brown's Private War," in *The Burden of Southern History* (Baton Rouge: Louisiana State University Press, 1960), 41–68.
39 "Kansas Troubles Thickening," *Daily Picayune* (New Orleans, LA), June 18, 1856.

40 Some newspapers that printed such headlines were *Boston* (MA) *Daily Atlas*, June 7, 1856; *New York Herald*, August 28, 1856; *Fayetteville* (NC) *Observer*, September 4, 1856; *Milwaukee* (WI) *Daily Sentinel*, September 6, 1856; *Bangor* (ME) *Daily Whig & Courier*, October 8, 1856; and *New York Herald*, February 27, 1857.

41 *Macon* (GA) *Weekly Telegraph*, June 17, 1856.

42 "Three Acts in the Life of Millard Fillmore. By His Acts Shall He Be Judged," *Macon* (GA) *Weekly Telegraph*, September 30, 1856.

43 "From Kansas," *Charleston* (SC) *Mercury*, June 11, 1856.

11

"LIKE SO MANY BLACK SKELETONS"

The Slave Trade through American and British Newspapers, 1808–1865

Thomas C. Terry and Donald L. Shaw

"A story is going the rounds of a slave ship chased, a year or two ago, by two British cruisers," the *Baltimore Sun* told its readers on October 13, 1846.[1] "She passed her 300 slaves up from the hold, tied shot to their feet, and slipped them overboard— then [hid] all the extra water, provisions, lumber, &c. and when overhauled by the cruiser exhibited regular papers, and was released."[2]

This chapter examines newspaper coverage in British and American newspapers from 1808 to 1865 to determine whether the transatlantic slave trade did in fact disappear after 1808, by which time both Congress and the British Parliament had officially banned it. There were nearly one hundred slave ships clearly identifiable in the study period, evidence of a thriving global trade, with nearly thirty of the ships likely bound for or from the United States. More than thirty-five British and American warships were mentioned by name.

Some historians believe very few slave ships plied the Atlantic after 1808 when Congress prohibited the slave trade. Their research is largely based on the Trans-Atlantic Slave Trade Database, which documents slave ship trips from 1501 to 1875 and is widely considered by historians to be the definitive source.[3] Over that span, according to the database, as many as 12.5 million Africans were forcibly taken to the Western Hemisphere from Africa and sold into slavery.[4] The first slaves were brought to Virginia in 1619, a year before the Pilgrims landed at Plymouth Rock and just a dozen years after the founding of Jamestown. The vast majority went into brutal slavery in South America and the Caribbean to work on sugar plantations. Five million alone were transported to Brazil. According to historian Henry Louis Gates, Jr., based on the Trans-Atlantic Database figures, 388,000 Africans were exported straight to the US, while another 70,000 initially landed in the Caribbean.[5]

The British West Africa Squadron captured 1,600 slave ships during the fifty-two years after the official cessation of the transatlantic slave trade.[6] Moreover, those are just the ships caught; it is impossible to speculate accurately about the number that escaped detection and/or interception entirely or that completed an unknown number of transits before being captured. Submitting documentation, shipping manifests, and other paperwork to authorities, for obvious reasons, would not be high on the agenda for captains and ship owners engaged in a highly profitable, but illicit trade.

Britain in the eighteenth century was the greatest and largest slave-trading nation in history, its navy defending, expanding, and enabling slavery. And the prime destinations were its American and Caribbean colonies. In 1785, Member of Parliament William Wilberforce became Britain's most vocal abolition advocate, spearheading efforts to eliminate the transatlantic slave trade. Prime Minster William Grenville, Lord Grenville, and Foreign Secretary Charles James Fox were also leaders of the campaign in Parliament. The Abolition of the Slave Trade Act (1807) passed the House of Lords on February 5, 1807 and the House of Commons on February 23. George III signed it (the Royal Assent) on March 25. The law abolished the slave trade throughout the British Empire, but it was not until the Slavery Abolition Act of 1833 that slavery itself was prohibited.

In the immediate aftermath of the American Revolution, slavery was eliminated in Massachusetts and New Hampshire. In Pennsylvania, Rhode Island, Connecticut, and New York, gradual abolition was legislated.[7] Delegates to the Constitutional Convention in 1787 established a twenty-year time limit on the transatlantic slave trade that allowed for its eventual elimination in 1808. After 1800, Georgia and South Carolina were the only states that officially allowed international slaving. Congress put some limitations on the trade in the Slave Trade Acts of 1794, 1800, and 1803 that addressed slave ships and the involvement of American citizens with them, while dodging the issue of the slave trade itself. None of these laws prohibited the landing of slaves or gave them their freedom. They also did not ban the trade outright. That was left to the Abolition of the Slave Trade Act of 1807. On March 2, 1807, Congress banned "the importation of slaves into any port or place within the jurisdiction of the United States," effective January 1, 1808. President Thomas Jefferson signed the bill the same day. In 1818 and 1820, Congress would amend the 1807 law. Anyone engaged in the foreign slave trade was "adjudged a pirate" and "shall suffer death" if convicted. There were few if any prosecutions.

The US Navy's Africa Squadron, created and extended by treaties with Britain in 1819 and 1842, plied the north Atlantic from 1819 to 1861. In all, the US Navy captured only thirty-six ships from 1844 to 1861.[8] Numbers prior to that are minuscule. On the other hand, between 1808 and 1860, the Royal Navy's West Africa Squadron captured 1,600 slave ships and liberated 150,000 Africans.[9]

Though the possible penalties for being caught were horribly high, it was worth the high profits for some. The average price of a Texas slave, regardless of age, gender, or health, was approximately $800 in 1860,[10] amounting to $21,795 in 2014 money.[11] During the late 1850s, male field slaves, ages 18–30, cost $1,200 ($32,693), while a skilled slave, such as a blacksmith, could cost more than $2,000 ($54,488).[12] By contrast, top Texas cotton land could be purchased for $6 an acre ($163) at the same time.[13] By one estimate, the Ibo tribe in 1859 and the King of Dahomey (modern Benin) in 1860 sold their fellow Africans into slavery for $50 each ($1,362).[14] The cost to the slaver, then, for a 400-slave "shipment" would be about $20,000 in 1860 (or $544,800 in 2014) to realize perhaps $320,000 ($8.8 million) in gross revenue.[15] Slavers typically carried between 200 and 900 slaves, though one made in Liverpool, England, the *Nightingale*, was designed to transport as many as 1,800.[16] Approximately 80 percent of the slave ships were built and fitted in New York.[17] In 1859, there were as many as seven slave ships being fitted out at any given time in New York City with more being built in other major ports.[18] According to Carl Cutler, naval design was spurred by the continuing slave trade, creating fast clipper ships capable of outrunning warships.[19]

In the mid-1830s, most slave ships were built in Baltimore, according to the Kingston, Pennsylvania, *Wyoming Republican and Herald*, and were fitted out for the slave trade in either Baltimore or New York City.[20] The ships were then crewed, many with Americans, and if the newspaper is to be believed, some of the crew did not even know the nature of the voyage until they appeared off the African coast. "The slavers sail from our ports [under] an American flag with American papers and appears like a regular trader," the newspaper added.[21] Sometimes they put into ports in Cuba and Puerto Rico and exchanged their American papers for Spanish ones. The shackles that would eventually chain slaves below decks "are put up in barrels and shipped as merchandise."[22] "When a slaver is expected" on the coast of Africa, the *New York Times* explained, "certain flags are displayed, denoting whether the coast is clear or not, and whether she can anchor with safety" from warships.[23] A slave ship could be loaded with hundreds of slaves in two hours.[24] Slaves, initially captured by local chiefs and leaders, were sold to the slavers. The King of Dahomey made at least 14 slave-hunting expeditions to fulfill the demand, according to the *Times*.[25]

Coverage of the Transatlantic Slave Trade

News coverage of the slave trade in the first few decades of the nineteenth century was characterized by stories clipped from other newspapers in the US and Europe that emphasized debates and resolutions in Congress and state legislatures and the petitions of various civic organizations, such as abolitionist groups. The same story might appear in many newspapers. For example, a story about an American slave ship off the coast of Africa first appeared in the *Boston Daily Evening Transcript* on January 21, 1843. Within seven days, the article had spread to Vermont, Washington, DC, South Carolina, and Georgia, eventually appearing in at least eleven other newspapers.[26]

One limiting factor of this project was the difficulty sometimes in determining whether coverage refers to slave trade to the American South or to the Caribbean and Brazil. Ships docking in Cuba, for instance, could head to any of these three main destinations. In addition, Texas and Florida, before they became states, were outside the prohibition of the foreign slave trade and coverage clearly establishes that both were slave transit points into the wider US. Some historians contend that as many as 50,000 slaves were brought into the US through Spanish Florida and Texas before they became states in 1845.[27] Broadly speaking, when a slave ship contained a large number of women, it was most likely headed for the US, rather than the Caribbean and its insatiable demand for men to harvest sugar.

In January 1818, members of the House of Representatives debated the situation of Amelia Island, near the Georgia–Florida border.[28] Ostensibly under Spanish colonial rule and outside American jurisdiction, the island had become a base for privateers and slave ships. Its location was a "powerful encouragement" for the flourishing of the slave trade, according to the *New York Evening Post*, and a convenient way to avoid US legal prohibition.[29] In the opinion of an unnamed congressman, it was the "duty of the government" to utilize all in its power "to restore the security" of American commerce by eliminating the sanctuary provided by Amelia Island.[30]

The Cape Verde Islands are in the east central Atlantic, 350 miles off the West African coast, and since the sixteenth century, a way station for slavers. A May 1820 letter from a resident of the archipelago to a friend in Boston (which was published in the *Boston Patriot*) described the "alarming extent" of the slave trade there.[31] An "important check" on the slavers was the arrival of the USS *Cyane*, captained by a Captain Trenchard, which "entirely cleared the coast of every slaver."[32] The *Cyane* captured "a great number of slave vessels," most sailing with Spanish papers.[33] Captain Trenchard was forced to release most of them, but only after removing the Americans, according to the *Patriot*. However, members of his crew boarded four American-owned slave ships, ordered them impounded, and sailed them to New York.[34] The British Royal Navy had been prowling the vicinity "endeavoring with redoubled energy to suppress the horrible traffic" and "rendered great service," but "nothing has had such an effect on the slavers as the arrival of a single American ship of war," the letter writer claimed.[35] The writer added that the trade was "too notorious" and that "scarce a vessel proceeds to the coast for slaves, but what is wholly or in part owned in America."[36]

> The manner in which most of the slavers carry on the trade is this: they sail from the United States to some port in Cuba, with cargo of blue and white cottons, India checks, pankin, powder, tobacco…where they make a sham sale of the vessel for the purpose of procuring a set of Spanish papers, and the officers make oath that the cargo…has been landed, and procure the requisite certificates, which every article remains untouched on board. They then take on board a Spaniard, who passes for the captain, but perhaps [on]

his first voyage…hoist the Spanish flag and proceed to the coast of Africa… keeping three log books, two in Spanish, one true and the other false, and one in English; on arrival the supercargo lands with the goods, under cover of the guns of the vessel, on the beach in huts erected for the purpose, and sends circulars to all the neighboring kings, acquaints them of his arrival, and that he has a handsome assortment of goods, which he wishes to dispose of for slaves in a given number of days. They immediately flock to his depot with their slaves, which they exchange for goods…. [I]n the mean time the vessel is [loading] rice, wood and water, and when the slaves are collected they are all embarked in one day, and the same night the vessel puts to sea….[37]

"When overhauled by English [warships, the slavers] exhibit American papers and when by the Americans, Spanish papers," the writer concluded, "by which means many escape capture and condemnation."[38] During the crossing, slave ship captains routinely doctored their logbooks.

In its September 30, 1823 issue, the *Connecticut Courant* reprinted a *London Morning Herald* article quoting Captain (later Admiral) Sir W. Mends's report to the British Admiralty "respecting the progress of efforts to suppress the Slave Trade."[39] Sailing up the River Benny in command of an unnamed Royal Navy warship, he encountered nine smaller slave ships flying French and Spanish flags that began firing on his vessel. After twenty minutes, Mends silenced all the guns opposing him and boarded the ships.[40] Brief firefights ensued and sixteen slaves were killed on at least one ship, used by the slavers as distractions and perhaps as human shields.[41] Some slaves were even armed with muskets and fired up the hatchways at the British sailors.[42] Another slave ship was abandoned by its crew to avoid capture. They booby-trapped it, however, leaving "a lighted match hanging over the magazine," which would have "blown up 300 poor fellows ironed in the hold" if the British boarding party had not extinguished it.[43] In all, Mends captured five of the ships, carrying a total of 1,485 slaves, ranging from 218–380 per ship.[44]

The *USS Erie* pursued a slave ship in early 1828 after it was chased onto the reefs near Key West, Florida. Onboard the unnamed slaver were 700 slaves and "a large quantity of gold dust and ivory."[45] The slaves escaped onto the island, but the slavers did not give up so easily; they rounded up four hundred of the slaves, reclaimed the gold dust, and boarded another vessel to escape to Havana, where the slaves were sold.[46] The 300 remaining were "to be sent for," according to the *New-York Daily Advertiser* account.[47] The escaped slaves were "men, women, and children; some families consisting of father, mother, brothers, and sisters." The slaves were very skittish and when "approached, they walk off as if afraid."[48]

"Our country has, we fear," the *Freeman and Messenger* lamented in July 1832, "given the enemies too much cause to charge it with hypocrisy…[involving] the slave trade."[49] The newspaper claimed American citizens were "engaged in the nefarious trade at least to as great an extent as the people of any other country."[50]

The newspaper also excoriated the country for its internal slave trade as well, though it stopped short of condemning the institution of slavery itself. Quoting a *Baltimore Republican* source, Captain McDonald of the *USS North*, the *Freeman and Messenger* reported that thirty-one slave ships had been captured near Sierra Leone between January and April 1832. According to a British captain, "a considerable number of vessels fitted out as slavers" sailed under the US flag. The Royal Navy was "obligated to let them pass, tho' having no doubt…[of] their cargoes [and the]…human suffering on board."[51]

During the middle decades of the study, Texas, as either Mexican territory or independent, and Florida before statehood were outside US jurisdiction and wide-open territories for slavers. Six weeks after Texas declared its independence from Mexico in March 1836, its role in the slave trade to America was being discussed by Northern newspapers, mainly because things were getting too risky for slavers in New York City. Investors in the slave trade were looking further afield and the newly independent Texas provided fertile fields for safer expansion. "The fitting out of two or three or a dozen slave ships is a small matter," wrote the *Pittsburgh Gazette*, "compared with the scale [of] which some capitalists of [New York City] and New Orleans are preparing to carry on the traffic."[52] Speculators owned large tracts of land in Texas, according to the newspaper, and were leveraging their money to expand the slave trade. Large-scale cotton farming required slave labor, the *Gazette* stated. "If slaves can be had from Africa for 100 dollars a head…they can be set to work on the fertile soil of Texas" at great profit.[53] "This is the grand scheme of which the Texas rebellion and Texas independence are to be the instrument…under the banner of 'liberty.'"[54] External slave trading was prohibited by every American state and most European countries…but not Texas. "This odious traffic is to be taken under the exclusive protection of 'free and independent' Texas and should anyone object, the Texas flag will be waved" as protection for slavers.[55] Slave trade agents soon appeared in New York, Boston, and New Orleans to buy ships outfitted for the slave trade, according to the *Gazette*. Havana was a major transit point for slaves and Texans went there to purchase "and then take cargoes of slaves to Texas."[56]

The *Adams Sentinel* of Gettysburg, Pennsylvania, published a British estimate that the "western coast of Africa…swarms with slavers."[57] The *Pittsburgh Gazette* confirmed that view, quoting English newspapers describing a marked increase in the slave trade. Most slavers used American-built vessels, the *Gazette* added, mainly from northern states. Some were adapted for the trade, while others were purpose-built.[58]

News Coverage of the Coastal Slave Trade

Further complicating the identification of foreign slave trading into the United States was the coastwise slave trade. To facilitate the American internal slave trade, vessels were legally allowed to carry slaves and navigate US coasts and rivers among states where slavery was legal. They were allowed only to transport slaves bought and sold within the United States. The Act of March 2, 1807, outlawing the foreign slave trade imposed some regulations on the domestic coastal slave trade. Ships under 40 tons were prohibited from transporting slaves, while those over that limit were required to provide a shipping manifest of slave cargo at the ports of departure and arrival.[59] Hundreds of vessels were engaged in this trade. The unpredictability of storms and navigation errors (intentional and unintentional) would occasionally blow coastal slave ships into ports in Bermuda, Cuba, and the British West Indies.

Once in Texas or Florida, it was relatively easy to smuggle slaves across poorly patrolled borders with Southern slave states or to load them onto ships sailing along the coasts and rivers. The *New Orleans Courier* reported in mid-1839, "It is believed that African negroes have been repeatedly introduced into the United States" because the "proximity of Florida ports to the island of Cuba make it no difficult matter."[60] The newspaper added that human smuggling "in all countries and ages...has afforded hopes of high profits."[61]

In June 1812, Congress debated the colonization movement as one remedy to slavery.[62] Slaves would be freed and sent to Africa to establish colonies and countries under the aegis of the American Colonization Society (ACS). The first colony, in what was to become the country of Liberia, was purchased for the project and in 1822 began to be settled by free African Americans. Several representatives claimed that smuggling slaves into the US was thriving, and that for colonization to be successful, continued slave trading into the US would have to be choked off.[63]

Between 1822 and 1867, approximately 13,000–15,000 African Americans settled in Liberia on land purchased by the ACS. Liberia declared its independence in July 1847 and modeled its political system on that of the US. James Monroe, Henry Clay, and Abraham Lincoln were supporters of the ACS. Many abolitionists opposed colonization, since it siphoned resources and intellectual energy from abolition and was regarded as an alternative to widespread emancipation.[64]

Coverage: The 1850s

In 1858, a slave ship called the *Echo* was intercepted by an American warship off the coast of Cuba and forced to surrender. The slaver was filled with a half-American, half-Spanish crew under the command of an American citizen. The *Echo* and its captives were taken to Charleston, North Carolina.

The *Chicago Tribune* reported on this incident by quoting a pro-slavery reporter for the *New York Herald*, who wrote, "There was great joy in Charleston" when "the recaptured Africans found on board the slave ship Echo" landed there, eliciting "a perfect blaze of enthusiasm" among all residents.[65] "[A]dvocates of the reopening of the slave-trade looked upon them as the first installment of a perennial stream of black...humanity that, as they hope, is hereafter to gladden their sight...."[66] However, the slaves were not destined for the fields, but were to be repatriated under federal law. Charleston authorities "furiously and pertinaciously telegraph[ed]" back and forth with officials in Washington, DC for instructions.[67] The Africans were quarantined in a never-occupied "vacant fort in our harbor, erected when Uncle Sam's purse was fuller than it is now."[68]

The crew of the *Echo* was marched "handcuffed through the streets of a slaveholding city," wrote the outraged *Herald* reporter, who characterized slavery "in a spiritual as well as utilitarian" light.[69] The correspondent was aghast that the crew was arrested "[f]or rescuing undying souls from the night of heathen barbarism and transporting them to the full blaze of Christianity...for redeeming [them] from the life of indolence and uselessness, and making [them] an active energetic unit among philanthropists."[70] The reporter claimed many of the liberated Africans did not want to return to Africa, including a chief's daughter, who "positively refused to go back, and says her followers will not go either."[71] They "much rejoiced at their arrival" in Charleston and were "singing songs, dancing, and testifying in every conceivable manner their attainment of increased happiness."[72]

The London *Observer* interpreted the situation very differently, calling the *Echo* one of "a dozen American vessels...kidnapping negroes on the African coast," and reporting that "the vessel was crowded with native Africans...and as flagrant" a case of "slave-trading as could be found."[73] The newspaper further reported that the "seaboard States" were in "great excitement" as to whether the "pirates" would be punished.[74] Not unexpectedly, a Charleston grand jury, despite what the *Chicago Tribune* called "confessedly convincing evidence," refused to indict any of the *Echo*'s crew.[75]

President Buchanan ordered the *USS Niagara* to Charleston to return the Africans home. The *Observer* believed this incident might even trigger a long-anticipated civil war:

> We could wish that this were all. We would thankfully close the subject here, but there is more to be told—disheartening, from the mere fact of the controversy.... Thus there seemed to be every probability that the collision which is hourly expected, and on which the fate of the Union hangs, would be brought on by this first apparition of a captured slaver on the American coast.
>
> The speedy announcement from Washington that the Government intended to adhere to its treaty obligations has produced a powerful and instantaneous political effect. The Northern citizens...proclaimed their strength, as certified by this act of the Cabinet. The South and its abettors in the North...are saying now that it is a small thing for the Government to send away the first batch of captives, under the full gaze of the world, while the trade itself is carried on vigorously by ships belonging to New York, Baltimore, and other ports of their coast.

The absurdity of returning a single cargo with much demonstration, while a dozen American vessels are kidnapping negroes on the African coast, will keep up an outcry which will compel one of the two difficult courses— abolishing the slave trade completely, or giving up the whole question to the South.[76]

The *Times* reported on two more slave ships, the *Eliza Jane* and the *William Clarke*, both American ships, though without flags and papers, that had been captured off West Africa.[77]

The slave ship *Wanderer* was a rarity, a fully documented slaver transporting slaves to the United States, landing its cargo of 350 Africans on Jekyll Island off Georgia in December 1858. Its slaves were sold to planters in the area. To the *Chicago Tribune* at the time, this represented the inauguration of the slave trade, not its concluding stages.[78] "The African Slave Trade is again an established fact [and] one of the branches of Southern commerce; and although it will not be as openly carried on as the trade in cotton or sugar, it will nevertheless be followed in the future with as much energy as its returns are larger in proportion to the capital employed."[79] Efforts to equate the slave trade with piracy had notably failed, according to the *Tribune*. The newspaper recalled that the *New Orleans Delta* had published a story the previous summer of slaves being landed on the Pearl River in Louisiana. The report had been dismissed as a "peculiar half-sardonic half-serious…clever canard" designed to "stir up the bile of Abolitionists…and 'black' Republicans."[80] Now, the *Tribune* believed, the "truthfulness of the story" seemed likely.[81] "That purchasers were found…for the two pioneer cargoes," the newspaper scolded, "is not more truly an indication of the deterioration of moral sentiment among the people of the slaveholding States, on this subject, than a thousand other well known facts."[82]

On February 16, 1859, the *Chicago Tribune* observed, "true bills" were brought against Selvas & Mares, a partnership, and the captains of the bark *Angelica*, which had been "seized here [Savannah] on suspicion of being fitted out for the slave trade."[83] Earlier in the month "privy advices from Jacksonville, Fla." indicated that a vessel had left "that port a few weeks ago…to take in a cargo of slaves," and that a "brig sailed" the previous day to meet the other ship and "transfer the cargo."[84]

A *New York Times* correspondent surveyed the status of the slave trade in late 1859:

> I learn that the trade was never more thriving than now…. Not an English cruiser can touch them as long as the American Flag is flying, even if they have a full cargo of slaves about their decks, and the American cruisers are chary of their seizures, unless the negroes are on board, since no matter what fittings they may have for the trade, they are sure to escape in the courts, and then the U.S. officers must endure suit without a government to back them up.[85]

The *Times* quoted the *Salem* [MA] *Register* regarding an American slaver that set sail near Zanzibar, headed for Cuba. The ship "took off 1,200 negroes from the coast a few days since," purchased from members of the Ibo tribe for $60,000 ($50 each).[86] If 1,000 slaves survived the voyage, a $27,244 "investment" yielded a $400,000 return (or $10.9 million in inflation-adjusted 2014 dollars).[87] The *Times* reported, "If people send ships on this coast for slaves they can get them," though there were British ships "watching a chance to pounce."[88]

Coverage: The 1860s

Two slave ships, one the *Stephen H. Townsend* of New Orleans and the other the *Tavernier*, were captured by British cruisers off St. Helena.[89] Both ships were condemned in vice-admiralty court. The captain of the *Tavernier* said he had been boarded at least twice previously, but eluded capture by somehow hiding the slave cargo and denying the British any pretext to search the vessel. The *New York Journal of Commerce* reported that seventy-four slaves died during an eighteen-day period on the *Tavernier* and the remaining ninety-four "were miserable specimens of humanity crawling out of their filthy stowage, like so many black skeletons, their very bones almost rattling as they totter in weakness."[90]

In October 1860, the *New York Times* reported the apprehension of the slaver *Erie* off the coast of Africa with 897 slaves onboard.[91] Seventeen months later, in March 1862, the London *Observer* reported that the *Erie*'s captain, Nathaniel P. Gordon, was executed in New York City on April 29 after being found guilty for captaining a slave ship.[92] It was the only capital punishment found during the study period.

The coverage in newspapers of the slave trade does not definitively answer the question of how widespread the slave trade was post-1808. Nor does it conclusively contradict the Trans-Atlantic Slave Trade Database that put the official numbers of slaves in the hundreds. However, there is certainly substantial and convincing coverage in this study that puts the numbers far greater than the number claimed by the database. With a typical slave ship carrying 200–400 or more slaves generating high profits despite high risk, and given the number of stories and mentions of slave ships, the database's numbers simply do not seem plausible. The trade was so profitable, according to the *Baltimore Sun*, that one group of slavers that landed on Anguila Island in the Bahamas could afford to burn their ship "to escape detection" and send the 600 slaves onboard to Cuba via two schooners.[93] The very nature of an illegal and therefore secret enterprise is the dearth of evidence that could bring retribution or prosecution.

The capacity of the slave trade ships alone mentioned in this study, figuring 200–400 on average, would account for as many as 12,000 African slaves. Even if many or most were not destined for the American South, the numbers of those identifiable as owned by American companies or captained by Americans far exceed the figures of the Atlantic Slave-Trade Database. The accumulated mass of newspaper coverage provides tantalizing evidence that the slave trade continued and did so robustly. Does it seem believable that in the five-year period before and just after the 1808 ban on the slave trade nearly 48,000 slaves would be transported to the American South, and then in the ensuing five years the number would plummet to scarcely more than a fraction of that at 1,230, which would account for just two or three slave ships? Did the demand and profit motive disappear? Was there a sudden rush to fulfill the demand for slaves before it became illegal? Or was a report laid before the British House of Commons in 1845 more accurate?[94] That report stated, "the total number of slaves landed in America from the West Indies and South America, from 1815 to 1843" was 638,145.[95] It seems likely the slave trade escaped into the mists of history and the fog banks of the Atlantic as fast clipper ships outran and outfoxed their naval pursuers.

Notes

1 "A Horrible Story," *Baltimore Sun*, October 13, 1846, 4.
2 Ibid.
3 "Voyages," The Trans-Atlantic Slave Trade Database, accessed June 15, 2017, www.slavevoyages.org/assessment/estimates.
4 Ibid.
5 Henry Louis Gates, Jr., "How Many Slaves Landed in the U.S.?" *The Root*, January 6, 2014, www.theroot.com/articles/history/2014/01/how_many_slaves_came_to_america_fact_vs_fiction/.
6 "Voyages," Trans-Atlantic Slave Trade Database.
7 Ibid.
8 Donald L. Canney, *Africa Squadron: The U.S. Navy and the Slave Trade, 1842–1861* (Lincoln, NE: Potomac Books, 2006), 233–34.
9 "Cashing Freedom Information Sheet," Royal Naval Museum.
10 "Slavery," Texas State Historical Association, accessed June 24, 2017, www.tshaonline.org/handbook/online/articles/yps01.
11 "Inflation Calculator," accessed June 24, 2017, www.in2013dollars.com/1860-dollars-to-2014-dollars. Inflation was roughly 2.17% annually from 1825–2014.
12 "Slavery," Texas State Historical Association.
13 Ibid.
14 David Pilgrim, "Question of the Month: Cudjo Lewis, Last African Slave in the U.S.," Ferris University, accessed June 26, 2017, www.ferris.edu/jimcrow/question/july05; "Inflation Calculator."
15 "Slavery," Texas State Historical Association; "Inflation Calculator."
16 "Capture of a Slave Ship from Liverpool," *The Observer* (London, England), June 24, 1861, 2.
17 "The Slave Trade as It Is," *Manchester* (England) *Guardian*, April 12, 1860, 3.
18 Carl C. Cutler, *Greyhounds of the Sea: The Story of the American Clipper Ship*, 3rd ed. (Annapolis, MD: Naval Institute Press, 1984), 39.
19 Ibid.
20 "Untitled," *Wyoming Republican and Herald* (Kingston, PA), January 6, 1836, 2.
21 Ibid.

22 Ibid.
23 "Enormities of the Slave-Trade," *New York Times*, November 28, 1860, 2.
24 Ibid.
25 Ibid.
26 "An American Slaver," *Daily Evening Transcript* (Boston, MA), January 21, 1843, 2. The other newspapers were *Boston Courier*, January 23; *Daily National Intelligencer* (Washington, DC), January 23; *Centinel of Freedom* (Newark, NJ), January 24; *Southern Patriot* (Charleston, SC), January 25; *New Bedford* (MA) *Mercury*, January 27; *Vermont Phoenix* (Brattleboro), January 27; *Easton* (MD) *Gazette*, January 28; *Augusta* (GA) *Chronicle*, January 28; *Maine Cultivator and Hallowell Weekly Gazette*, January 28; and *Boston Gazette*, January 28.
27 Randy J. Sparks, *Africans in the Old South: Mapping Exceptional Lives Across the Atlantic World* (Cambridge, MA: Harvard University Press, 2016), 80.
28 "Congress-House of Representatives: Saturday, January 10," *New York Evening Post*, January 14, 1818, 2.
29 Ibid.
30 Ibid.
31 "From Africa: Extract of a letter from the Cape-de-Verde Islands, to a gentleman in this town, received the by brig *Rebecca*, dated 'Villa-da-Prays, St. Jago, May 29th, 1820,'" *Boston Patriot*, July 10, 1820, reprinted in *Connecticut Courant*, July 18, 1820, 2.
32 Ibid.
33 Ibid.
34 Ibid.
35 Ibid.
36 Ibid.
37 Ibid.
38 Ibid.
39 "Slave Trade–Starvation of Negroes," *London Morning Herald*, August 12, 1823, reprinted in *Connecticut Courant*, September 30, 1823, 2.
40 Ibid.
41 Ibid.
42 Ibid.
43 Ibid.
44 Ibid.
45 "Extract from a letter dated Malangas, January 30th," *New York Daily Advertiser*, reprinted in *Connecticut Courant*, February 18, 1828, 3.
46 Ibid.
47 Ibid.
48 Ibid.
49 "The Slave Trade," *Freeman and Messenger*, July 11, 1832, 2.
50 Ibid.
51 Ibid.
52 "Texas and American Slave Trade," *Pittsburgh* (PA) *Gazette*, April 26, 1836, 2.
53 Ibid.
54 Ibid.
55 Ibid.
56 Ibid.
57 "Untitled," *Adams Sentinel*, May 9, 1836, 1.
58 "The Slave Trade," *Pittsburgh* (PA) *Gazette*, October 19, 1836, 2.

59 For insights into the coastal trade, see *Slave Manifests of Coastwise Vessels Filed at New Orleans, Louisiana, 1807–1860*, National Archives, www.archives.gov/research/african-americans/slave-ship-manifests.html; "Slave Manifests, 1819–52, 1860–61," in Forrest R. Holdcamper, comp. *Preliminary Inventory of the Records of Bureau of the Customs* (Washington, DC: National Archives and Records Service, 1968).

60 "Arguments for Slave Smuggling," *New Orleans Courier*, reprinted in *Rochester* (NY) *Freeman*, June 12, 1839, 2.

61 Ibid.

62 "Smuggling Slaves into the U.S.," *New York American*, reprinted in *Rochester* (NY) *Freeman*, June 12, 1839, 2.

63 Ibid.

64 During the Civil War, Lincoln met with African American leaders before issuing the Emancipation Proclamation and suggested that Central America might be an option for colonization.

65 "The Congo Fever in Charleston," *Chicago Tribune*, September 6, 1858, 2.

66 Ibid.

67 Ibid.

68 Ibid. The unnamed fort may have been Fort Sumter, site of the first shots of the Civil War.

69 Ibid.

70 Ibid.

71 Ibid.

72 Ibid.

73 "Negro Slavery in the United States," *The Observer* (London, England), September 27, 1858, 3, based on a *New York Daily News* story.

74 Ibid.

75 "The African Slave Trade Reopened," *Chicago Tribune*, December 17, 1858, 2.

76 Ibid.

77 "From Sierra Leone: The Emperor of the French in the Slave-Trade," *New York Times*, November 18, 1857, 6.

78 "The African Slave Trade Reopened," *Chicago Tribune*, December 17, 1858, 2. *Clotilde* was the last documented slave ship, landing as many as 160 Africans in the vicinity of Mobile, AL, in 1859. There was no mention of the *Clotilde* in any of the newspapers in this study.

79 Ibid.

80 Ibid.

81 Ibid.

82 Ibid.

83 *Chicago Tribune*, February 16, 1859, 4.

84 *Chicago Tribune*, February 11, 1859, 1.

85 "The Slaver," *New York Times*, November 14, 1859, 5.

86 "The Slave Trade: Captured Slavers at St. Helena—The Trade on the African Coast," *New York Times*, November 9, 1859, 2.

87 "Inflation Calculator."

88 "The Slave Trade: Captured Slavers at St. Helena," *New York Times*, November 9, 1859, 2.

89 Ibid.; and "The Slave Trade Flourishing," *Chicago Tribune*, February 6, 1860, 2, quoting the *New York Journal of Commerce*.

90 Ibid.
91 "The Slave Trade," *New York Times*, October 4, 1860, 5.
92 "America: Reuter's Express," *The Observer* (London, England), March 9, 1862, 13.
93 "The Slave Trade," *New York Sun*, October 5, 1861, 4.
94 "Foreign News," *Alexandria* (VA) *Gazette*, May 22, 1845, 2.
95 Ibid.

The Election of 1856, Dred Scott, and the Lincoln-Douglas Debates

PART II
The Election of 1860,
Dred Scott, and the
Lincoln-Douglas Debates

12

1856

A Year of Volatile Political Reckoning

Dianne M. Bragg

On May 22, 1856, South Carolina Representative Preston Brooks strode into the Senate Chamber and proceeded to beat Massachusetts Senator Charles Sumner, an antislavery Republican, about the head with a gutta-percha cane. As Sumner, who had been seated at his desk, tried to escape, Brooks hit him again and again, even after the cane had broken, until Sumner lay unconscious and bleeding on the floor. Throughout the summer and on toward the presidential election in November, Northern and Southern newspapers continued to report on the attack, which only exacerbated the bitter political campaign. In the end, Democrat James Buchanan was elected president, but the Democratic Party appeared to be splintering and the antislavery ranks were growing.

This chapter offers a glimpse of how newspapers, North and South, covered the Sumner attack and the Buchanan election, and how this coverage can be seen as an indicator of the increasingly hardened sectional positions being taken in the country's divided political arena.

The impetus for Brooks's attack began when Sumner delivered a fiery antislavery speech on May 19, 1856. Since coming to the Senate in 1851, the Massachusetts senator had often been critical of slavery in the South and had not endeared himself to the Southern delegation. Indicative of such feeling, in June 1854, Georgia's *Daily Morning News* had carried an item that noted the malice felt toward Sumner. "Considerable feeling exists here among the Northern members in consequence of an article in the Star of this evening," the *News* stated, "which is construed to bear an intimation of personal violence to Mr. Sumner and other anti-Nebraska members of Congress." The *News* had printed a portion of the *Star* article in question, one that noted the difficulty Southern slave owners were having enforcing the Fugitive Slave Law in the North. In response, the *News* stated:

> If Southern gentlemen are to be threatened and assaulted while legally
> seeking to obtain a possession of property, for the use of which they have
> a constitutional guarantee; if legal measures can only be sought for and
> established at the bayonet's point, certain men now in our midst will have
> to evince a little more circumspection than they ever evinced, in their walk,
> talk and acts.[1]

In September 1854, Massachusetts newspapers had covered a Free Soil Convention
where Sumner had appeared and spoken strongly against "the two great outrages
of the season to wit: the passage of the Kansas-Nebraska bill, and the rendition of
Anthony Burns."[2] The forced return of Anthony Burns from freedom in Boston
to slavery in Virginia in June 1854 had garnered a tremendous amount of publi-
city. In covering the Free Soil Convention, the *Pittsfield* (MA) *Sun* described the
Senator as follows:

> As he was always a Wilmot provisoist, and an ultra one, it is not strange that
> he spoke against the former; and as he, even in the senate, openly denied any
> obligation to support the constitution, as to the return of fugitive slaves, it
> is not strange that he denounced the latter. We suppose none will deny that
> this republican is a freesoiler! His speech was received with applause by the
> audience. It is due to this politician to say, that whatever opinion might have
> existed as to his tameness in his first congressional term, it exists no longer.[3]

Sumner's opposition to slavery had only increased since 1854, and when he rose
on May 19, 1856, to give what would famously become known as his "Crime
Against Kansas" speech, his words would surpass anything he had ever dared
before to speak.

A few accounts of Sumner's remarks quickly appeared in newspapers' regular
summaries of events in Congress. The *Daily National Intelligencer* (Washington,
DC) reported that Sumner began his remarks by reviewing some of the history
of Kansas in a speech that was "remarkable for lucidity of statement and beauty of
diction."[4] Sumner decried recent events in Kansas and stated what he believed to
be the foundation of it all. "That crime was the forcible infliction of slavery on a
reluctant people," Sumner said, "accomplished at the risk of intestine war and at
the bidding of what he termed the 'slave power of the Republic.'" Sumner charged
Illinois Senator Stephen Douglas and South Carolina Senator Andrew Butler, co-
authors of the Kansas-Nebraska Act, for their complicity in allowing Kansas to be
"overrun and subjugated by the pro-slavery propagandists of Missouri."[5]

A *Boston* (MA) *Daily Atlas* article carried a portion of Sumner's speech that
included the following criticism of Douglas and Butler:

> As the Senator from South Carolina is the Don Quixote, the Senator from
> Illinois [Mr. Douglas] is the squire of Slavery, its very Sancho Panza, ready

to do all its humiliating offices. This Senator, in his labored address, vindicating his labored report—piling one mass of elaborate error upon another
mass—constrained himself, as you will remember, to unfamiliar decencies
of speech.[6]

Although Sumner's speech was filled with many other similarly inflammatory
accusations, one of the most well-remembered paragraphs was quoted by the
Charleston (SC) *Mercury* on May 26. It read as follows:

> The Senator from South Carolina has read many books of chivalry, and
> believes himself a chivalrous knight, with sentiments of honor and courage.
> Of course, he has chosen a mistress, to whom he has made his vows, and
> who, though ugly to others, is always lovely to him; though polluted in the
> sight of the world, is chaste in his sight—I mean the harlot Slavery. For her,
> his tongue is always profuse in words.[7]

Just as the *Mercury* printed this passage to incite outrage over Sumner's words, several Northern newspapers praised Sumner's speech and his mastery of language,
a fact several Southern newspapers noticed. One Southern newspaper snidely
reported that a *New York Tribune* correspondent was "in raptures over the speech
of Mr. Sumner." To prove the point, the *Macon* (GA) *Weekly Telegraph* reprinted the
following from a *Tribune* article that lauded Sumner's words:

> Senator Sumner's Kansas speech is the most masterly, striking and scathing
> production of the session. The galleries were crowded with intellect, beauty
> and fashion, and the ante-rooms were also thronged. His excoriation of
> Douglas was scornfully withering and scorching. He was animated and
> glowing throughout, hurling defiance among the opposition, and bravely
> denouncing the Kansas swindle from first to last. Some passages quite elec-
> trified the Chamber, and gave a new conception of the man. Finer effect has
> rarely been produced.[8]

Over the next day or so, several newspapers reported that the speech was so filled
with vitriol that several senators immediately rebuked it for violating Senate
decorum.[9] South Carolina's *Charleston Mercury* initially published only a few lines,
noting senators Lewis Cass (Massachusetts), James Mason (Virginia), and Douglas
had "denounced Mr. Sumner's speech in severe terms, and characterised [*sic*] it as
being destitute of the truth."[10] Several Northern newspapers such as the *Vermont
Patriot & State Gazette* also noted Senator Cass's denunciation of the speech. "Mr.
Cass said he had listened to Mr. Sumner's speech with equal regret and surprise,"
the paper reported. "It was the most un-American and unpatriotic speech he had
ever heard on this floor. He hoped he might never hear such a speech again, here
or elsewhere."[11]

The *Boston* (MA) *Daily Atlas* also printed some critics' reactions to Sumner's speech. "Messrs. Cass, Douglas and Mason assailed him with the fury of ban-dogs," the paper stated. "Mr. Sumner answered them with vigor, firmness and spirit, and all three retired rather the worse for the contest."[12] The *Vermont Patriot & State Gazette* gave a more detailed report of the verbal sparring between Douglas and Sumner. "That speech was written and committed to memory," Douglas had asserted, "practised [sic] before a glass, a negro boy holding a candle and watching the gestures." Douglas's remarks infuriated Sumner, who responded in kind that the "poisome [sic], nameless animal, whose nature it is to discharge venom, is not a proper model for a senator." The Vermont newspaper also reported the response of Senator Mason, whom Sumner had also verbally assailed: "The senator is certainly *non compus mentis.*—(Laughter.) Here the war of words ended."[13] But this was not to be the end of the affair.

Newspapers did not print Sumner's speech in its entirety, probably because of its length. The speech was so long that it lasted into the next day's session. In general, considering its length and inflammatory nature, newspapers normally might have published more of the speech's content in subsequent issues. But the sacking of Lawrence, Kansas, occurred on May 21, and then, on May 22, an unexpected turn of events on the Senate floor quickly captured newspapers' attention.

When Sumner's speech singled out Andrew Butler for criticism, the South Carolina senator was not in the chamber to respond. As it turned out, though, he did not need to be present. South Carolina Representative Preston Brooks took it upon himself to defend his relation's honor.[14] On May 22, Brooks entered the Senate Chamber and after a few words about libel upon his state and kin, struck Sumner with his cane. Brooks continued to hit him until the Massachusetts senator lay unconscious on the floor. Newspaper coverage of the attack spread quickly, fanning sectional tensions.

The morning after the attack, Georgia's *Daily Morning News* carried a telegraphic report with the headline, "Mr. Sumner assaulted in the Senate Chamber." The short article, which was fairly accurate, read as follows:

> Washington, May 22.—Mr. Brooks of South Carolina assaulted Mr. Sumner of Massachusetts in the Senate Chamber to-day while the latter was seated at his desk writing. Mr. Sumner received a stunning blow over the head with a heavy cane, cutting and contusing him badly. Mr. Brooks repeated the blows frequently. Mr. Keitt attempted to prevent interference. Mr. Crittenden, who was present pronounces the affair a shameful outrage. The provocation for the attack was given by Senator Sumner in his speech in the Senate against Senator Butler. [Sumner in his speech denounced Senator Butler in a most gross and insulting manner. Mr. Brooks is a relative of Senator Butler.][15]

The *Boston* (MA) *Daily Atlas* also published an article on the incident. It carried the headline "Dastardly Assault on Senator Sumner" and reported that a "most

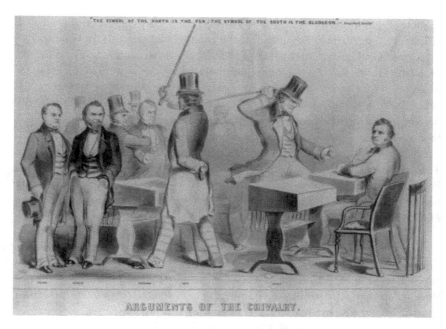

FIGURE 12.1 "Arguments of the chivalry," lithograph by Winslow Homer, 1856. This image illustrates the May 22, 1856, attack on Massachusetts Senator Charles Sumner by South Carolina Representative Preston S. Brooks, motivated by Sumner's public derogatory remarks about senators Andrew Pickens Butler and Stephen A. Douglas. An unwary Sumner is shown sitting at his desk while an infuriated Brooks raises his cane above his head to strike a blow. To Brooks's left, fellow South Carolina Representative Lawrence M. Keitt raises his own cane threateningly towards other legislators, deterring intervention on Sumner's behalf. Behind Keitt's back, he holds a pistol in his left hand. Senators Douglas and Robert Toombs stand with hands in pockets, unperturbed. Above the scene, a quote from abolitionist Henry Ward Beecher reads, "The symbol of the North is the pen; the symbol of the South is the bludgeon." (From the Library of Congress Prints and Photographs Division Washington, DC. <www.loc.gov/pictures/item/2008661576/> Reproduction Number: LC-USZ62-38851 (b&w film copy neg.). Winslow Homer, "Arguments of the chivalry.")

brutal assault has just been made upon Senator Sumner by Brooks, of South Carolina." Sumner was bleeding badly, the article stated, and "is now in a state of partial stupor."[16] As with most articles on the attack, the *Atlas* noted that Sumner was seated in the Senate Chamber, which was not in session, and was taken by surprise. The *Daily Scioto* (OH) *Gazette* said that the *New York Mirror* believed the assault would assure Sumner's reelection to the Senate, which it did. "Personal and political differences are swallowed up," the *Gazette* declared, "in the overwhelming feeling that Massachusetts has been grossly outraged in the person of her Representative."[17]

James Gordon Bennett's *New York Herald* carried a blatantly sensational article headlined: "An Authentic Account of the Fracas—Colonel Brooks Held to Bail to Answer—Indignation of the Nigger Worshippers—Movement to Expel Brooks From the House, Etc."[18] Despite the headline, the article began with a defense of its report on the attack. The *Herald* probably did so because of its reputation for being pro-South and anti-abolition. "The following will be found to be a strictly correct and impartial account of the attack on Mr. Sumner, in the Senate chamber, to-day," the *Herald* stated. "Colonel Preston S. Brooks, of South Carolina, took exception to the following language used by Senator Sumner in his speech on Tuesday last." Then the article published portions of Sumner's speech followed by an account of the attack, during which the *Herald* gave as favorable a view of Brooks as possible.[19] The *Charleston* (SC) *Mercury* followed suit and published the *Herald's* account on May 26, but with the headline "Messrs. Brooks and Sumner."[20]

Following the attack, the House of Representatives attempted to expel Brooks, and although the resolution passed with a vote of 121 to 95, it did not receive the required two-thirds majority.[21] Instead, the House took action to censure Brooks and South Carolina Representative Laurence Keitt, who had accompanied Brooks, but they resigned in protest, only to be returned to their seats in the next election.[22] In the South, both men received favorable press attention, with Brooks being especially praised. North Carolina's *Weekly Raleigh Register* printed a short article noting that several Charleston gentlemen had given Brooks a new cane to show their appreciation for what he had done. The article said the governor of South Carolina (James Hopkins Adams) was leading an effort to raise money to present "Mr. Brooks with a 'silver pitcher and goblet.'" Most strikingly, though, was the assertion that the slaves in Charleston were going to present their own token of regard to Brooks, "who has made the first practical issue for their preservation and protection in their rights and enjoyments as the happiest laborers on the face of the globe." The article implied that slaves wanted slavery to continue because they benefited from the institution, a stance often taken by Southern slaveholders to defend the practice.[23]

A *Weekly Raleigh* (NC) *Register* article on June 4 took note of the North's opinion of Brooks's actions. "The Northern papers are all condemning and denouncing Mr. Brooks," the paper stated, "for his assault on Senator Sumner, in the severest terms." The Senate should not forget, the article added, that Brooks was motivated by "foul language, abuse, taunts, and opprobrious epithets," and the affair should be cause for the Senate to prevent such kinds of debates. "One evil leads necessarily to another," the paper declared. "The Senate must preserve its own dignity, in order to command the respect of the public."[24]

Northern newspapers took note of the South's admiration for Brooks and responded accordingly. That fall, Philadelphia's *North American and United States Gazette* noted that the Columbia *South Carolinian* had reported on a dinner given in Brooks's honor. The *Gazette* reported on some of the speeches from the dinner as follows:

Brooks made one of his usual violent speeches—full of disunion and egotism. Senator Toombs of Georgia, also spoke for two hours in favor of
Southern sectionalism and disunion, and said that he approved most heartily
of Brooks' assault on Sumner—that he "saw it done, and saw it well done."[25]

Similarly, in November, a Massachusetts newspaper called for a comparison of
a speech Sumner had given at a reception to one Brooks made to some of his
constituents. "Mark in the one the accomplished scholar, the liberal, highminded
statesman, the generous patriot, the Christian gentleman," the paper stated. "Mark
the coarse brutality of the other, the egotism, the insolence, the contempt of
authority, of order, and the open demand for the dissolution of the Union."[26]

Newspaper responses to Brooks's attack on Sumner show how divided the
North and South had become and is indicative of each region's contempt for
the other and its politicians. Northerners were horrified that such an attack had
taken place in the Senate, and Northern newspapers tended to gloss over the
vitriol in Sumner's speech. In the South, Sumner, who was already considered
an enemy of slavery, was vilified for his speech while Brooks was praised for
protecting the honor of Butler, South Carolina, and, indeed, the entire South.
But, even as Brooks was being hailed as a hero, his actions created a dilemma for
some Southerners. By attacking Sumner without warning, Brooks had violated
the South's honor code that would have had Brooks challenge Sumner to a duel.
Some Southerners felt Brooks was justified because Sumner was not worthy of
being asked to a duel.[27] The North's use of the incident to attack Southern honor
and chivalry infuriated Southerners.[28] One article, "Congressional Courtesy,"
appeared in an Ohio newspaper and noted the South's predicament in supporting
Brooks, who had so clearly violated the South's own code. Massachusetts Senator
Henry Wilson, the article reported, had said in the Senate that "his colleague,
Charles Sumner, had been stricken down by a murderous, brutal and cowardly
assault." Wilson's remarks, the article said, "called out Senator Butler, of the chivalrous State of South Carolina."[29]

The widely varying tone of newspaper coverage on the character and actions
of the politicians involved in this incident attests to how intense the rancor had
become between the North and South. The incident raised awareness that some
sort of reckoning between the two regions surely was at hand. A Vermont paper
plaintively said:

Never have our hopes of the prosperity of the Union been so shaken as
by the events of the last few weeks. We have long hoped with trembling,
but hope is fast yeilding [sic] to despair—and we are ready to say that if it
can only be preserved by subserviency to Slavery, and continued, constant
yeilding [sic] to Southern dictation and injustice, better far, by a mutual
agreement, dissolve the copartnership, and let the South alone bear the evil
and its responsibilities. In view of the crisis of our nation, with Jefferson, we

tremble to remember that God is just, and only hope in view of his great forbearance and compassion. May Heaven avert the harm that threatens us![30]

The aftermath of the incident left in its wake much indignation throughout the country. It would take Sumner more than three years to recover enough, physically and mentally, to return to the Senate, and his constituents considered his empty Senate chair to be a reminder of what had occurred. After his reelection, Brooks returned to the House, but his tenure, like his life, was short. He died on January 27, 1857, of croup, and failed to see the outcome of a saga in which his actions played a crucial role.

The presidential election of 1856 would give some idea of how Brooks's actions and the rise of the Republican Party would serve to change America's political power structure. As November 1856 approached, it became clear the country's fractured nature would be a deciding factor in the election. The Democrats had come to despise President Franklin Pierce, and he failed to receive the party's nomination for a second term. As the Democratic Convention in Cincinnati, Ohio, neared, newspapers reported on the various names being considered for nomination. "The acrimony which this feeling has in part engendered," the *Macon* (GA) *Weekly Telegraph* reported, "between the '*friends*' of Messrs. Douglas, Buchanan, and Pierce is already, among the 'knowing ones,' considered a fatal bar to the nomination of either of the three."[31] James Buchanan, though, did secure the nomination, and although the Democrats appeared confident, the *Telegraph* alluded to some concerns. "Political relations are so loose and unsettled now," the paper stated, "that the only ground of absolute confidence is actual success."[32]

The Democratic ticket won the election, but newspapers were quick to note how the votes for Buchanan and the Republican candidate, John C. Frémont, were divided along regional lines, with Frémont carrying the northernmost states and none in the South. Although Buchanan received 58.8 percent of the electoral votes, he only won 45.3 percent of the popular vote, while Frémont claimed 33.1 percent, and Millard Fillmore, the Whig-American candidate, 21.5 percent. The votes indicated that if Fillmore, who only carried Maryland in the election, had not run, Frémont might very well have won Illinois, Indiana, New Jersey, and Pennsylvania.[33] Newspapers noted that the Republican Party's loss, in many ways, had the air of victory about it, while Buchanan's win was shrouded in division.

Newspapers offered various explanations, including the *Weekly Herald* (NY), which said Buchanan had won the election "by the divisions of the opposition forces and a series of accidents." The paper then warned:

Now the work among the various managers, cliques and coteries of the democracy, North and South, for the first seats at the table begins. Mr. Buchanan will soon discover what Mr. Jefferson Davis and his controlling secession faction expect and demand; he will perceive, on the returns of the election, however, that some steps of conciliation in behalf of the North are

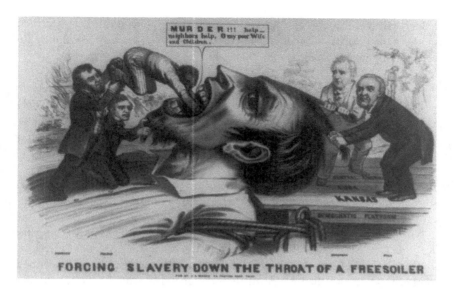

FIGURE 12.2 "Forcing Slavery Down the Throat of a Freesoiler," lithograph by John L. Magee, 1856. A major issue in the 1856 presidential election was the "Crime Against Kansas." This anti-Democratic political cartoon depicts a Kansas Free Soiler being tied down to the "Democratic Platform" by two Lilliputian politicians, presidential nominee James Buchanan, and Democratic Senator Lewis Cass. The Free Soiler's head rests on the platform, which has written on it "Kansas," "Cuba," and "Central America," likely in reference to the Democratic view of expanding slavery into these regions. A black man depicting slavery is forced into the Free Soiler's mouth by Democratic Senator Stephen A. Douglas and President Franklin Pierce. In the background, a scene of chaos, pillaging, fire, and a man hanging from a tree further depicts the widespread violence in Kansas. The Free Soiler pleas: "MURDER!!! help— neighbors help, O my poor Wife and Children." (From the Library of Congress Prints and Photographs Division Washington, DC. <www.loc.gov/item/2008661578/> Reproduction Number: LC-USZ62-92043 (b&w film copy neg.). John L. Magee, "Forcing Slavery Down the Throat of a Freesoiler.")

indispensable to the continued existence of the democratic party. Inevitably he must disappoint the North or the South.[34]

The *Charleston* (SC) *Mercury* asserted that if the election had been held just a few months before, when Northern sentiment against Preston Brooks was at its height, Frémont might have won. Also, Brooks's actions against Sumner, the paper stated, had caused Northerners to unite against the Democrats, even though the paper reaffirmed its position that Brooks was in the right. "The merited caning of Sumner by Brooks," the *Mercury* stated, "was perverted into a powerful lever to act upon the northern fanatical mind. All seemed lost. Know Nothingism subsided into a mere ripple upon the wave of Black Republicanism."[35]

Recognizing that the election easily could have gone the other way, most Southern newspapers somberly assessed the Democrats' victory. As the results were coming in, the *Charleston* (SC) *Mercury* stated:

> We make no particular comments upon this result until we get full returns of the election; but what seems beyond dispute is, that the Northern people, in this Presidential election, have declared themselves a distinct people, with principles and purposes essentially and permanently at war with our safety and equality in the Union.[36]

Although Buchanan's election was a victory, the results gave Southern newspapers pause as they were forced to acknowledge that any momentum in the South's favor seemed to be dissipating.

Northern newspapers kept readers apprised of the South's concerns after the election. The *New York Herald* published a *Charleston* (SC) *Mercury* article on Southerners' belief that the election proved that the country was now divided. Southerners said they had put forth the only qualified candidate, a Northerner, and the North instead chose to support a less qualified candidate. The headline, "What Will Mr. Buchanan Do?" became a frequent query. The *Mercury* wrote:

> The hope of peace, of good understanding, has all passed away. Henceforth we are necessarily two peoples—the North and the South. The democratic party have presented to the country a Northern man, no way identified with peculiar Southern interests; a man who could raise against himself no prejudices on the part of the North; and yet it has depended upon the almost unanimous vote of the South whether this man, great in talents, famous by a long life of noble statesmanship, irreproachable in morals and manners, should be elected to the Presidency over an adventurer, without experience in politics, with a doubtful reputation even as an explorer, and a still more doubtful one as a man, whose sole qualifications as a candidate was that he was willing to embody the sentiment of the hostility of the North to the South. What conclusion can we draw from this result except that we are on the verge of revolution or destruction? For ourselves, we prefer the former.[37]

The *Boston* (MA) *Daily Atlas* asked the same question about Buchanan's agenda for the future, but framed it in light of Franklin Pierce's support of slavery and his disappointing term as president. "We owe all the agitation which has shaken the country to the inexplicable fatuity of Franklin Pierce," the *Atlas* stated, "who has thought slavery and dreamed slavery ever since he took an oath only to break it! Will Mr. Buchanan be mindful of aught else?"[38] The *Bangor* (ME) *Daily Whig & Courier* reprinted a *New Orleans Delta* article acknowledging that Buchanan's election ensured that the Democrats would have "possession of the government" for eight consecutive years and would be "open to all the disadvantages of being in power." More importantly, though, the *Delta* took note of the Republicans' first run at the presidency. "Mr. Fillmore is laid on the shelf," the *Delta* stated. "Fremont has served the purpose of laying a broad foundation for a party, the essential character of which is to be aggressive, and its object to control the country and subject the South to the despotism of a sectional majority."[39] Such newspaper comments indicate that the South now looked with great concern toward its political future—in particular, to what would happen in the next presidential election.

The votes had barely been tabulated when Northern and Southern presses began speculating about the 1860 contest. The *Charleston Mercury* printed a *New Orleans Delta* article that stated that even though the election was over, the issues it had raised were not settled. "These are yet in the womb of the future," the article predicted, "and what the next four years may bring forth, we must wait to see, hoping for the best while we should be forearmed against she [*sic*] worst." The end of the article posed a question that the *Delta* said required an answer: "Whether this Union shall be Northern and Sectional, (to use a seeming contradiction in terms), or Southern and National?"[40]

The *Milwaukee* (WI) *Daily Sentinel* wrote that "…the contest just past is but the pastime of children. Close up the ranks again, then, and stand ready for the day of need."[41] The *Weekly Herald* (NY) said it was not surprised to learn that Democrats did not consider Buchanan's election to be a true victory. "The heavy vote for Fremont—the enthusiasm of his friends—," the paper said, "the deliberately expressed determination of the Northern people, evidenced by the popular vote, to overthrow the present democratic party—have caused a deep and powerful impression throughout the South."[42] The *New Hampshire Statesman* noted that although Buchanan had won, "it is very clear that the candidate of the Republican Party has made a capital run." The paper then looked forward to the next four years and predicted:

> Agitation of the Slavery Question, perhaps a War for conquests in behalf of the Slave System, will in all probability engross the public mind for the four succeeding years. Already the ablest of the Southern presses are advocating the revival of the Slave Trade and complete extension of black servitude over the entire country.[43]

Such articles would help to ensure that slavery would increasingly become the center of debate. Suggestions that the South would only be satisfied with the revival of the international slave trade and the extension of slavery across the entire country exacerbated Northern anxiety.

The *Daily Cleveland* (OH) *Herald* carried an article from the *Richmond* (VA) *Enquirer* arguing that slavery should continue. The *Herald* printed the article to show its readers the depth of Southern devotion to slavery. The article, "Slavery The Strongest Bond of Union," was divided into several sub-paragraphs with headings such as "Without Slavery We Should All Sink Down Together," "Slavery Gives Character To Civilization," and "Slavery Is Conservative And Elevating." A closing paragraph, apparently from the *New York Courier*, commented on the *Enquirer* article. "We have no comment to make on the above," the paragraph declared, "except to say that those are the political principles which have elected James Buchanan to the office of President of the United States." The article concluded by asking if it was upon these principles that the Republic was founded. "The North, with like voice," the article asserted, "has just answered 'No,' and will continue thus to answer, until that answer is heard and heeded."[44]

Southerners, however, were just as emphatic with their answer, as an article in the *Charleston* (SC) *Mercury* clearly showed:

> Let the South persevere in this new policy. Let her ask nothing (as she never has) but what she is entitled to under the Constitution, and resist all infringements upon her rights, and all will be well. Let the North see and know that the South is united in the defence [*sic*] of what she is entitled to, and the North will respect her rights. The way to maintain the Union is for the North to be taught that she must attend to her own business, and let that of the South alone. That is all the latter asks. And that she will have.[45]

The country's political environment sank to an abysmal point with the Brooks and Sumner incident, and Buchanan's election would do nothing to ease the situation. His March 4, 1857, inauguration became little more than the starting point for the election of 1860.

On the eve of the Civil War, a North Carolina newspaper would mention Brooks and his claim to infamy, a legacy few would wish to own:

> It may be doubted if the country will for many years feel the last of the evils resulting from the attack of Brooks on Sumner in the Senate Chamber. But for that attack…the country might never have arrived at that deplorable state of enmity that now exists. We have no doubt, also, that Lincoln's election is directly traceable to that assault, for it alone made half a million of Republicans—so a friend at the North assured us last summer.[46]

Notes

1   "Excitement in Washington," *Daily Morning News* (Savannah, GA), June 3, 1854.
2   "The Fugitive Slave Case. Burns Returned to Slavery," *Boston* (MA) *Daily Atlas*, June 3, 1854. When Sumner gave his speech at the Free Soil Convention in September 1854, Massachusetts was still in an uproar over the Anthony Burns case.
3   "The Freesoil Convention," *Pittsfield* (MA) *Sun*, September 14, 1854.
4   "Congress—Yesterday," *Daily National Intelligencer* (Washington, DC), May 20, 1846.
5   Ibid.

6 "Mr. Sumner's Speech. The Crime Against Kansas," *Boston* (MA) *Daily Atlas*, May 23, 1856. Brackets in original.

7 "Senator Sumner's Speech," *Charleston* (SC) *Mercury*, May 26, 1856. Original text found in "Kansas Affairs. Speech of Hon. C. Sumner, of Massachusetts, In the Senate, May 19, 1856," *Congressional Globe*, 34th Cong., 1st sess., appendix, 530, https://memory.loc. gov/cgi-bin/ampage?collId=llcg&fileName=042/llcg042.db&recNum=541.

8 "Senator Sumner's Speech," *Macon* (GA) *Weekly Telegraph*, May 27, 1856.

9 "Kansas Affairs," *Congressional Globe*, 34th Cong., 1st sess., appendix, 529–47, https:// memory.loc. gov/cgi-bin/ampage?collId=llcg&fileName=042/llcg042.db&recNum= 540. The text of Sumner's speech spreads over 15 pages of the *Congressional Globe*, followed by three pages of comments.

10 "Telegraphic Intelligence," *Charleston* (SC) *Mercury*, May 22, 1856.

11 "34th Congress—1st Session. Senate, Washington, May 20," *Vermont Patriot & State Gazette* (Montpelier), May 23, 1856. Original text found in "Mr. Cass," *Congressional Globe*, 34th Cong., 1st sess., appendix, 544, https:// memory.loc.gov/cgi-bin/ampage? collId=llcg&fileName=042/llcg042.db&recNum=555.

12 *Boston* (MA) *Daily Atlas*, May 22, 1856.

13 "34th Congress—1st Session," *Vermont Patriot & State Gazette* (Montpelier), May 23, 1856. The phrase "*non compus mentis*" means "not of sound mind."

14 Many sources say that Brooks was a relative of Butler's. Some say he was a cousin; others say he was a nephew.

15 "By Magnetic Telegraph. *Mr. Sumner assaulted in the Senate Chamber*," *Daily Morning News* (Savannah, GA), May 23, 1856. Keitt was a US Representative from South Carolina along with Brooks. It was reported that Keitt pulled a gun to stop anyone from helping Senator Sumner. Brackets in original.

16 "Dastardly Assault on Senator Sumner," *Boston* (MA) *Daily Atlas*, May 23, 1856.

17 "How It Promises," *Daily Scioto Gazette* (OH), May 31, 1856.

18 Editor's Note: This is a work of history, a major focus of which was the racial hatred that existed in the United States. This racial hatred was openly reflected in the language of the newspapers of the time. That language, when provided in direct quotations, is generally retained in this book in order to accurately reflect the historical context.

19 "An Authentic Account of the Fracas—Colonel Brooks Held to Bail to Answer— Indignation of the Nigger Worshippers—Movement to Expel Brooks From the House, Etc. Washington, May 22," *New York Herald*, May 23, 1856.

20 "Messrs. Brooks and Sumner," *Charleston* (SC) *Mercury*, May 26, 1856.

21 *Congressional Globe*, 34th Cong., 1st sess., July 14, 1856, 1628.

22 See Allan Nevins, *Ordeal of the Union, II: A House Dividing, 1852–1857* (New York: Charles Scribner's Sons, 1947), 443–48.

23 *Weekly Raleigh* (NC) *Register*, June 4, 1856.

24 Ibid.

25 "A Complimentary Dinner," *North American and United States Gazette* (Philadelphia, PA), October 10, 1856.

26 *Lowell* (MA) *Daily Citizen and News*, November 7, 1856.

27 Ken Deitreich, "'Ever Able, Manly, Just and Heroic': Preston Smith Brooks and the Myth of Southern Manhood," *Proceedings of The South Carolina Historical Association*, May 2011, 27–38, https://dc.statelibrary .sc.gov/handle/10827/23862.

28 David H. Donald, *Charles Sumner and the Coming of the Civil War* (New York: Alfred A. Knopf, 1961).

29 "Congressional Courtesy," *Daily Scioto Gazette* (OH), May 31, 1856.

30 "To What Are We Coming?" *Farmers' Cabinet* (Amherst, MA), June 5, 1856.

31 "A Party in Peril," *Macon* (GA) *Weekly Telegraph*, May 13, 1856.

32 Ibid.

33 John Wooley and Gerhard Peters, "Election of 1856," *The American Presidency Project*, accessed June 2013, www.presidency.ucsb.edu/showelection.php?year=1856.

34 "Election of Mr. Buchanan and a Democratic Congress—Shoals and Breakers Ahead," *Weekly Herald* (NY), November 8, 1856.

35 "The True Cause of the Victory," *Charleston* (SC) *Mercury*, November 18, 1856.

36 "The Presidential Election," *Charleston* (SC) *Mercury*, November 10, 1856.

37 "A Loud Blast from South Carolina. First Results of Mr. Buchanan's Election. A New Constitution demanded, or a Southern Confederacy. What Will Mr. Buchanan Do? [From the Charleston Mercury] Mr. Rhett's Letter to Gov. Adams," *New York Herald*, November 11, 1856.

38 "What Will Mr. Buchanan Do?" *Boston* (MA) *Daily Atlas*, November 11, 1856.

39 "From the New Orleans Delta, Nov. 11th," *Bangor* (ME) *Daily Whig & Courier*, November 26, 1856.

40 "The Presidential Election in 1860—The Policy of the South," *Charleston* (SC) *Mercury*, November 17, 1856.

41 "What is Expected of Democracy," *Milwaukee* (WI) *Daily Sentinel*, November 27, 1856.

42 "Affairs in Washington. Our Washington Correspondence. Washington, Nov. 25, 1856," *Weekly Herald* (NY), November 29, 1856. As the name suggests, the *Weekly Herald* was a weekly version of the *New York Herald* owned by James Gordon Bennett.

43 "The Presidential Election," *New Hampshire Statesman* (Concord), November 8, 1849.

44 "Gems of Democracy," *Daily Cleveland* (OH) *Herald*, November 14, 1856. The article began by saying the slavery text was from the *Richmond* (VA) *Enquirer*, but the *N.Y. Courier* was cited after the concluding commentary. It is possible the entire article, including the introduction to the *Enquirer* text and the final paragraph, were all republished from the *Courier*. In any event, the opening and closing paragraphs gave a Northern perspective on the *Richmond* (VA) *Enquirer* text.

45 "The True Cause of Victory," *Charleston* (SC) *Mercury*, November 18, 1856.

46 "Brooks and Sumner," *Fayetteville* (NC) *Observer*, December 10, 1860.

13

DOUGHFACE DEMOCRATS, JAMES BUCHANAN, AND MANLINESS IN NORTHERN PRINT AND POLITICAL CULTURE

Brie Swenson Arnold

In the decades preceding the Civil War, Northern Democrats who sided with the South on slavery were commonly known as "doughfaces." When political debate over slavery intensified in the late 1850s as a result of controversy over the Kansas-Nebraska Act and the westward expansion of slavery, Northerners mercilessly attacked and lampooned the doughfaces—the most prominent of whom included senators Stephen Douglas (Illinois) and Lewis Cass (Michigan), Attorney General Caleb Cushing, and presidents Franklin Pierce and James Buchanan.[1] These and many other doughface politicians were portrayed in Northern print and political culture as physically unattractive, as possessing undesirable personality traits, and, most unflatteringly of all, as being spineless, soft, and cold-blooded—imagery and euphemisms associated with sexual impotency and sterility in Antebellum popular culture. Such gendered and sexualized representations of the doughfaces' less-than-ideal masculinity circulated widely across the North, appearing in mass-market publications as well as in private letters, diaries, and scrapbooks.

Despite the prevalence of the pejorative portrayals of the doughfaces in Northern print culture, historians have not yet examined the gendered and sexualized discourse surrounding the doughfaces or its role in the slavery politics of the late 1850s.[2] In Northern books, newspapers, magazines, pamphlets, plays, political cartoons, campaign banners, scrapbooks, diaries, and more, doughfaces like Buchanan were depicted as cowardly, spineless, and "soft" on the political issue of slavery and in terms of their own manliness. By examining the prevalence of gendered and sexualized depictions of doughfaces like Buchanan, scholars can better understand the interplay between the press, the public, and politics during the Civil War era. Antebellum print culture and cultural understandings about gender and sexuality played an important role in shaping Northern political sentiment during the decisive elections and party realignments that precipitated the

Civil War. By linking manhood with political persuasion, the stories and satire associated with being doughfaced contributed to a growing sense that ideal Northern white men supported the anti-extension-of-slavery Republican Party and not the soft-on-slavery, doughfaced Democratic Party. The gendered and sexualized discussion about the doughfaces in Northern print culture helped to undermine the authority of the Democratic Party in the 1850s. It encouraged Northern men who had yet to take a decisive stance on slavery to identify with the Republicans and to stand firm, "shew [sic] more backbone," and no longer be soft on slavery or soft as men.[3]

Since the 1820s, the word *doughface* had designated Northern politicians sympathetic to the South on slavery, but it was not until after the passage of the Kansas-Nebraska Act of 1854 that "the term 'doughface' [was] so commonly applied to Northern men with Southern principles."[4] Over the course of the late 1850s, *doughface* became "a contemptuous nickname" widely applied to members of Congress from the North who demonstrated a willingness to support the views and demands of the South with regard to slavery.[5] Because doughfaces submitted to the political demands of proslavery Southern men, Northern newspapers, novels, pamphlets, and handbills consistently described them as "subservient," "servile," "weak," "cowardly," "spineless," and "soft."[6] Northerner John Walden clipped articles from Northern newspapers that described doughfaces like James Buchanan in "emphatic language" as "*show[ing] a want of moral courage.*"[7] Such articles clearly made an impression on Walden, who pasted them in his personal scrapbook where he also generated a handwritten list of Buchanan's "characteristics"—which prominently included "lacks moral courage."[8] New York City diarist George Templeton Strong concurred, writing in his private journal that Buchanan was "devoid of all decision and courage" and "weak, of course."[9]

One of the most common ways of talking about the doughfaces was to describe them as "weak-backed," "spineless," or having "no back-bone."[10] Antebellum Northerners commonly used the words *spine* and *backbone* to describe, as one dictionary put it, a man's overall "moral stamina, strength of will, [and] firmness of purpose."[11] Yet, in the late 1850s, Northerners came to use and think of the terms more specifically in relation to slavery politics. Two late 1850s editions of Bartlett's *Dictionary of Americanisms* noted that the term *backbone* was "a figurative expression recently much used in political writings," and an 1859 *Harper's Weekly* article listed "weak-backed" and "dough-faced" as synonyms.[12] Just days after the passage of the Kansas-Nebraska Act, when abolitionist Henry Ward Beecher was asked if he thought the North would stand firm on the question of slavery's expansion, he talked about "spine" and declared he was "afraid that the North will not stand [firm]." But Beecher held out hope Northern resolve would stiffen such that he would "see inscribed on the banner of [the] 1856 [presidential election], 'Death to doughfaces.'"[13] That same year, another famous abolitionist, William Wells Brown, wrote a satirical antislavery play titled *Experience, or How to Give a Northern Man a Backbone*, which became so popular he had to "give up his agency" with the American Anti-Slavery Society to "devote his time to…the reading of his drama."[14]

Brown's play was one of many Northern pop culture sources produced during the 1856 election year that prominently discussed political backbone. The cartoon "The Great Presidential Race of 1856," for example, featured two everyday Northern white men discussing the traits of doughface presidential candidates Millard Fillmore and James Buchanan. When asked about Fillmore's suitability for the presidency, one of the men commented that Fillmore had "a decided Curvature of the Spine, no Back Bone Sir, 'All Dough' Sir, ha! ha! ha!"[15] Northerners attending political rallies in the summer and fall of 1856 were known to shout out comments about politicians' "backbone." During one speech denouncing proslavery atrocities in Kansas, the large crowd gathered on New York City's Broadway let out "Three groans for BUCHANAN," and a loud "VOICE" yelled out that only Republicans were men with "a backbone."[16] By the election of 1860, Republican political organizers from New York and Iowa asserted in their private correspondence that Northern voters "revere, admire, worship, adore pluck. A stiff backbone is worth all the rest of the human anatomy. …Brain is nothing compared to the dorsal column."[17]

While newspapers, plays, and letters remarked on the "backbone" and "human anatomy" of many politicians, none was so widely discussed as that of James Buchanan. Throughout his political career, explained one Northern newspaper, "Mr. Buchanan seems to have been troubled with a *weak spine*, which has grown weaker with each ascending year."[18] By the 1856 election, Democratic papers like the *Brooklyn Daily Eagle*, as well as Republican and Know-Nothing/American Party newspapers, noted that "all knew him [Buchanan] to be a statesman without backbone."[19] The *New York Herald* asserted that Buchanan's tacit support of the expansion of slavery into Kansas and his refusal to denounce the hostilities and sham elections going on there confirmed "the want of pluck and backbone… of the Buchanan democrats," which factored into that long-time Democratic paper's decision to instead endorse Republican John C. Frémont in 1856.[20] Even proslavery Southern Democrats who benefitted from Buchanan's soft-on-slavery policies admitted that "Mr. BUCHANAN is a little weak in the back."[21]

As Northerners watched Buchanan bungle the slavery crisis during his presidency, they exasperatedly concluded that he now lacked "a backbone" altogether.[22] Buchanan and other doughfaces were increasingly depicted as "invertebrates"— that is, creatures entirely devoid of a backbone—including spiders, crabs, jellyfish, eels, and mollusks. One diarist alone called Buchanan "invertebrate," a "jellyfish," and an "old mollusk," and referred to members of the Northern Democratic Party as "star-fishes."[23] An 1860 *Vanity Fair* cartoon titled "What is it?" even superimposed James Buchanan's head on the body of some sort of starfish or spider.[24]

Not only were these creatures invertebrate, they were also cold-blooded— which further implied that the doughfaces exhibited questionable manliness. As historian Helen Horowitz has demonstrated, mid-nineteenth century popular understandings of sexual desire and virility were still rooted in humoral theory.[25] Antebellum Americans believed the body was regulated by four humors—blood, phlegm, yellow bile, and black bile—which correlated to four states—hot, cold, dry, and moist. In this formulation, heat and blood were the sources of sexual desire. Sexual desire, explains Horowitz, was "imagined as springing from heated blood" and men were considered "the hotter, drier, and lustier sex," while women were "cooler" and less passionate.[26] Though such understandings were being debunked by medical professionals, these centuries-old ideas about the relationship between blood, heat, sex, and masculine virility still held sway in Antebellum America.[27]

Presenting doughfaces as cold-blooded creatures, therefore, did more than simply caricature them as animals: it attacked their manliness and raised questions about their virility. Such ideas melded perfectly with long-standing public perceptions of James Buchanan as "cold," "chilly," "bloodless," and "cold-blooded." The Democratic newspaper *The Weekly Portage Sentinel* of Ravenna, Ohio, explained that "political enemies" had "call[ed] Mr. Buchanan cold and heartless" throughout his decades-long political career.[28] During the run-up to the 1856 election, Buchanan was often called "cold" and "too cool"; even his campaign biographer had to admit that many writings and speeches "represented him as cold."[29] Democratic newspapers asserted that "Black Republicans" promoted the use of the "stereotyped phrase…that Mr. Buchanan is 'so very *cold*'" simply to drum up greater support for Frémont, and Republican papers did routinely describe Buchanan "with all his coolness" as "frigid enough to make ice-creams."[30]

Northern newspaper articles, speeches, and political parade banners called Buchanan a "cold-blooded bachelor" and bluntly referred to "his cold celibacy" and "HIS COLD-BLOODED BACHELOR HEART," thereby linking his coolness with his single, never-married status.[31] Perhaps Northerners so frequently commented on Buchanan's status as a "cold-blooded bachelor" because never marrying or having children was considered unusual during this period.[32] Historian Michael Pierson and literary scholar David Grant have explained that during the 1856 election, Buchanan's never-married, childless status presented a stark contrast to his married, father-of-five Republican rival John C. Frémont—who had first become nationally famous in the 1840s for his elopement with Jessie Benton, the daughter of Democratic Senator Thomas Hart Benton.[33] During the 1856 campaign, pro-Frémont newspapers, poems, songs, banners, and cartoons frequently featured both John and Jessie Frémont and emphasized the couple's romantic love and fecundity. Buchanan, on the other hand, was the celibate, childless "Old Bachelor." Although Democratic papers attempted to circulate stories about "Mr. Buchanan's First Love" to "[propagate] the myth that he maintained his single status as a measure of devotion" to a woman to whom he had once been engaged, most Northerners dismissed the story as "Buchanan's excuse for old bachelordom."[34] Even the *Pennsylvanian*—"Old Buck's leading organ in that State"—"knock[ed] the love story [as] 'higher than a kite.'"[35]

For many Northerners, Buchanan's status as the "Old Bachelor"—or even more derisively the "maiden lady," "spinster aunt," or "old maid"—factored into their assessment of his masculinity and fitness for political office.[36] "I hold," a speaker at a Republican mass meeting told his audience, "that no man who has not had the courage to marry a wife (laughter, and cries of 'Three groans for BUCHANAN' [from the crowd]) has courage to fill a responsible office."[37] Banners created by individual Northerners—which were displayed before crowds of five, ten, and twenty thousand people in hundreds of political parades across the North—featured slogans like "No Bachelors for Us," "Down with 'old Buck' and up with a married man for President," and "The Yankee nation spurns the man / That cannot win a woman."[38] When making the case for Frémont at a political rally, one speaker argued that "it is by no means a small item in the qualifications he possesses for being the President" that he had eloped with Jessie Benton, a comment met with loud "cheers" from the large crowd.[39] Scrapbook-maker John Walden agreed, as evidenced by the fact that he saved several articles about "Buchanan's bachelorship" and directly contrasted them with Frémont's "rascal style [with] Jessie Benton."[40]

Buchanan's "cold celibacy" was especially questionable in the eyes of many Northerners because he, "in all his coolness," had never "begotten" a child.[41] "Spicy communications" in Northern newspapers explained that the people "think the idea…that an old withered up bachelor like Buchanan should ever preside over the destinies of the American people" was "preposterous."[42] "A man without babies a President!" the article concluded, "Just think of it!"[43] The *New York Daily Times* reported that "over the ladies' car" in one Northern political parade hung "a banner on which was [printed] 'We want no old Buck who has no little *Dears!*'"[44]

Some sources plainly stated that doughfaces like Buchanan were sexually sterile. In an 1850 speech on "the Slavery Question," abolitionist Thaddeus Stevens—who, like Buchanan, was a career politician from Pennsylvania—"described the 'race of doughfaces' as 'an unmanly, unvirile race, incapable, according to the laws of nature, of reproduction."[45] The lyrics of a popular 1856 song proclaimed, "Old Bachelors are low in rate" as "They'd never populate a state."[46] An 1856 Know-Nothing political pamphlet similarly attacked the "statesmanship and prowess" of doughfaces like Buchanan, Pierce, and Douglas who had supported the Kansas-Nebraska Act by referring to their inability to reproduce:

> For from North and from South
> Is the stuff that composes doughfaces;
> Thank Heaven! like mules,
> By nature's stern rules,
> They never can increase their species.[47]

An article that appeared in multiple Northern newspapers was even more blunt, noting the reason some Northerners opted not to support Buchanan was "his total barrenness…physically."[48]

Referencing Buchanan's "barrenness" and lack of biological children was also a way to allude to his sexual impotency. As historian Kevin Mumford has explained, Antebellum Americans considered impotence in men comparable with barrenness in women.[49] The fact that so many Northern popular press sources linked "backbone," "coolness," and "softness" further indicates that these critiques of the doughfaces had sexualized as well as gendered connotations. Sources that discussed the need for "firmness" and "backbone" among Northern men also referred to the fact that "soft," "unfirm," "chilly" doughfaces needed to be "heated up," "warmed," and "hardened by heat"—language that almost exactly mirrors gendered and sexualized rhetoric deployed in other moments of political upheaval such as the Spanish-American War and the Cold War.[50] Antebellum doughfaces were called "men of wax," as they "softened" and "melted" instead of "stiffened" under heat.[51] Cartoons frequently depicted doughfaces like Buchanan as candles and icebergs that melted under heat.[52] The author of the anti-Kansas-Nebraska Act pamphlet *A Bake-pan for the Dough-faces* concluded that politicians who were "pale-blooded, chicken-livered, and dough-faced, or what in vulgar parlance is termed slack-baked" were "soft" and "unmanly" and needed to be "hardened by the heat."[53]

While they most often spoke euphemistically, Northerners sometimes directly discussed Buchanan's sexual anatomy. In the same breath in which he referenced the fact that Buchanan "had no backbone" or "blood in his veins," a speaker at a large Republican rally in Brooklyn proclaimed that Buchanan had "lost a limb"— not "a hand, an arm, or a leg…Mr. BUCHANAN's infirmity surpassed all these."[54] Responding to "loud cheers" from the crowd, the speaker went on to connect Buchanan's "infirmity" with his unmarried and ostensibly celibate status. The speaker disingenuously pleaded with the audience not to make too much of his remark, as "he did not mean to refer to anything peculiar in Mr. BUCHANAN's physical relations in life."[55] After waiting for laughter from the crowd to subside, the speaker asked the audience if they would "trust a man like that" for the presidency, to which the people cried "No, never."[56]

Other sources called Buchanan a eunuch or hermaphrodite. A satirical *Vanity Fair* article implied that Buchanan, like the "Chief Eunuch" servant who worked for him, was asexual and physically emasculated.[57] A lengthy article by "a lady correspondent" in the *New York Evening Post* proclaimed "THE OLD BUCK BUT A HALF A MAN."[58] "An old Bachelor," she continued, "is at most but a Half Man; and how can such a person make more than a Half President?"[59] Political cartoons like "Letting the Last 'Democratic Drop'" and "The Candidates. Young America, …and… Old Fogyism" depicted Buchanan with male and female attire—a mid-nineteenth-century representation of ambiguous gender or sexual identity.[60] An article titled "Under which *Sex*, Buchanan?" in the *Lowell* (MA) *Daily Citizen and News* frankly observed that "the democrats are fond of calling their presidential candidate 'old Buck.' The people, better versed in natural history, think he is too *doe*-faced for a genuine *buck*, and put him down as an hermaphrodite."[61]

Historians like Richard Hofstadter, Kevin P. Murphy, and Kristin Hoganson have written about the figure of "the political hermaphrodite" that appeared in the contexts of the Gilded Age and the Spanish-American War.[62] This figure, they argue, was used to call popular attention to the ambiguous gender or sexuality and the ambiguous politics of a particular candidate. Historians have surprisingly overlooked "the political hermaphrodite" in Antebellum America—a society that, as historians have thoroughly documented, "thought nearly exclusively in terms of absolute sexual [i.e., gender] dichotomies,"[63] and that increasingly thought of the politics of slavery in dichotomous terms.

Throughout the late 1850s, Antebellum Northerners called attention to the "double-sexed, double-gendered, and hermaphroditic" personal and political identity of the doughfaces.[64] In contrast, Republican politicians like John

LETTING THE LAST "DEMOCRATIC DROP"

FIGURE 13.1 "Letting the Last '*Democratic Drop*,'" lithograph by unknown author and publisher, 1856. In this political cartoon, Democratic presidential candidate and prominent doughface James Buchanan is depicted wearing female attire. Another politician on the left bleeds Buchanan into a bowl held by Republican candidate John C. Frémont on the right. (From the Library of Congress Rare Book and Special Collections Division Washington, DC. <www.loc.gov/pictures/item/2008661582/> Reproduction Number: LC-USZ62-92022 (b&w film copy neg.). "Letting the Last '*Democratic Drop*'")

C. Frémont—and later Abraham Lincoln—who exhibited a clear political stance on slavery were "perfect men" of "heat," "passion," "firmness," and "backbone."[65] Over the course of the late 1850s, more and more Northern men became convinced of the need to "shew [sic] backbone," "assert…manhood," and "especially never vote…for a *dough-face*."[66] After years of political compromise with the South on slavery, George Templeton Strong concluded in his diary that it was increasingly "hard for any of us [Northern men] possessing even the rudiments of a backbone to tolerate talk about conciliation and concession."[67] In 1860, a majority of Northern voters would elect "*a man*" with enough "backbone" and "firmness" to stand up to the South.[68] Under the leadership of president-elect Lincoln, "the Northern backbone [was] much stiffened" and "the Great Iceberg" of doughface rule and submission to the Slave Power was "melting away."[69]

Notes

1   For an example list of politicians considered to be doughfaces, see "The Vote in the Senate on the Nebraska Bill," *Daily Missouri Democrat*, March 9, 1854.
2   Some studies have explored gendered characterizations of particular politicians like James Buchanan: for example, Andrea Foroughi, "Father Abraham, Mammy Lincoln, and Aunty Abe: Gender in the Civil War Cartoons of Abraham Lincoln," in *A Press Divided: Newspaper Coverage of the Civil War*, David Sachsman, ed. (New Brunswick, NJ: Transaction Publishers, 2014): 75–90; Michael D. Pierson, *Free Hearts and Free Homes: Gender and American Antislavery Politics* (Chapel Hill: The University of North Carolina Press, 2003); and Brie Swenson Arnold, "Competition for the Virgin Soil of Kansas: Gendered and Sexualized Discourse about the Kansas Crisis in Northern Popular Print and Political Culture, 1854–1860" (PhD dissertation, University of Minnesota, 2008).
3   George Templeton Strong, diary entry, December 9, 1859, in *The Diary of George Templeton Strong*, vol. 2, Allan Nevins and Milton Halsey Thomas, eds. (New York: The Macmillian Company, 1952), 476.
4   "Origin of the Term Doughface," *New York Daily Times*, April 27, 1854.
5   John Russell Bartlett, *Dictionary of Americanisms. A Glossary of Words and Phrases Usually Regarded as Peculiar to the United States, Second Edition Greatly Improved and Enlarged* (Boston: Little, Brown and Company, 1859), 128.
6   "Sketches in Washington. Number Four," *New York Daily Times*, June 30, 1854; *New York Daily Times* June 3, 1856; *New York Daily Times*, March 10, 1856; [Darius Lyman?], *Leaven for Doughfaces; or Threescore and Ten Parables Touching Slavery* (Cincinnati: Bangs and Company, 1856); [Leonard Marsh?], *A Bake-pan for the Dough-faces. By One of Them. Try it* (Burlington, VT: Charles Goodrich, 1854); Lewis Cass, *Address of General Cass to the Democracy of Detroit, Delivered at the City Hall, November 4, 1854* (Detroit: 1854), 8; *Nebraska, A Poem Personal and Political* (Boston: John P. Jewett and Co., 1854), 17; Henry Ward Beecher, *Defence of Kansas* (Washington, DC: Buell & Blanchard, 1856), 7; and Charles H. A. Bulkley, *Removal of Ancient Landmarks: Or, The Causes and Consequences of Slavery Extension* (Hartford, CT: Case, Tiffany and Company, 1854), 15.
7   J. M. Walden, scrapbook, vol. 1 [1850–1856], 30, www.kansasmemory.org/item/221235. Emphasis in original.

8 Ibid.

9 George Templeton Strong, diary entry, January 4, 1858, in *The Diary of George Templeton Strong: The Turbulent Fifties, 1850–1859*, Allan Nevins and Milton Halsey Thomas, eds. (New York: Macmillan Company, 1952), 382; George Templeton Strong, diary entry, December 4, 1860, in *Diary of the Civil War, 1860–1865*, Allan Nevins, ed. (New York: The Macmillan Company, 1962), 69.

10 "Mr. Senator Brown of Mississippi," *Bradford Reporter* (Towanda, PA), October 14, 1858; "Buchanan's Slavery Record," *Bedford* (PA) *Inquirer and Chronicle*, August 29, 1856.

11 Bartlett, 17. For an example of Northerners using the terms to generally comment on one's manliness, see Strong, diary entry, October 4, 1854, *The Diary of George Templeton Strong*, vol. 2, 185, which noted that certain acquaintances he disapproved of were men "without backbone, unvertebrated [sic]."

12 Bartlett, 17; "Butter and Eggs," *Harper's Weekly*, January 22, 1859.

13 "Judge Culver, in his speech," *Meigs County Telegraph* (Pomeroy, OH), September 11, 1855.

14 As cited in John Ernest, ed., introduction to *The Escape; or, a Leap for Freedom* by William Wells Brown, 1858 (Knoxville, TN: University of Tennessee Press, 2001), x.

15 John L. Magee, "The Great Presidential Race of 1856" (Philadelphia, PA: 1856), Library of Congress, www.loc.gov/pictures/item/2008661583/. See also "Grand Rally of the 9th Ward Democracy; 5,000 Democrats in the Field," *Brooklyn* (NY) *Daily Eagle*, August 21, 1856.

16 "Republicanism; Grand Ratification Meeting," *New York Daily Times*, June 26, 1856.

17 Letter from Fitz-Henry Warren of Burlington, Iowa, to James Shepard Pike (associate editor of the *New York Times*), February 25, 1860, in *First Blows of the Civil War: The Ten Years of Preliminary Conflict in the United States. From 1850 to 1860. … Progress of the Struggle Shown by Public Records and Private Correspondence* (New York: American News Company, 1879), 526.

18 "Buchanan's Slavery Record," *Bedford* (PA) *Inquirer and Chronicle*, August 29, 1856. Emphasis in original.

19 "The Rocky-Mountain Club Re-organized," *Brooklyn* (NY) *Daily Eagle*, November 25, 1856.

20 "The Appropriation Bill in Congress—Bringing the Kansas Question to Bear," *New York Herald*, July 26, 1856.

21 "Mr. Senator Brown of Mississippi," *Bradford Reporter* (Towanda, PA), October 14, 1858; "Mr. Senator Brown of Mississippi," *The Intelligencer* (Wheeling, VA), October 13, 1858.

22 Strong, diary entry, January 5, 1858, in *The Diary of George Templeton Strong*, vol. 2, 382.

23 Ibid.; Strong, diary entries, January 15, 1861, and June 21, 1860, in *Diary of the Civil War, 1860–1865*, 89, 35.

24 "What Is It?" *Vanity Fair*, March 10, 1860, http://gettysburg.cdmhost.com/cdm/singleitem/collection/p4016coll2/id/138/rec/10. Antebellum Northerners would have instantly been familiar with the racial connotations that also went along with the "What is it?" title of the cartoon. From 1859 onward, the "great what is it" exhibition of an African American man at PT Barnum's Museum was widely known across the North. This phrase was also featured in other political cartoons like "An heir to the throne, or the next Republican candidate" (New York: Courier and Ives, 1860).

25 Helen Lefkowitz Horowitz, *Rereading Sex: Battles over Sexual Knowledge and Suppression in Nineteenth-Century America* (New York: Alfred A. Knopf, 2002).

26 Horowitz, *Rereading Sex*, 6, 26. See also Winthrop Jordan, *White Over Black: American Attitudes toward the Negro, 1550–1812* (Chapel Hill, NC: University of North Carolina Press, 1968), 18; Elizabeth Stephens, "Pathologizing Leaky Male Bodies: Spermatorrhea in Nineteenth-Century British Medical and Popular Anatomical Museums," *Journal of the History of Sexuality* 17, no. 3 (Sept. 2008), 424–26.

27 Horowitz, *Rereading Sex*, 5, 26, 88.

28 "Monument of James Buchanan," *The Weekly Portage Sentinel* (Ravenna, OH), August 28, 1856.

29 R. G. Horton, *The Life and Public Services of James Buchanan. Late Minister to England and Formerly Minister to Russia, Senator and Representative in Congress, and Secretary of State; Including the Most Important of his State Papers* (New York: Derby & Jackson, 1856), 420.

30 "Mr. Buchanan's Social Character," *Eaton* (OH) *Democrat*, August 22, 1856 (emphasis in original); "Close of the Constitutional Convention…Wyandotte, K.T.," *New York Daily Times*, August 6, 1859.

31 "Plain Dealer on Buchanan in 1851," *The Daily Cleveland* (OH) *Herald*, October 23, 1856; "The Plain Dealer's Unbiased Opinion of James Buchanan," *Holmes County* (OH) *Republican*, September 25, 1856; "What was said of Buchanan," *The Bradford Reporter* (Towanda, PA), October 23, 1856; "The Plain Dealer's Unbiased Opinion of James Buchanan," *Western Reserve Chronicle* (Warren, OH), October 1, 1856; "A Very Pretty Quarrel as it Stands," *Chicago Daily Tribune*, December 31, 1857.

32 "Only three of every hundred American men stayed single" in this period, according to Jean H. Baker in *James Buchanan* (New York: Times Books/Henry Holt, 2004), 21–22. While Buchanan remains America's first and only bachelor president, surprisingly little scholarship exists on his bachelor status. Buchanan's unmarried status as well as other evidence of his long-term close relationship with Southern Senator William Rufus King has also raised speculation about whether he may have been "America's first homosexual president" (Baker, *James Buchanan*, 25). Though mid-nineteenth century Americans would not have identified themselves or others as homosexual, some Antebellum evidence of Buchanan and King's relationship and the popular discussion of Buchanan's "coolness" and "hermaphodism" may suggest that Buchanan was involved in or perceived to have been involved in a same-sex intimate relationship. For brief references to Buchanan's sexuality and/or possible homosexuality, see Baker, *James Buchanan*, 20–22 and 25–26; Etcheson, "General Jackson Is Dead: James Buchanan, Stephen A. Douglas, and Kansas Policy" in *James Buchanan and the Coming of the Civil War*, John W. Quist and Michael J. Birkner, eds. (Gainesville, FL: University of Florida Press, 2013), 93; John Gilbert McCurdy, *Citizen Bachelors: Manhood and the Creation of the United States* (Ithaca: Cornell University Press, 2009), 199; Michael J. Birkner, "Introduction" in *James Buchanan and the Political Crisis of the 1850s* (Selinsgrove, PA: Susquehanna University Press, 1996), 21. For recent popular press discussion of this, see Bill Kaufman, "For President Buchanan," *American Conservative* 12, no. 3 (May/June 2013): 38; Andrew Romano, "The Bachelor," *Newsweek Global* 161, no. 31 (August 30, 2013): 1. On mid-nineteenth century meanings and understandings of hermaphrodism as being tied to homosexuality, see Christina Matta, "Ambiguous Bodies and Deviant Sexualities: Hermaphrodites, Homosexuality, and Surgery in the United States, 1850–1904," *Perspectives in Biology and Medicine* 48, no. 1 (Winter 2005), 78.

33 Pierson, *Free Hearts, Free Homes*, Chapters 5–6; David Grant, "'Our Nation's Greatest Hope is She': The Cult of Jessie Fremont in the Republican Campaign Poetry," *Journal of American Studies* 42, no. 2 (August 2008): 187–213.

34 "Mr. Buchanan's First Love," *New York Herald*, July 23, 1856; "A Romantic Story Spoiled," *Wyandot Pioneer* (Upper Sandusky, OH), August 14, 1856.

35 "A Romantic Story Spoiled," *Wyandot Pioneer* (Upper Sandusky, OH), August 14, 1856.

36 "Faces of the Candidates," *New Hampshire Statesman* (Concord), August 18, 1856.

37 "Republicanism: Grand Ratification Meeting," *New York Daily Times*, June 26, 1856.

38 "Grant Fremont Demonstration in Poughkeepsie," *New York Herald*, October 17, 1856; "A Shot at the Nominee," *Meigs County Telegraph* (Pomeroy, OH), July 8, 1856; "Our Trophy," *Dixon* (IL) *Telegraph*, August 2, 1856, cited in Grant, "'Our Nation's Greatest Hope is She,'" 200.

39 "Republicanism: Grand Ratification Meeting," *New York Daily Times*, June 26, 1856.

40 J. M. Walden, scrapbook, vol. 1 [1850–1856], 28.

41 "James Buchanan," *Fremont* (OH) *Journal*, July 11, 1856.

42 "We clip the following…," *Meigs County Telegraph* (Pomeroy, OH), July 8, 1856.

43 Ibid.

44 "Political Miscellany," *New York Daily Times*, October 1, 1856. Emphasis in original.

45 Cited in Grant, "'Our Nation's Greatest Hope is She,'" 199.

46 "There is the White House Yonder," *Rocky Mountain Song Book* (Providence, RI: Du Dah & Co, 1856), 35, cited in Grant, "'Our Nation's Greatest Hope is She,'" 200.

47 [Henry Chase], *The Life and Times of Sam / Written by Himself* (Claremont, NH: Tracy and Sanford, 1855), 28.

48 *Cincinnati* (OH) *Evening Post*, quoted in *New York Evening Post*, June 23, 1856, as cited in Pierson, *Free Hearts, Free Homes*, 124.

49 Kevin J. Mumford, "'Lost Manhood' Found: Male Sexual Impotence and Victorian Culture in the United States," *Journal of the History of Sexuality* 3, no. 1 (July 1992), 34–36.

50 See Frank Costigliola, "'Unceasing Pressure for Penetration': Gender, Pathology, and Emotion in George Kennan's Formation of the Cold War," *Journal of American History* 83, no. 4 (March 1997), 1309–39; K. A. Cuordileone, "'Politics in an Age of Anxiety': Cold War Political Culture and the Crisis in American Masculinity, 1949–1960," *The Journal of American History* 87, no. 2 (September 2000), 515–45; K. A. Cuordileone, *Manhood and American Political Culture in the Cold War* (New York: Routledge, 2005); Robert Dean, *Imperial Brotherhood: Gender and the Making of Cold War Foreign Policy* (Amherst, MA: University of Massachusetts Press, 2001); Kristin Hoganson, *Fighting for American Manhood: How Gender Politics Provoked the Spanish-American and Philippine-American Wars* (New Haven: Yale University Press, 1998); Kevin P. Murphy, *Political Manhood: Red Bloods, Mollycoddles, and the Politics of Progressive Era Reform* (New York: Columbia University Press, 2008).

51 [Leonard Marsh?], *A Bake-pan for the Dough-faces. By One of Them. Try it* (Burlington, Vermont: Charles Goodrich, 1854).

52 "Good Night," *Vanity Fair*, March 17, 1860, http://gettysburg.cdmhost.com/cdm/singleitem/collection/p4016coll2/id/147/rec/11; "Head of the Nation," *Vanity Fair*, March 31, 1860, https://gettysburg.contentdm.oclc.org/digital/collection/p4016coll2/id/166/.

53 *A Bake-pan for the Dough-faces*, 1, 4, 25.

54 "The Presidency: Grand Republican Ratification Meeting in Brooklyn," *New York Daily Times*, June 28, 1856. Also in "Black Spirits and White! Collision between the Republicans and Know Nothings," *Brooklyn* (NY) *Daily Eagle*, June 28, 1856.

55 "The Presidency: Grand Republican Ratification Meeting in Brooklyn," *New York Daily Times*, June 28, 1856.

56 Ibid.

57 "Letter to King James, of America," *Vanity Fair*, January 21, 1860, 59. Baker also refers to Buchanan as "eunuchlike" and "asexual" (Baker, *James Buchanan*, 26).

58 "No Chance for Bachelors. The Old Buck but a Half a Man," *Western Reserve Chronicle* (Warren, OH), June 25, 1856. Originally published in the *New York Post*.

59 "No Chance for Bachelors," *Western Reserve Chronicle*, June 25, 1856.

60 "Letting the Last 'Democratic Drop'" (1856), Library of Congress, www.loc.gov/pictures/item/2008661582/; "The Candidates. Young America, ... and ... Old Fogyism" (1856), Wisconsin Historical Society, www.wisconsinhistory.org/museum/exhibits/elections/1856.asp.

61 "Under which *Sex*, Buchanan?" *Lowell Daily Citizen and News*, June 26, 1856. This same "doe-face" pun appeared in *Prenticeana*, a compendium of "wit and humor": "One of the northern papers calls Mr. Buchanan a dough-face. Who would expect such an Old Buck to have a doe-face?" George D. Prentice, *Prenticeana; Or, Wit and Humor in Paragraphs* (New York: Derby and Jackson, 1860), 205.

62 Richard Hofstadter, *Anti-Intellectualism in American Life* (New York: Knopf, 1962), 187–91; Murphy, *Political Manhood*, 14–15, 27–31; and Hoganson, *Fighting for American Manhood*, 23.

63 Matta, "Ambiguous Bodies and Deviant Sexualities," 75.

64 The "double-sexed, doubled-gendered, hermaphroditic" phrase was specifically used in reference to Northern doughface Caleb Cushing. This discussion of Cushing appeared in a lengthy and widely reprinted article that generally critiqued doughfaces like Buchanan and Pierce. "The Presidency," *New York Herald*, June 27, 1856; "Benton on the State of Politics," *The Nashville* (TN) *Patriot*, July 19, 1856; "Benton on the Cincinnati Convention," *Holmes County* (OH) *Republican*, October 9, 1856; "Benton in the Field," *The Sudbury* (PA) *American*, August 2, 1856.

65 "Fremont at Home—Description of a Call upon him at his House," *New York Daily Times*, August 13, 1856.

66 Strong, diary entry, December 9, 1859, in *The Diary of George Templeton Strong*, vol. 2, 476; "Just Judgment from Missouri-Let Every Buchanan Democrat and Fillmore Union Saver-in the Free States 'Read, Ponder, and Inwardly Digest," *The Daily Cleveland* (OH) *Herald*, September 10, 1856.

67 Strong, diary entry, February 6, 1861, in *Diary of the Civil War, 1860–1865*, 98.

68 Strong, diary entry, March 5, 1861, in *Diary of the Civil War, 1860–1865*, 106 (emphasis in original); "Nomination of Abraham Lincoln," *Cleveland* (OH) *Morning Leader*, May 21, 1860.

69 Strong, diary entry, April 13, 1861, in *Diary of the Civil War, 1860–1865*, 119; "Great Iceberg Melting Away," *Vanity Fair*, May 9, 1861.

14

"FREE MEN, FREE SPEECH, FREE PRESS, FREE TERRITORY, AND FRÉMONT"

Gregory A. Borchard

With news that Democratic candidate James Buchanan had just won the 1856 election, *The New York Tribune* featured a column titled "Frémont for 1860," which evaluated John C. Frémont's 1856 campaign as a long-term victory for Republicans. Horace Greeley, editor of the *Tribune*, responded to election results by using Frémont's trailblazing campaign as synonymous with a growing antislavery platform. The November 11, 1856, column listed a string of freedoms that Republicans had helped to advance. According to the *Tribune*, "Free Men, Free Speech, Free Press, Free Territory, and Frémont" produced "prodigious results," and Frémont was "the man of the hour."[1]

This chapter profiles Frémont and his bid for the presidency in 1856 as the first Republican candidate for the office, citing commentary from *The New York Tribune* and leading Republican newspapers as voices of the newly formed party. It focuses specifically on the way editors—Greeley among them—addressed claims from the Democratic press about Frémont's past that included his alleged ties to Catholicism, which at the time would have brought into doubt his legitimacy as a candidate among a portion of the electorate. Exploring issues of political and cultural importance, this study describes one of the ways the press in the years leading to the Civil War made personal lives political issues. It features Frémont, dubbed "the pathfinder of the West" for his trailblazing explorations, as both a controversial and essential character in helping launch Abraham Lincoln's campaign for the presidency in the subsequent election of 1860.

Well before the 1856 election, Frémont's family had created a trajectory for his later years that entailed drama and intrigue. As a young boy, Frémont grew from roots rich in aristocracy and scandal. His mother's family had ties to Virginia's governing body, and she, Anne Beverly Whiting, at the age of seventeen, married a much older man in Richmond to maintain the family's wealth. Anne later fell in love with Charles Fremon, who identified himself as a royalist who had fled the French Revolution. In 1811, Anne deserted her husband and eloped with Fremon. Moving repeatedly, the couple evidently never wed but had several children. When Anne's husband died, she and Charles left for Savannah, GA, where their child John was born in 1813. Charles died when John was young, and thereafter, the family added a "t" and the accent to their last name.

Young Frémont worked in a law office and then studied at the College of Charleston from 1829 until his expulsion in 1831, shortly before graduation, for "incorrigible negligence." The school, however, bestowed on him a bachelor's degree five years later. Joel Poinsett, a South Carolina politician and botanist, became Frémont's patron. In the late 1830s, Poinsett arranged to have Frémont travel with Joseph Nicollet, a French-born scientist and explorer, on surveys of the Mississippi and Missouri Rivers. After returning, Frémont eloped with Jessie Benton, daughter of US Senator Thomas Hart Benton of Missouri. In 1842, Senator Benton secured congressional authorization for an expedition, headed by his new son-in-law, to explore, survey, and map the Oregon Trail. Ignoring the governmental directive to return via the same path, Frémont and his party traveled into Nevada, over the Sierra Nevada, and into California. In all, the expedition covered almost 6,500 miles.

Frémont returned to Washington, DC, to write reports on his travels. One of his accounts put "Las Vegas" literally on the map. On May 3, 1844, after a day's journey of 18 miles, Frémont wrote, "we encamped in the midst of another very large basin, at a camping ground called Las Vegas—a term which the Spaniards use to signify fertile or marshy plains, in contradistinction to llanos, which they apply to dry and sterile plains."[2]

Poetically phrased—and likely written by his articulate wife—his reports filled newspapers with content read by audiences eager for details about sensational locations and events. Years later, Jesse Benton Frémont helped commemorate her husband's travels through the West with the legendary verse: "Cities have risen on the ashes of his lonely campfires."[3]

In 1847, Captain Frémont was appointed to serve as the military governor of the territory of California. However, Frémont was recalled to Washington, DC, for a court-martial. He was found guilty, pardoned by President James K. Polk, and would later return to California as a private citizen. In 1850, he was selected to be one of California's first two senators. He voted against the Fugitive Slave Act and for the ban on the slave trade in Washington, DC. In 1851, the state legislature denied him a second term and chose a proslavery Democrat instead. Frémont's antislavery record would make him an attractive recruit for the newly formed Republican Party.

Democrat Franklin Pierce was elected president in 1852—a time when he could expect support from a predominantly Democratic Congress. The climate in Washington reflected a national mood of optimism that the Compromise of 1850 had solved the crises over land in the West acquired in the war with Mexico.

However, the compromise would fail under Pierce when the Western territories exploded in violence after settlers in the Nebraska Territory petitioned for statehood.[4] The hostilities in the West were a prelude to the conflict that would sweep the nation—in the prophetic words of New York editor Thurlow Weed, "The great battle was yet to come."[5] Popular outrage over the Kansas-Nebraska Act, which officially allowed popular sovereignty in the territory and included a provision that abrogated the Missouri Compromise, found expression at meetings and in newspapers, leading, in sections of the North, to a new organization dedicated to opposing the expansion of slavery. When Asahel N. Cole, editor of *The Genesee Valley Free Press* in Allegany County, New York, organized a convention for members of the movement, he contacted Horace Greeley, editor of the *New York Tribune*, about naming the group. Greeley suggested "Republican," and in May 1854, the *Free Press* became the first newspaper to display the name of the party in its masthead.[6]

The Republican Party sprang to life in meetings, conventions, and rallies with the help of editors and affiliates in the Northern press, who issued columns devoted to the destruction of slavery. The Republicans grew on a state-by-state level, holding their first convention in Michigan on July 6, 1854. Kinsley S. Bingham was nominated for governor and elected later that year. Ohio also held a Republican state convention and nominated a ticket that won in the fall. After consistent growth throughout the North, party leaders organized in June 1856 the first Republican National Convention in Philadelphia's Music Fund Hall. About six hundred delegates attended, attracting more than one hundred reporters, who sat at front tables. On the first ballot, the Republican Party selected John C. Frémont as its presidential candidate, who received and won the nomination. The conventioneers matched him with William L. Dayton, a former senator from New Jersey, as a running mate. The *Philadelphia Public Ledger* reported that a delegate from Illinois, Abraham Lincoln, "a good fellow, a firm friend of freedom and an old line Whig," had received 110 votes for vice president, but Dayton later secured the spot.[7]

Frémont's acceptance speech on July 8, 1856, indicated that he believed in the need to prevent the extension of slavery. "The influence of the small but compact and powerful class of men interested in Slavery, who command one section of the country and wield a vast political control as a consequence in the other," he said, "is now directed to turn back this impulse of the Revolution and reverse its principles."[8] Republicans praised his speech. They noted that he had produced more public service than elder statesmen twice his age. Editors who published the speech described him as "the saviour [sic] and regenerator of a great nation!"[9]

The Republican platform advocated both free labor and a ban on the expansion of slavery into the West.[10] Having extensively navigated settlement routes in the West, Frémont's image as "The Pathfinder" was his greatest asset. The fact that the voting public knew little else about him made his nomination safe, as Republicans needed a contender with a fresh and uncompromised political record. Frémont's role in promoting a free California in 1849 and his endorsement of a free Kansas helped to build his image. In accordance with popular expectations of a presidential candidate at the time, Frémont did little active campaigning himself; however, an aspiring Republican from Illinois named Abraham Lincoln made dozens of speeches on Frémont's behalf.[11]

The chaos in the West that emerged from the ambiguities of popular sovereignty in the 1854 Kansas-Nebraska Act created tremendous difficulties for the Democrats who had supported it, namely incumbent President Franklin Pierce and one of its chief architects, Illinois Senator Stephen Douglas. Both of these potential presidential nominees for the Democratic Party in 1856 eventually opted out of the race and instead lent their support to James Buchanan, a long-time member of Congress and diplomat. Buchanan's position on the issue of Western settlement asserted that individual states and territories should decide on their own the future of slavery within their borders, a position held by most Democrats at the time. Upon receiving his party's nomination, Buchanan chose the traditional approach to presidential campaigning, making no public appearances and saying nothing to the press.

By 1856, the Whig Party had collapsed over the slavery issue and a secretive group of nativists, described as Know-Nothings, had grown in influence, stoking fears of perceived threats from Irish and German immigrants who had fled economic and political failures in Europe.[12] Whigs who reluctantly supported the Republicans had urged Frémont to seek votes from nativists, who were especially strong in Pennsylvania and Indiana. However, Greeley, who had established himself as a standard-bearer for the Republicans, insisted that "no tangible data" warranted the compromise. He believed Frémont's "adventurous, dashing career" had given him sufficient popularity among young voters, and he had faith "in the practicability of our winning many votes from those 'Americans' who were not heartily Republicans."[13]

Greeley's assessment was not entirely accurate. By mid-decade, those associated with the Know-Nothings included eight governors, five senators, and forty-three members of Congress nationwide; and in 1855, the party held its first and only national convention in Philadelphia in hopes of bringing together the many local variants of the movement. Its platform reflected the anti-immigrant and nativist positions anchoring the party. The items on the Know-Nothing platform drafted for public circulation in pamphlets and broadsides included a call for resistance to "the aggressive policy and corruption tendencies of the Roman Catholic Church" and staunch adherence to the maxim that "Americans only Shall Govern America."[14] In time for the 1856 election, former President Millard Fillmore (1850–1853) headed a nativist ticket as the American Party candidate, an organization backed by the Know-Nothings.[15] The American Party, as the public face of the Know-Nothings, appealed to a demographic that resisted the influence of Roman Catholicism and promoted the interests of native-born Americans.[16]

Throughout the summer of 1856, *The New York Tribune* advocated a belief that Frémont would bring "free men, free speech, free press, and free territory."

The Democrats meanwhile remained well-organized and hopeful, despite the failings of the Pierce presidency. But precisely because Frémont's limited political record provided such little fodder for Democratic attacks, the campaign of his rival James Buchanan instead focused on the Republican's personal life. While Democrats pointed to questionable business deals Frémont had made in the past, as well as his conduct as territorial governor of California, he also had to endure smears of being born a "Frenchman's bastard" and of having direct ties to Roman Catholicism.[17] The press often exaggerated details to describe his background. For example, Frémont's father was indeed French and by extension may have had an indirect tie to Catholicism (synonymous at that time with holding Papal loyalties over American ones), but Frémont himself practiced as a Protestant and had his oldest son baptized in the Episcopal Church.

FIGURE 14.1 "The Great Republican Reform Party, Calling on their candidate," lithograph published by Nathaniel Currier, 1856. This political cartoon presents a negative view of Republican presidential nominee John C. Frémont during the 1856 election. Frémont (far right) stands before an assorted group of constituents (including a Catholic priest and a free black) and responds to their demands: "You shall all have what you desire, and be sure that the glorious Principles of Popery, Fourierism, Free Love, Woman's Rights, the Maine Law, & above all the Equality of our Colored brethren, shall be maintained; If I get into the Presidential chair." (From the Library of Congress Prints and Photographs Division Washington, DC. <www.loc.gov/pictures/item/2003656588/> Reproduction Number: LC-USZ62-10370 (b&w film copy neg.). Nathaniel Currier, "The Great Republican Reform Party, Calling on their Candidate.")

Republicans countered the smears with remarks about Buchanan's age and his bachelorhood, using the nickname "Ten-Cent Jimmy" for him after he unwisely said in public that he considered ten cents a day a fair wage for manual laborers. They also smeared Buchanan with innuendos that as a bachelor, he could not serve in full capacity. A report from the *Times* described a Republican campaigner who delivered an oft-repeated sentiment: "No man who has not had the courage to marry a wife," he said, "has courage to fill a responsible office."[18]

Frémont had recognized that a denial of Catholicism would imply there was something wrong with its practice, and he refused throughout the campaign to dignify or even address the allegation.[19] As a result, pro-Democratic newspapers exploited the silence, using unanswered questions to their advantage by adding additional speculation to the mix. His silence opened the opportunity for opponents to portray him as the leader of a collection of radicals and reformers, and they associated him with drunks, utopians, free-love advocates, priests, and freed slaves.

The orchestrated smear against Frémont eventually made its way to Catholic publications as well, and, perhaps surprisingly, publications associated with the Roman Catholic Church began to distance themselves from Frémont by suggesting that he was indeed a Catholic and that his election should be opposed for this very reason. While *The St. Louis Pilot* and *The St. Louis Leader* (both connected to the Catholic Church) claimed Frémont took communion and went to confession during his stays in the city, they also told readers not to vote for him because he associated with Know-Nothings.[20]

Orestes Augustus Brownson, a New England intellectual and noted Catholic convert and writer, took the suggestion further and wrote that it was in the best interest of Catholics to oppose a Catholic nominee. Brownson's *Quarterly Review*, which he used to promote religiously based reform, avoided describing Frémont as a Catholic, but it insinuated that Catholic voters should vote for the Democratic Party's candidate because the success of a Catholic would, he thought, promote anti-Catholic sentiment. "We care not by whom the offices are filled, whether by Catholics or non-Catholics, provided they are well filled," he wrote. "If we as Catholics forget this, and ask office because we are Catholics, we make ourselves a Catholic party in politics, and necessarily provoke a non-Catholic party which will oppose us."[21]

One of the nation's leading Roman Catholic newspapers, *The Boston Pilot*, took the opportunity to question how Frémont's background would influence the United States. "We shall see the strange sight of this Anti-Catholic, dark-lantern, oath-bound party, nominating a Catholic, and a son of a foreigner, for the highest office in the gift of the people," the editorial stated. "For a Catholic to be put into the presidential chair, by the party whose only creed is hatred and persecution of Catholicity, will be in the world's history like England, restoring Pope Pious VII to Rome, a proof that governments and parties are unwilling instruments in God's hands to use as he will."[22]

Recognizing the salience of the issue, Democrats intensified their attacks on the Republican candidate. *The Brooklyn Daily Eagle*, *The Auburn American*, and

Democratic newspapers across the nation circulated claims that Frémont received his education from St. Mary's College in Baltimore. Furthermore, according to these press accounts, he was raised a Catholic and subscribed to Catholicism, and even though he married a Presbyterian and a Presbyterian minister conducted the service, "a Catholic priest made a fuss about it as being null, void, and heretical, and the ceremony was re-performed by him!"[23] J. A. McMaster, editor of the *Freeman's Journal*, claimed Frémont himself had "professed to be a Catholic, and nothing but a Catholic."[24]

While Frémont remained silent on the Catholic issue, the allegations about him often found quick rebuttals in Republican newspapers—for each account of

THE 'MUSTANG' TEAM

FIGURE 14.2 "The 'Mustang' Team," lithograph published by Nathaniel Currier, 1856. This cartoon mocks the abolitionist Republican Party, and jabs at the newspaper editors who supported Republican presidential candidate John C. Frémont. Seated in a wagon pulled by the "wooly nag" of abolitionism, Frémont holds a large cross, referring to the rumor that he was Catholic. On the horse (from left to right) are New York *Tribune* editor Horace Greeley, New York *Herald* editor James Gordon Bennett, and *New York Times* editor Henry J. Raymond. Clinging to the back of the wagon on the far right is James Watson Webb, editor of the New York *Courier & Enquirer*. In the back of the wagon, Frémont's wife Jessie reclines against a sack labeled "Bleeding Kansas Fund," an allusion to the unrest in Kansas between pro- and antislavery advocates. Frémont's wagon is blocked by the "Union Tollgate," manned by Brother Jonathan (a fictional personification of New England popular in the Antebellum period). (From the Library of Congress Prints and Photographs Division Washington, DC. <www.loc. gov/item/2008661587/> Reproduction Number: LC-USZ62-1998 (b&w film copy neg.). Nathaniel Currier, "The 'Mustang' Team.")

Frémont's religious history, another account directly contradicting it would appear in the press supporting him.[25]

The Frémont controversy attracted the attention of voters with nativist beliefs, who would instead support Buchanan or Fillmore. William Brownlow, editor of *Brownlow's Knoxville Whig*, summarized the sentiments of Know-Nothings at the time, expressing horror at the prospects of such a threat to the interests of native-born Americans. "The placing of such a candidate before the public," he wrote, "seems especially designed to defy public sentiment, and mock the Protestant American feeling of the country!"[26]

Initially, the insinuations and smears about Frémont's past had not troubled Republicans, as most assumed that the issue was simply beneath recognition. As the campaigns dragged on throughout the summer, however, Democrats continued to publish questions about Frémont's past and Republicans could no longer ignore them. Frémont's supposed ties to Catholicism even drew a response from Henry Raymond, editor of *The New York Times*, who at the Republican organizing convention in February had delivered a keynote address. He published a letter widely circulated in newspapers attempting to dismiss the issue. "But so far as these rumors assert, or imply, that he is a Roman Catholic, they are without the slightest foundation in fact," Raymond wrote. "Col. Frémont is not now, nor has he ever been, a Roman Catholic."[27] But despite Raymond's efforts, by the end of the summer the incessant innuendo of the press brought abolitionist Thaddeus Stevens to express a concern held by many Republicans that allegations about Frémont's Catholicism—regardless of their truth or lack thereof—had grown to the point that the perception of them would cost the new party the election.[28]

Greeley, for his part, used his *New York Tribune* to urge voters to look at issues in a much larger scope and cast their votes as a referendum on slavery. On Election Day, November 4, 1856, the *Tribune* pitched the decision as one about freedom. "The eventful day has dawned at last. To-day gives a tremendous triumph to Freedom or Slavery," Greeley wrote. "If this country is already wise enough, good enough, to throw off the domination of the Slave Power, then to-day's vote insures and proclaims that result."[29] Despite the negative attacks and his relatively scant political résumé, Frémont surpassed the expectations of many Republicans by winning eleven out of thirty-one states and 114 electoral votes to Buchanan's 174. Buchanan carried the eleven slave states and Pennsylvania, New Jersey, Indiana, Illinois, and California with 1,838,169 votes. Fillmore won 874,534 popular votes (21.6 percent) and electoral votes from the state of Maryland, the most substantial returns for a third-party candidate to that date.[30]

Republicans focused on the fact that Frémont had finished respectably and established the Republican Party as a real political force and the main rival to the Democratic Party. Columns published in the *Tribune* in the days following the election illustrated sentiments popularized by Greeley among his wide readership—quotes that developed a narrative anticipating a victory in 1860, a sentiment no doubt shared by Republican editors elsewhere.

On Monday, November 10, Greeley declared a victory of sorts, noting that the Republicans had failed to elect Frémont, but that none of the party's supporters had complained. "On the contrary," according to the *Tribune*, "the universal greeting has been—'We have done nobly for a beginning'—'we can surely beat them next time.' And this confidence is based on facts which every one can appreciate. We have carried more States than our party existed in barely one year ago."[31]

Of course, the victory described by the *Tribune* was more figurative than literal, and while Republican newspapers nationally celebrated a symbolic victory for the party, few Republicans expected that Frémont himself would win the next election. Instead, they eventually turned to other prospects, among them Abraham Lincoln, the former Whig lawyer who would run against Stephen Douglas for a Senate position from Illinois.

In 1856, Lincoln had campaigned extensively for Frémont. He criticized the Democrats over issues of Western settlement, and said little publically about the Know-Nothings, preferring to focus on the threat of the extension of slavery. However, in an 1855 letter to his friend Joshua Speed, Lincoln had written a scathing rebuke of the Know-Nothings, suggesting, "As a nation, we began by declaring that 'all men are created equal.' We now practically read it 'all men are created equal, except negroes.'"

> When the Know-Nothings get control, it will read "all men are created equals, except negroes and foreigners and Catholics." When it comes to that I should prefer emigrating to some country where they make no pretense of loving liberty—to Russia, for instance, where despotism can be taken pure, and without the base alloy of hypocrisy.[32]

Lincoln would encourage supporters to look forward to 1860. "Let us reinaugurate the good old 'central ideas' of the Republic. We can do it," Lincoln said. "The human heart *is* with us—God is with us. We shall again be able not to declare, that 'all States as States, are equal,' or yet that 'all citizens are equal,' but to renew the broader better declaration, including these and much more, that 'all *men* are created equal.'"[33]

While Republicans began gearing up for the next election, President James Buchanan and Vice President John Breckinridge ("Buck and Breck") entered the White House with less than half the votes cast and presided over one of the nation's most disastrous administrations. Two days after Buchanan assumed office, Supreme Court Chief Justice Roger Taney delivered the *Dred Scott* decision that held, according to a popular interpretation, "negroes have no rights which white men are bound to respect."[34] For the next four years, national troubles only worsened as the new administration failed to address the crisis that had begun in Kansas. (Indeed, reporters in the territory could later argue that the first shots of the Civil War did not take place at Sumter, but rather in fights that included the Sack of Lawrence and the Battle of Osawatomie—fights that ultimately spilled east.[35])

Between 1856 and 1860, Republican Party numbers grew, and members resisted attempts by Democrats to weaken their resolve. New York Senator William H. Seward and Abraham Lincoln suggested the same solutions to the slavery issue. Seward warned that the nation was entrenched in an "Irrepressible Conflict," and Lincoln responded by saying to him, "I have been thinking about what you said in your speech. I reckon you're right. We have got to deal with this slavery question, and got to give much more attention to it hereafter than we have been doing."[36]

While the Republican press continued to heap accolades on Frémont for another decade, Lincoln inherited momentum from the party's first presidential campaign and used it to navigate areas that had caused trouble for Frémont in the 1856 campaign. First in his 1858 campaign for Senate and then in his 1860 campaign for president, Lincoln addressed—and in some cases avoided—issues of slavery, the settlement of the West, and nativist interests. In November 1860, Lincoln captured a substantial number of votes from those who had not voted Republican in 1856. He won the presidency with an estimated 40 percent of the popular vote against a divided field.

The war that ensued after Lincoln's election brought a new role for Frémont, as the new president appointed the would-be president to a position of General in the Union Army. In a new chapter of Frémont's colorful career, the General, as commander of the Department of the West, issued an emancipation proclamation in August 1861 that applied only in Missouri. Lincoln disapproved of the measure, as it preempted the administration's war efforts and the president's own designs on eventual abolition. Lincoln scaled back the scope of Frémont's order and eventually recalled him to the East where he was no match for the generalship of Stonewall Jackson. Although Frémont again considered a presidential bid in 1864, he declined to run because he felt his candidacy would cause division within the party.

Frémont received an appointment as governor of Arizona Territory from 1878 until 1881, when he moved back to New York to seek financing for various ideas to develop the West. His last home was in Los Angeles. During a visit to New York City on July 13, 1890, Frémont fell ill with peritonitis, possibly resulting from a ruptured appendix. He died at the age of 77, and his body, initially interred at Trinity Cemetery in Manhattan and buried at his own orders in a plain coffin and a civilian suit, was reinterred at Rockland Cemetery in Nyack, New York. Both sites claim no denominational affiliation. History adopted as his epitaph the short phrase attributed to his wife, "From the ashes of his campfire have sprung cities." For his efforts in the 1856 election, he could claim credit in finding a path for Lincoln's 1860 victory.[37]

Notes

1 "Frémont for 1860," *New York Tribune*, November 11, 1856.

2 John C. Frémont, *Report of the Exploring Expedition to the Rocky Mountains*, vol. 2 (Washington: Blair and Rives, 1845), 266.

3 Frémont, *The Expeditions of John Charles Frémont*, vol. 1, Donald Dean Jackson and Mary Lee Spence, eds. (Urbana, IL: University of Illinois Press, 1970), xxi.

4 Thurlow Weed, *Life of Thurlow Weed Including Autobiography and a Memoir*, vol. 2 (Boston: Houghton, Mifflin, 1884), 184, 219, 241.

5 Ibid., 239.

6 Ibid., 241.

7 Ibid., 246; *Philadelphia Public Ledger*, June 20, 1856.

8 Charles Wentworth Upham, *The Life, Explorations, and Public Services of John Charles Frémont* (New York: Livermore & Rudd, 1856), 110–14.

9 Upham, "James C. Ashley, Anthony J. Blecker, Joseph C. Hornblower, E. R. Hoar, Thaddeus Stevens, Kingsley S. Bingham, John A. Wills, C. F. Cleveland, Cyrus Aldrich, &c.," *The Life, Explorations, and Public Services of John Charles Frémont*, 115.

10 Weed, *Life of Thurlow Weed*, 246; David W. Bulla and Gregory A. Borchard, *Journalism in the Civil War Era* (New York: Peter Lang), 28.

11 Roy P. Basler's *The Collected Works of Abraham Lincoln*, 8 vols. (New Brunswick, NJ: Rutgers, 1953–1955) includes the following speeches of Lincoln on behalf of Frémont in volume 2: Belleville, Illinois, October 18, 1856 (379, 380); Vandalia, Illinois, September 23, 1856 (377, 378); Jacksonville, Illinois, September 6, 1856 (369–73); Petersburg, Illinois, August 30, 1856 (367, 368); Kalamazoo, Michigan, August 27, 1856 (361–66); Galena, Illinois, July 23, 1856 (354, 355).

12 Gregory A. Borchard, "Revolutions Incomplete: Horace Greeley and the Forty-Eighters at Home and Abroad," *American Journalism*, 27:1 (Winter 2010): 7–36.

13 Horace Greeley, *Recollections of a Busy Life* (New York: J. B. Ford, 1868), 354.

14 *Young Sam, or, Native American's Own Book, Containing the Principles and Platform on which the Order Stands* (New York: American Family, 1855), 3.

15 Weed, *Life of Thurlow Weed*, 245.

16 Greeley, *Recollections of a Busy Life*, 290.

17 Henry A. Wise, "1856 Speech in Richmond," in Alan Nevis, *Frémont, Pathmarker of the West* (Lincoln and London: University of Nebraska Press, 1983), 451.

18 "Republicanism: Grand Ratification Meeting," *New York Times*, June 26, 1856.

19 "To Correspondents," *New York Tribune*, November 8, 1856.

20 "The Bishop of St. Louis on the Stand!" *The St. Louis Pilot*, September 16, 1856; *The St. Louis Leader*, September 16, 1856, in *Frémont's Romanism Established* (n.p., 1856), 12–14.

21 "The Presidential Election," *Brownson's Quarterly Review*, 1:4 (New York: E. Dunigan, 1856): 509–13.

22 *The Boston Pilot*, in "The Duty of Native Americans in the Present Crisis," *Frémont Campaign Literature* (New York: Democratic National Committee, 1856), 198.

23 *Brooklyn* (NY) *Daily Eagle*, June 25, 1856; William Brownlow, *Americanism Contrasted with Foreignism, Romanism, and Bogus Democracy* (Nashville, TN: Brownlow, 1856), 181.

24 J. A. McMaster, *Capital City Fact* (Columbus, OH), September 26, 1856, in *Frémont's Romanism Established* (n.p., 1856), 12–14.

25 Rev. Michael Olivetti, "To the Editor of *The Whitehall Chronicle*," *The Whitehall* (NY) *Chronicle*, August 28, 1856, in *Col. Frémont's Religion* (n.p., 1856), 7.

26 Brownlow, *Americanism*, 181.

27 "Authorized Letter from Lt. Governor Raymond," *Cincinnati Gazette*, August 5, 1856, in *Col. Frémont's Religion* (n.p., 1856), 1.

28 Tyler Anbinder, *Nativism and Slavery: The Northern Know Nothings and the Politics of the 1850s* (New York: Oxford University Press, 1992), 225. Anbinder cites, among other primary sources, *The Hartford Courant*, July 17, 1856; Thurlow Weed to Edwin

D. Morgan, August 9, 1856, Morgan Papers, NYSL; James Walker to Salmon P. Chase, September 22, 1856, Chase Papers LC; Schuyler Colfax to John Bigelow, August 29, 1856, Bigelow Papers, NYPL; Colfax to Francis P. Blair, Sr., August 15, 1856.

29 "To Correspondents," *New York Tribune*, November 4, 1856.

30 Weed, *Life of Thurlow Weed*, 248.

31 *New York Tribune*, November 10, 1856.

32 Abraham Lincoln to Joshua Speed, August 24, 1855, in Basler, *The Collected Works of Abraham Lincoln*, vol. 2, 323.

33 Lincoln, "Speech at a Republican Banquet, Chicago Illinois," December 10, 1856, in Basler, *The Collected Works of Abraham Lincoln*, vol. 2, 383–85.

34 Weed, *Life of Thurlow Weed*, 251.

35 Borchard, *Abraham Lincoln and Horace Greeley* (Carbondale, IL: Southern Illinois University Press, 2011), 51–53; Borchard, *The Firm of Greeley, Weed, and Seward: New York Partisanship and the Press, 1840–1860*, PhD diss. (Gainesville: University of Florida, 2003), 160–62.

36 Thorton Kirkland Lothrop, *William Henry Seward* (Boston: Houghton Mifflin, 1896), 56; Frederic Bancroft, *The Life of William H. Seward*, vol. 1 (New York: Harper and Brothers, 1900), 518.

37 John Bicknell, *Lincoln's Pathfinder: John C. Frémont and the Violent Election of 1856* (Chicago: Chicago Review Press, 2017), 165–200.

15

NEWSPAPER COVERAGE OF DRED SCOTT INFLAMES A DIVIDED NATION

William E. Huntzicker

On March 6, 1857, two days after the inauguration of Democrat James Buchanan as the fifteenth US president, Supreme Court Chief Justice Roger B. Taney announced that the court had resolved the Dred Scott case. The court ruled that Scott, a slave who had sued for his freedom because he had lived in a free state and a free territory, could not be freed after returning to the slave state of Missouri. The Washington *Union* reported that the court had made slavery constitutional and thus removed it from legislative control, party politics, and "public prejudice."[1]

The case (called *Dred Scott v. Sandford* because the court misspelled John F. A. Sanford's name) had already become a cause célèbre when Taney read his decision that spoke for a majority of the court, even though each justice wrote his own opinion in this unprecedentedly controversial case. In a 7–2 vote, the justices disagreed on many aspects of the case, but Taney spoke for the majority when he led off the readings.

Not only did Taney refuse Scott his freedom, he declared the Missouri Compromise unconstitutional, stating that the people in the territories had no right to decide the slavery issue because the Constitution protected slavery everywhere. Taney denied slaves, former slaves, descendants of slaves, and people with African heritage access to the courts. Thus, Scott had no right to sue. Taney also upheld the Fugitive Slave Law, protecting slaves like other property owned by people visiting free states.

Ironically, Democrats had hoped that the Dred Scott case would settle the major slavery disputes. In his inaugural address on March 4, 1857, President Buchanan said he expected the Supreme Court to decide the divisive question of slavery in the territories. Like his fellow Democrat, Senator Stephen A. Douglas, he celebrated popular sovereignty—the policy of allowing each territory to vote whether to be admitted as a slave state or free state. At the time, Kansas was literally at war over the issue.

The 1856 election, Buchanan said, was "a Presidential contest in which the passions of our fellow-citizens were excited to the highest degree by questions of deep and vital importance; but when the people proclaimed their will the tempest at once subsided and all was calm." Such calm, he predicted, would also follow the pending Supreme Court decision to work out the details of slavery in the territories.[2]

However, Buchanan's speech ignored many of the major issues of the case, focusing solely on the power of territorial legislatures on slavery. Only one justice, Taney, addressed that issue. The Dred Scott decision was so splintered and complex that each of the nine justices wrote his own opinion, which some of them read aloud to a crowded courtroom over several days.[3]

Clearly, the Dred Scott decision failed to achieve Buchanan's hope that the Supreme Court would settle the question of slavery's expansion, unite the nation, and calm political turmoil. In fact, the decision further polarized and stirred up sectional and racial strife.

Republican Horace Greeley's *New York Tribune* editorialized almost immediately: "This decision, we need hardly say, is entitled to just so much moral weight as would be the judgment of a majority of those congregated in any Washington bar-room.... It is a *dictum* prescribed by the stump to the bench—the Bowie-knife sticking in the stump ready for instant use if needed." The *Tribune* accused Taney of delaying the decision and announcement "over from last year in order not too flagrantly to alarm and exasperate the Free States on the eve of an important Presidential election."[4]

Its cardinal points are reported as follows:

1. A negro, because of his color, is denied the rights of a citizen of the United States—even the right to sue in our Courts for the redress of the most flagrant wrongs.
2. A slave, being taken by his master into a Free State and thence returning under his master's sway, is not therefore entitled to his freedom.
3. *Congress has no rightful power to prohibit Slavery in the Territories:* hence the Missouri Restriction was unconstitutional.[5]

"This judgment annihilates all Compromises and brings us face to face with the great issue in the right shape," the *Tribune* wrote.[6]

The *New York Evening Post*, whose editor William Cullen Bryant had begun as a Democrat and converted to abolitionism and the Republican Party, warned of encroaching governmental power.

> The dangers apprehended from the organic tendencies of the Supreme Court to engross the legislative power of the federal government, which Jefferson foresaw and so often warned his countrymen against, are no longer imaginary. They are upon us. The decision rendered by that body yesterday, in

the case of a Missouri negro who had appealed to it for assistance in asserting his right to share in the promises of the Declaration of Independence, had struck at the very roots of the past legislative policy of this country in reference to slavery. It has changed the very blood of the constitution, from which we derive our political existence, and has given to our government a direction and a purpose as novel as it is barbarous and humiliating.[7]

The *Richmond* (VA) *Enquirer*, by contrast, said abolitionists ignored laws and made treasonous arguments based on exaggerated views of the South. "Uplifting their hand in holy horror, its designing demagogues have been wont to shriek and wail, rave and storm, and impiously appeal to Heaven by turns, as they impose upon the popular mind of the North with their perverted portraitures of Southern slavery, contemptible caricatures of Southern society, and harrowing calumnies upon Southern character." Opponents of slavery were "the maligners of Washington and repudiators of the Bible."[8]

In a June speech in the Illinois legislative chamber, Senator Stephen A. Douglas defended Taney's decision, mostly on racial grounds. He warned that "violent resistance to the final decision of the highest judicial tribunal on earth" would be futile and that all people must put aside past political differences. Douglas said: "the main proposition decided by the court" was that "a negro, descended from slave parents imported from Africa, is not and cannot be a citizen of the United States." Douglas mentioned racial mixing—a topic he repeatedly used in his debates with Lincoln, who challenged him for his Senate seat the following year. "If the principle of negro equality be true," Douglas said, "we shall certainly be compelled, as conscientious and just men, to go one step further—repeal all laws making any distinction whatever on account of race and color, and authorize negroes to marry white women on an equality with white men."[9]

In the audience, country lawyer and former congressman Abraham Lincoln began his plan to challenge Douglas for his Senate seat. Lincoln took two weeks to prepare a response that he delivered in the same place to a smaller audience. Lincoln agreed with Douglas who was "especially horrified at the thought of the mixing blood by the white and black races." However, slavery encouraged racial amalgamation, Lincoln said. "In 1850 there were in the United States, 405,751, mulattoes. Very few of these are the offspring of whites and free blacks; nearly all have sprung from black slaves and white masters." Lincoln said that Republicans believed "the negro is a man; that his bondage is cruelly wrong, and that the field of his oppression ought not to be enlarged." Challenging Taney's history, Lincoln said that the writers of the Declaration of Independence would not recognize the use of the document "to aid in making the bondage of the negro universal and eternal."[10] Even though he would narrowly lose the Senate seat to Douglas in 1858, Lincoln discovered that Dred Scott resonated in the North and West, using his name several times during his famous debates with Douglas.[11]

On opposite sides, both Buchanan and Lincoln invoked Dred Scott's name, but they increasingly separated the people involved from the issues and symbolism of their case. News coverage reflected this same phenomenon.

The *New York Herald* on March 31 compared the attitudes of two newspapers— the *Charleston Mercury* and the St. Louis *Democrat*—representing two sides of the slavery issue.[12] The *Democrat* favored gradual emancipation in Missouri, and the *Mercury* represented "the incipient elements of a Southern ultra defection against Mr. Buchanan's administration." The *Herald* said John C. Fremont's turnout in the 1856 presidential election did not bode well for the Democrats in 1860. "The meaning of all this is that neither Mr. Buchanan's Cabinet, nor his inaugural, nor the Dred Scott decision, nor the harmonized democracy over the spoils, can any longer be relied upon by the secession faction of the South." The St. Louis *Democrat* attacked the Democratic Party that "placed slavery on its banner before the world." The *Herald* concluded: "Now, here we have a democratic faction which aided in the election of Mr. Buchanan, separating from party upon a local issue, and practically joining hands with the republican party, because the 'national democracy' ask too much in behalf of slavery." The solution would be to support Fremont in 1860.[13]

Supporting the Dred Scott decision, the *Baltimore Sun* praised it as "probably the most important that ever emanated from that highest tribunal of our country." The *Sun* continued: "The decision, we are glad to say, seems to be welcomed in most quarters. There are indiscreet and suicidal ravings among some of those who know no love except that of their own violent self-will and passions."[14] Shortly thereafter, the paper wrote, "If it is acquiesced in it will afford a peaceful solution to the only question which has for twenty-five years, disturbed the tranquility of the country."[15] The Raleigh *Standard* of North Carolina went even further: "The idea of the abolitionists, that a slave is free as soon as he touches the soil of a free State, is again exploded; for it is declared that he remains a slave, though sojourning in a free State, and the right of his owner to his body and to his services cannot be affected."[16] Some newspapers, like the New Orleans *Daily Delta*, defended their positions by attacking the New York press where "no end of incomprehensible legal profundity is employed to mystify the few intelligible points of constitution-ality and law contained in the decision."[17]

Three of the major New York newspapers heading into the Civil War were the *New York Herald* founded by James Gordon Bennett in 1835, the *New York Tribune* founded by Horace Greeley in 1841, and the *New York Daily Times* founded by Henry J. Raymond in 1851. The first major penny newspaper in New York had started in 1833 using a London circulation model. Instead of relying on strong par-tisan support, like the elite press, they sold for a penny on the street and relied heavily on advertisers. Nonetheless, partisan instincts died hard. Bennett, an immigrant who had received a Jesuit education in Scotland, was the most accomplished at using sen-sationalism to sell newspapers. He leaned Democratic and mocked reformers, such as suffragists and abolitionists. Greeley, by contrast, was a passionate reformer on temperance, abolition and, early on, women's suffrage. Raymond, who had worked with Greeley in both journalism and politics, created the *Times* as an alternative that did not get into a passion over issues. The fiercely competitive *Herald* and the *Tribune* would make by far the greatest investments in covering the Civil War.[18]

The *Daily Delta* cited the *New York Times* as saying that an apparent peace would follow the court's decision, but "it has laid the only solid foundation which has ever yet existed for an Abolition party." The *Delta* said the *New York Herald* foreshadowed increased fighting in Kansas: "Had the partisans of anti-slavery principles hired the United States Supreme Court to give them help and comfort, they could not have been more faithfully, more dexterously or more opportunely served. No sooner does the fire threaten to go out for want of fuel than this Supreme Court appears, and loads the embers with dry combustible material." The *Delta* continued: "The New York Tribune is of course rebellious, truculent and blatant, and will not fail to lash abolition fanaticism into greater violence than ever, while such conservative and soberly behaved sheets as the Times will organize the more solid materials of anti-slavery upon a well-seeming national sort of platform. The organization for the contest in 1860 commenced as soon as the last was decided."[19]

The Albany *Evening Journal* in upstate New York said the court followed "the lead of the Slavery Extension party, to which most of its members belong. Five of the Judges are slaveholders, and two of the other four owe their appointments to their facile ingenuity in making State laws bend to Federal demands in behalf of 'the Southern institution.'"[20] The *Evening Journal* had been created by Thurlow Weed in 1830. Weed helped elect Seward to the New York governorship and helped found the Republican Party in New York.[21] The Albany newspaper saw Taney's decision as part of a conspiracy to advance slavery and constrict freedom. Slave owners controlled the Senate, the presidency, and now the courts. The *Journal* encouraged readers "who love Republican institutions and who hate Aristocracy" to resist the growing slave power.[22]

"Judge Taney requests the American people to believe that the framers of the Constitution did not know their own minds," the *Evening Journal* said. "For the same Statesmen who drew up the Constitution, (which he says forbids Congress to prohibit Slavery in the Territories,) adopted the Ordinance of '87, which prohibited it in all the Territories we then had." Without mentioning the word slave, the Dred Scott decision legalized slavery everywhere.[23] The *Evening Journal* said the new president had conspired with other politicians to stop states and territories from freeing slaves, and the Dred Scott case had been designed to conquer all subordinate courts.[24]

Few articles actually covered Dred and Harriet Scott as people. *Frank Leslie's Illustrated Newspaper* on June 27, 1857, reported on a visit with the Scott family in St. Louis, where they were living in freedom. Scott, Harriet, and their two daughters went to Fitzgibbon's gallery for their only known photographs, which were reproduced on the cover. According to historian Lee VanderVelde, the *Leslie's* article provides "the only existing first-person description of [Harriet] by anyone who ever

FIGURE 15.1 "Visit to Dred Scott—his family—incidents of his life—decision of the Supreme Court," *Frank Leslie's Illustrated Newspaper*, June 27, 1857. These woodcuts were printed on the front page of *Frank Leslie's Illustrated Newspaper*. Dred Scott and his wife, Harriet, were interviewed following their final court appearance in May 1857,

actually met her," and "that picture, subsequently engraved for the newspaper, is the only remaining likeness."[25] *Leslie's* concluded with its own analysis of the case, saying the Taney's decision restored the Constitution's "true meaning and intention when it was formed and adopted" by removing Scott's citizenship because of his race. After the case ended, *Leslie's* reported, his owner filed a quitclaim deed for the Scotts and their daughters and transferred ownership to Taylor Blow, who then freed them.[26]

Although journalism historians have tended to criticize one another for placing too much emphasis on the New York newspapers, it is clear the metropolitan papers were strongly influential, helping set the agenda and frame the national debate. Many newspapers around the nation reprinted and quoted from the New York press. New York papers, in turn, depended on local papers for stories and analysis. National publications could aggregate news from around the nation to synthesize and follow trends.

Major newspapers announced that they received summary bulletins of breaking news by telegraph, with details, like the justices' opinions, to follow by mail. Some major papers sent reporters into the field, but none followed the lives of the Scott family—most relied on shared stories.[27]

Fiery rhetoric split the nation apart in a divide greater than any other in its history. "We hear much of the dangers of agitation," the *Tribune* wrote. "We admit them; but we know of another danger far greater, and that is the danger that our liberties may be subverted, our rights trampled upon; the spirit of our institutions utterly disregarded, and our great republican experiment turn out a disastrous failure. We confess that this danger particularly occupies our mind just about this time."[28]

In 1860, one cartoonist depicted Dred Scott as playing the tune to which all four major candidates danced in the nation's most divisive presidential election. Southern Democrat John C. Breckinridge dances arm-in-arm with outgoing President James Buchanan, dressed as Buck, his nickname. Democrat Stephen A. Douglas dances with an Irish ruffian—a common anti-Catholic, anti-immigrant stereotype. John Bell, the Constitution Union candidate, dances with a Native American for reasons that may have to do with an earlier political position or as a symbol of nativism, which like slavery permeated politics. Abraham Lincoln dances with an African-American woman fulfilling the charges that he was an abolitionist and that he favored "amalgamation" of the races—two allegations raised by Douglas that Lincoln denied in their famous debates in 1858. Each of the candidates had to dance with his constituents while dealing with slavery as symbolized by Dred Scott.[29]

Caption for Figure 15.1 (Cont.)

after emancipation papers were officially presented to Scott. The top image shows Scott's children, Eliza and Lizzie. Below are portraits of Dred and Harriet Scott. (From the Library of Congress Prints and Photographs Division Washington, DC. <www.loc. gov/pictures/item/2002707034/> Reproduction Number: LC-USZ62-79305 (b&w film copy neg.). "Visit to Dred Scott—his family—incidents of his life—decision of the Supreme Court," *Frank Leslie's Illustrated Newspapers,* June 27, 1857.)

FIGURE 15.2 "The Political Quadrille. Music by Dred Scott," lithograph by unknown artist, 1860. This cartoon illustrates the importance and impact of the Dred Scott decision on the 1860 presidential campaign. In the center, Scott sits playing the violin. Around him, the four presidential contenders dance with stereotypical caricatures of their constituents. In the upper left, Southern Democratic candidate John C. Breckinridge is paired with incumbent James Buchanan, depicted as a buck. In the upper right, Republican nominee Abraham Lincoln is illustrated arm-in-arm with a black woman, demonstrating the Republican sympathies for abolition. At the lower right, Constitutional Union candidate John C. Bell dances with a Native American man, perhaps in reference to Bell's brief foray into indigenous people's interests. In the lower left, Northern Democratic candidate Stephen A. Douglas is paired with a raggedy Irishman, emphasizing the allegations of Douglas's Catholicism. (From the Library of Congress Prints and Photographs Division Washington, DC. <www.loc. gov/pictures/item/2008661605/> Reproduction Number: LC-USZ62-14827 (b&w film copy neg.). "The Political Quadrille. Music by Dred Scott.")

Notes

1 Paul Finkelman, ed., *Dred Scott v. Sandford: A Brief History with Documents*, 2nd ed. (Boston: Bedford/St. Martin's, 2017); *Union* (Washington, DC), March 12, 1857.

2 James Buchanan, "Inaugural Address," March 4, 1857, *The American Presidency Project*, www.presidency.ucsb.edu/ws/?pid=25817.

3 Don E. Fehrenbacher, *The Dred Scott Case: Its Significance in American Law and Politics* (New York: Oxford University Press, 1978), 1–7, 310–17.

4 *New York Tribune*, March 7, 1857. Many moderate publications merely reported the delays. The *New York Times* on December 20, 1856, in a small item at the bottom of

page 1: "The arguments in the Dred Scott case were concluded yesterday. The Court takes time for deliberation, and the decision will not be made for several weeks—probably not until the end of the term."

5 Ibid.

6 Ibid.

7 *New York Evening Post*, March 7, 1857.

8 *Richmond Enquirer*, March 13, 1857.

9 *Remarks of the Hon. Stephen A. Douglas, on Kansas, Utah, and the Dred Scott decision. Delivered at Springfield, Illinois, June 12th, 1857*, "Our Collection," The Gilder Lehrman Institute of American History, www.gilderlehrman.org/collections/b1a6a423-ba27-4400-9783-296089734550. Douglas spoke without notes but later published the speech in a pamphlet.

10 James F. Simon, *Lincoln and Chief Justice Taney: Slavery, Secession, and the President's War Powers* (New York: Simon & Schuster, 2006), 133–64.

11 Abraham Lincoln, speech at Springfield, IL, June 26, 1857, "Speech on the Dred Scott Decision," TeachingAmericanHistory.org, accessed August 15, 2017, http://teachingamericanhistory.org/library/document/speech-on-the-dred-scott-decision/.

12 *New York Herald*, March 31, 1857.

13 Ibid.

14 *Baltimore Sun*, March 9, 1857.

15 Ibid., March 11, 1857.

16 "Highly Important Decision," *Standard* (Raleigh, NC), March 11, 1857.

17 *Daily Delta* (New Orleans, LA), March 19, 1857.

18 William E. Huntzicker, *The Popular Press 1833–1865* (Westport, CT: Greenwood Press, 1999).

19 *Daily Delta* (New Orleans, LA), March 19, 1857.

20 *Evening Journal* (Albany, NY), March 7, 1857.

21 Huntzicker, *The Popular Press*, 37–39, 117.

22 *Evening Journal* (Albany, NY), March 9, 1857.

23 Ibid., March 10, 1857.

24 Ibid., March 11, 1857; March 19, 1859.

25 Lea VanderVelde, *Mrs. Dred Scott: A Life on Slavery's Frontier* (New York: Oxford University Press, 2009), 9–11. See also Lee VanderVelde and Sandhya Subramanian, "Mrs. Dred Scott," *The Yale Law Journal* 106, no. 4 (January 1997), 1033–122.

26 *Frank Leslie's Illustrated Newspaper*, June 27, 1857, 50.

27 The most detailed biographical information about the Scotts can be found in Walter Ehrlich, *They Have No Rights: Dred Scott's Struggle for Freedom* (Westport, CT: Greenwood Press, 1979); and VanderVelde, *Mrs. Dred Scott*.

28 *New York Tribune*, March 12, 1857.

29 The cartoon, "The Political Quadrille: Music by Dred Scott," was published by a maker of individually published prints when this business was at its peak in 1860, just as the major illustrated newspapers, like *Harper's Weekly* and *Frank Leslie's Illustrated Newspaper*, hired their own staff cartoonists to do satire. The quadrille invokes the popular four-couple dance of the early nineteenth century with the slave providing entertainment.

16

"MORE THAN A SKIRMISH"

Press Coverage of the Lincoln-Douglas Debates

David W. Bulla

In the late summer and fall of 1858, one of the greatest cultural spectacles of the Antebellum period occurred in Illinois, where Democratic incumbent Stephen Douglas debated Republican opponent Abraham Lincoln seven times during their US Senate race in that state. Debates during the political season in the small towns of the American prairie were common and offered a form of entertainment for citizens. It was also an opportunity for political opponents to state their views on various issues of the day—in this case, the major issue was slavery and its extension into the US territories. Douglas charged Lincoln with not being merely interested in keeping slavery out of the territories, but being an out-and-out abolitionist. Lincoln countered that Douglas did not really love liberty, as shown by his indifference to slavery. He also hammered Douglas on his favorite policy, popular sovereignty.

The Lincoln-Douglas Debates drew large crowds around the Prairie State beginning on August 21 in Ottawa and ending on October 15 in Alton. Coverage of the debate was intense, particularly in Illinois. The *Chicago Times* and *Chicago Press and Tribune* each sent reporting teams to chronicle every word of every debate. However, the quality of the debating and the issues at stake also brought relatively strong coverage nationally and even internationally. For example, the *New York Tribune* covered the debates as if they were just as important to New Yorkers as the political races in their own state.

This chapter examines how newspapers both in the United States and in Great Britain covered this important rhetorical exchange between the veteran Democratic politician Douglas and the upstart and relatively unknown member of the brand-new Republican Party, Abraham Lincoln. Of course, two years later, Lincoln would be the candidate for the Republicans and Douglas for the Democrats in the American presidential race. Lincoln would win the national election and thus precipitate the secession movement in the South. His election and the Civil War have given the debates even more significance historically. The Lincoln-Douglas Debates were a barometer of American journalism at mid-century, showing the mixture of reporting and interpreting that would be the hallmark of journalism during the Civil War.

There were four main approaches to press coverage of the debates: 1) description of the spectacle at each site; 2) transcription of the words spoken by each man; 3) description of the candidates; and 4) interpretation of what Douglas and Lincoln said.

In July of 1858, Douglas had begun to give speeches around Illinois, and Lincoln had attended each event. Each time, after Douglas spoke, Lincoln would give a talk as well. Several editors, including Joseph Medill of the *Chicago Press and Tribune* and Horace Greeley of the *New York Tribune*, suggested that the two men ought to speak in the same settings. Greeley wanted to see at least a dozen debates. Although Douglas was not sure at first that it was a good idea because he did not see how the debates would benefit him as the incumbent, eventually he agreed to a seven-debate tour of the state.[1]

In this era, democracy, at least in Illinois, was both entertainment and politics. The citizens of Illinois, as well as the press, highly anticipated the debates, and the largest of the seven debates drew twenty thousand attendees. The press of Illinois extensively covered the Lincoln-Douglas debates. Then newspapers outside Illinois picked up the transcripts and added their own analysis. The transcripts of each debate required several pages of an edition when they were published in their entirety. A correspondent named "Bayou," writing in the *Randolph County Journal* in Indiana, perhaps stated best what these seven debates meant to America when he called them "more than a skirmish," saying Douglas and Lincoln were two great titans representing two different approaches to government. "This battle is more than a skirmish for 1860," the correspondent wrote in a letter from Springfield. "It is the engagement of the vanguards of two great armies, each feeling that the main fight must depend very much upon the result now to be achieved."[2]

Northern Coverage of the Debates

The seven debates between Lincoln and Douglas occurred in August, September, and October of 1858. They took place in the Illinois cities of Ottawa, Freeport, Jonesboro, Charleston, Galesburg, Quincy, and Alton. Thousands attended each debate with the largest crowd at Knox College in Galesburg. The estimates for that debate were between fifteen and twenty thousand attendees. The assembled came from across the prairie, staying in hotels, churches, and private homes—and generally stretching local resources to the bursting point for a day or two. The format was an opening sixty minutes by one debater, followed by ninety minutes from his opponent, and then thirty minutes from the first speaker.

From the beginning, the pro-Republican *Chicago Press and Tribune* and the pro-Democratic *Chicago Times* bitterly competed for the best coverage of the debates. The *Press and Tribune* sent Horace White to cover them as the primary reporter, while Robert Hitt served as the transcriber, writing down everything the candidates said. For the *Times*, James B. Sheridan and Henry Binmore were the scribes. Once these four reporters had rendered a transcript, they would telegraph it back to their offices in Chicago. The two newspapers then printed the transcripts the day after a debate, and subsequently, other papers around the country picked up these reports and printed all or part of the text. The availability of the telegraph helped disseminate something like a public record of what Lincoln and Douglas had said at the seven debates and contributed to the significance of the debates as millions of American newspaper readers read the transcripts. The *Press and Tribune* transcripts sometimes polished what Lincoln said and tarnished what Douglas said. The *Times* was accused of the same thing, but the other way around—cleaning up Douglas's words and doctoring some of Lincoln's words in an unflattering way.

At the first debate, in Ottawa, the *Press and Tribune* claimed the *Times*'s reporters deliberately mangled Lincoln's words.[3] *Press and Tribune* editor Joseph Medill also felt Lincoln was too tame in the first debate and urged him not to be on "the defensive" in the second debate.[4] The *Times* was uncomplimentary of Lincoln, saying that he "seemed to be paralyzed" and had become sick during Douglas's talk. Because of Lincoln's apparent ill state, the *Times* called the first debate a "farce."[5]

The second debate, in Freeport, was "the best-remembered three hours" of the seven debates, according to Lincoln expert Harold Holzer.[6] Again, the *Chicago Times* described Lincoln in unflattering terms, stating that he was "uneasy" and "quaking" while Douglas spoke.[7] The pro-Democratic *Missouri Republican* commented in its coverage of the Freeport debate that the "contest is one to which the whole country is looking with intense anxiety" and that a Douglas victory in the Illinois election would send a stern message to the Republicans about "the odium attached to a sectional party."[8]

Lincoln spoke first at Freeport, in the Democratic-friendly southern part of Illinois, although he had to wait until the tardy journalist Hitt arrived. Lincoln responded to seven questions that Douglas had put to him from the previous debate. Lincoln then asked several questions of his own. Lincoln's second question would prove to be the most important of the seven debates. Overall, the feeling was that the second debate was a draw, except for Douglas's answer to that second question.

Lincoln asked Douglas to defend his popular sovereignty policy, under which the citizens of territories would decide democratically whether they would become free states or slave states. In 1857, the US Supreme Court, in *Dred Scott v. Sandford*, had ruled that slaves could be brought into any territories, thus refuting the concept of popular sovereignty. Under *Dred Scott*, Kansas's citizens had no right to determine if they would be free or slave. This court ruling had put Douglas in a sort of political no man's land by the time of the 1858 debates, and now Douglas was forced to explain his position. Douglas replied, "It matters not what way the Supreme Court may hereafter decide as to the abstract question whether slavery may or may not go into a Territory under the Constitution, the people have the lawful means to introduce it or exclude it as they please, for the reason that slavery cannot exist a day or an hour anywhere, unless it is supported by local police regulations."[9] This became known later as the Freeport Doctrine, and this segment of the debate got nasty. Douglas commented that one of Lincoln's supporters was the black abolitionist journalist Frederick Douglass, an attempt to paint Lincoln as an out-and-out abolitionist. Douglas sarcastically said that those in the crowd like Lincoln who believed in racial equality were certainly entitled to vote for the Republican candidate. Medill's *Press and Tribune* called Douglas "desperate" and "dishonest" for associating Lincoln with the abolitionist Douglass.[10]

Regarding the third debate in Jonesboro, the *Chicago Times* said that Lincoln supporters and those of "negro equality" were outnumbered one thousand to twelve.[11] In contrast, the *Press and Tribune* hailed Lincoln's "masterly argument."[12] About the Charleston debate, the *Times* said Lincoln had a muted greeting from the crowd and spoke as he had in the earlier debates. Of course, it said Douglas received "almost unanimous applause."[13]

In Galesburg, the local newspapers claimed twenty thousand attended the debate, and the *Galesburg Democrat* noted that the crowd was "immense" despite heavy rain the previous day. The *Democrat* described the many banners carried by supporters on both sides, including "Abe Lincoln Champion of Freedom" and "Knox College Goes for Lincoln."[14] Before Galesburg, the *Chicago Press and Tribune* had asked for "adequate accommodations for reporters" and that plenty of room be made for the press on the platform. Addressing the Committee for Arrangements, the paper had requested a table for six reporters and assurance that the folks on the speakers' platform would not interfere with the correspondents' work. The Chicago newspaper also asked for a security detail to "keep loafers out of the reporting corner."[15]

At Galesburg, Lincoln asked Douglas about his stance on *Dred Scott*. Douglas countered with his own question: If African American slaves were property, as the *Dred Scott* decision maintained, could property banned in one state be transported into that state from another state or territory? Indeed, could you transport liquor into a dry state like Maine?[16]

For the Quincy debate, the *Quincy Whig* reported that the town square "swarmed with people" and that Lincoln "gave Douglas one of the severest skinnings that he has received in the course of this canvass. In fact, he made the fur fly with every word."[17] The *Whig* also said that Douglas was hard to understand and that Democrats accordingly expressed their chagrin. The pro-Democratic *Quincy Herald* said twelve to fifteen thousand attended the debate and that the procession of candidates and their followers before the candidates' discussion was the largest ever seen in that city. For the final debate at Alton, Illinois, Republican papers criticized Douglas for using the same material he had on previous occasions, while Lincoln received praise for instead giving a proper summary of the law relative to both the facts and the will of the people.[18]

While the *Chicago Press and Tribune* consistently assessed the debates in Lincoln's favor, Democratic newspapers like the *Times* mainly saw Douglas ahead on points. However, some newspapers were more neutral. The *Springfield Republican* in Massachusetts conceded that Douglas had the upper hand and yet claimed that Lincoln overall was making inroads versus Illinois's Democratic senator.

Greeley's *New York Tribune* reported that "political excitement" in Illinois "was tremendous," and that the "Presidential contest of '56 was calm in comparison."[19] The *Tribune* ran the *Chicago Press and Tribune* transcripts verbatim. In addition to its news coverage, Greeley supplied interpretation on the editorial page. After the Ottawa debate, Greeley's newspaper noted that Lincoln's supporters were gleeful and that Douglas was a "used up man."[20] Since the *Tribune* was a national newspaper, its coverage was influential.

James Gordon Bennett was the rabidly conservative, though independent, editor of the *New York Herald* and the *Tribune*'s primary journalistic rival. The curmudgeonly Bennett took a dim view of the whole proceedings in Illinois. Bennett's newspaper called the oration on both sides "the merest twaddle upon quibbles," and stated that the debates had "descended into the dirty area of personalities."[21] The pedantic *Herald* editor told both candidates to go home and be quiet for the duration of the campaign (which was not unconventional in the mid-nineteenth century). When the debates concluded and Douglas edged out Lincoln for the senatorial seat, the *Herald* claimed that New York Republican Senator William H. Seward (a candidate for the Republican presidential nomination) had played a role in preventing Lincoln's election. "Indeed, we think it very likely that had not Mr. Seward desired the success of Mr. Douglas, Mr. Lincoln would have beaten him in the returns for the legislature," Bennett wrote.[22] At that time, state legislatures chose US senators.

There were nearly twenty newspapers in New York in 1858, and the *New York Times* was not as prominent as it would become. However, Henry J. Raymond, its editor, was a major player in the Republican Party and how his newspaper saw political matters was significant. The *Times* predicted Douglas would win the contest with Lincoln, and Raymond underemphasized the debates. Yet Raymond did on several occasions run critical *Washington Union* pieces about Douglas— editorials that usually saw little difference between Douglas and Lincoln, but seemed to top the scales slightly in favor of the Republican candidate. The *Times* commented that the schism between Douglas and the administration of President James Buchanan was "a great danger to the Democracy at the North."[23] Douglas, the champion of popular sovereignty, had disagreed with Buchanan's proslavery policy for Kansas, and in so doing, had alienated the vast majority of Southern Democrats. (Effectively, the Buchanan-Douglas schism over Kansas would lead to the fragmenting of the party and the rise of the Republicans as a national political force—which would lead to Lincoln's election as president in the fall of 1860.)

Typical of the coverage of the debates in the North was that of the *Holmes County Republican* in Millersburg, Ohio. The newspaper put debate coverage on its front page, saying "multitudes" had attended and that "excitement has run high with the masses." The *Republican* commented that Douglas won the early debates, but that Lincoln had rallied in the late stages and was driving "'the Giant' to the wall."[24] The Holmes County newspaper said the sixth debate, at Quincy, drew twelve thousand people, and it went on to quote both candidates with excerpts from the proceedings. Lincoln took the offensive on Douglas's contention that the nation could remain half-free, half-slave perpetually. At the end of his allotted time, Lincoln, the *Republican* noted, received a "deafening cheer" that "continued with unabated enthusiasm for several minutes."[25]

The *Weekly Portage Sentinel* in Ravenna, Ohio, called the campaign in Illinois the most important in the history of the federal government because of the clash of principles embodied in the two men and that no campaign in US history had "excited more general interest."[26]

In Lincoln's home state, the pro-Democratic *Ottawa Free Trader* hammered Lincoln over the issue of equality. The Ottawa editors criticized Lincoln for saying that black men, while inferior to white men politically, had some rights, and yet the senatorial candidate "dares not undertake to define" those rights. The Ottawa newspaper went on to call abolitionism an "odious" policy.[27]

Southern Coverage of the Debates

Southern newspapers generally sided with Douglas, although he was criticized for not being proslavery enough for many editors in the South. Lincoln was seen as a "black Republican," as an abolitionist or abolition sympathizer. No matter the viewpoints, most Southern newspaper realized that the debates were extremely important. For example, the *Louisville Journal* stated that Douglas vs. Lincoln was the "most remarkable" campaign in the history of the country.[28]

In Virginia, the framing of the issues in the Lincoln-Douglas Debates came down to the Constitution versus abolition. The *Alexandria Gazette* ran a report of Vice President John C. Breckinridge sympathizing with Douglas in his contest with "Black Republicanism" and abolition.[29] The *Richmond Enquirer* supported Douglas and ran a letter to the editor from a "Jackson Democrat" in which the writer defended Douglas's popular sovereignty policy on whether citizens of territories could exclude slavery even if the Supreme Court said the territories could not.

In South Carolina, the *Charleston Mercury* looked at Douglas as essentially being no different from the Republicans. The *Mercury* speculated Douglas would win the presidency in 1860, but that the Democrats would fall apart as a party. In the *Keowee Courier* of Pickens Court House, South Carolina, Editor Robert A. Thompson ran a transcript of a speech by Speaker of the US House of Representatives James L. Orr, who preferred Douglas. Orr characterized Lincoln as one who "proclaims the equality of the negro with the white man, and who avows to overthrow the Supreme Court of the United States, because the Dred Scott decision is distasteful to him."[30]

In Tennessee, the *Nashville Patriot* reprinted the Cincinnati *Commercial's* coverage, noting that the Cincinnati paper was pro-abolition. Before running the *Commercial's* copy, the Nashville newspaper termed the debates "highly entertaining and instructive."[31] During the debates, the pro-Democratic *Nashville Union and Patriot* printed Douglas's words one day and Lincoln's the next. The *Union and Patriot* took the excerpts from the *Chicago Tribune* and *New York Herald*.[32] The *Athens Post* counseled against underestimating Lincoln—that only New York's Seward was a more able man among the Republicans. The *Clarksville Chronicle* called Lincoln's stance that Congress had the power to abolish slavery in the territories "odious." But the Clarksville editor, R. W. Thomas, saw no reason to support Douglas either, because his popular sovereignty position gave the people of the territories powers that the Constitution and the Supreme Court had not given them. Thomas maintained that the policies of either man, if carried out, "would be fatal to the South."[33]

As the debates ended, Benjamin F. Deal, the editor of the *Memphis Daily Appeal*, ran a letter by a "WPF" wondering why any Southerner would oppose Douglas's reelection, even if the senator had had a single political squabble with President James Buchanan. Lincoln, he said, differed with the president "on all questions." Douglas was a man who played to the nation as a whole, whereas Lincoln was a sectionalist. The writer concluded: "If Democrats desire the unity and harmony and continued success of their party, they will not seek to strike down the champion of their party in Illinois."[34]

The proslavery *New Orleans Daily Crescent* actually defended Lincoln for not coming out categorically as pro-abolition. Editors James Nixon and William Adams said that after Lincoln denied being an abolitionist, they came to have a "much higher opinion of Mr. Douglas' opponent than formerly."[35] However, this nuanced understanding of Lincoln did not mean the *Crescent* editors had come to the position of supporting the Republican. The *Sugar Planter* of West Baton Rouge, Louisiana, reminded readers that Douglas said local authorities could nullify state law and ignore a new state government's democratically determined choice of slavery.

William W. Holden's pro-Democratic *Standard* in North Carolina looked ahead to the 1860 election, saying Democrats needed to "prepare for the storm which is impending." The Raleigh editor said the 1860 election would be the "most bitter and most important contest" in the nation's history because "the power is now in the free states."[36]

In Mississippi, the *Yazoo Democrat* ran a piece from the *New York Day Book* that stated that the contest between Douglas and Lincoln "threatens to shake the whole continent with its reverberations." The article praised Lincoln for "clearly and fairly" giving his position on the issues and defining "the purposes of his party." Yet Lincoln's party was "villainous," and clearly the Republican represented the party of revolution that would cause the "overthrow of our institutions," while Douglas stood for the "preservation of the existing condition."[37] Lincoln stood for racial equality; Douglas for black subordination to white men. However, Douglas had not performed well after having the upper hand in the early debates and had not put Lincoln away. He was too egotistical and lost his concentration.

John Richardson, editor of the *Prairie News* in Okolona, Mississippi, said he had no confidence in Douglas, even if he was a man of "giant intellect," and called Lincoln "a freedom shrieking, negro howling black Republican of the deepest dye." The *Prairie News* did not see Douglas as being proslavery enough, and concluded: "Verily there isn't so great a difference between Douglas and Lincoln after all."[38] The *Dallas* (TX) *Herald* noted that the debates were unusual for the national interest they were stimulating, pinpointing Douglas's national prominence as the reason for the attention. The Dallas newspaper also termed Lincoln's policies "dangerous, radical and incendiary."[39]

The Debates in the British and Canadian Press

The English magazine *The Spectator* used the fall 1858 political season to "cordially hail the American Republic to the anti-Slave policy" that Great Britain had already adopted. *The Spectator* continued: "When the Americans have carried an Anti-slavery President into the White House, they will carry the zeal of converts or antagonists into the suppression of the Slave-trade: and this will be a fair subject of rejoicing to England, and her faithful allies."[40] In 1858, *The Spectator* would rank Douglas second behind President Buchanan for the 1860 presidential election. However, the journal said there was "little to choose, indeed, between Mr. Buchanan and Mr. Douglas." *The Spectator* editors added: "The South expects the free states to throw out Mr. Buchanan, but believes that Mr. Douglas has a good chance, through the vastness of the Democratic party. The administration seems to be of the same opinion, by the virulence of their hostility to Douglas."[41]

The London *Morning Post* called the contest between Douglas and Lincoln "neck and neck" and remarked, "on the upshot depends in all human probability the next Presidency of the United States." The *Morning Post* said the "lank and ungainly" Lincoln was the "embodiment of principle." At stake was the perpetuation of slavery as represented by Douglas and the "opposite idea, put forth in the person of Lincoln." According to the newspaper, Lincoln was born poor in Kentucky and had risen to his "present position" as an Illinois attorney by "sheer industry." In assessing Douglas's political future beyond the Illinois election, the *Morning Post* suggested the 1860 presidential election might end in an Electoral College deadlock, in which case Congress would have to determine the next president.[42]

The *Kentish Gazette* in Canterbury, Kent, described Lincoln as being "tall and awkward" with a face of "grotesque ugliness." The *Gazette* added: "He presents the strongest possible contrast to the thickset burly bust and short legs of the judge." The *Gazette* also told a story, attributed to Lincoln, of a time when the lawyer was stopped by a hunter with his gun raised. The hunter said he was going to shoot Lincoln because he had been told if he ever met a man uglier than himself, he must shoot him. Lincoln replied: "'Well, shoot away, for if I am uglier than you, I don't want to live any longer!'"[43]

The *Glasgow Herald* noted that Democratic supporters of President James Buchanan wanted to crush Douglas "as a rebel." The *Herald* noted that Republicans had received a boost because of the Douglas-Buchanan rift, which had originated with the Illinois senator's loss to Buchanan in the 1856 Democratic convention and had culminated in the issue of popular sovereignty.

The *Jersey Independent and Daily Telegraph* stated that Douglas had much to rejoice about after he defeated Lincoln. The Douglas Democrats, the newspaper said, fought "under the banner of 'Popular Sovereignty.'" The paper termed Lincoln a "thorough anti-slavery Republican."[44]

In Canada, *The New Era* of Newmarket, Ontario, asserted that the Republicans had done well in the election of 1858, although Douglas had won in Illinois. The Republican gains were seen as a repudiation of President Buchanan. The *Daily Colonist* of Victoria noted that the election "ended favorably" for Douglas, and that in San Francisco there was a hundred-gun salute for the Illinois senator after his victory over Lincoln.[45]

Most coverage of the debates came from two types of reporting: 1) verbatim transcriptions of what the two candidates said during the seven debates, reported by the two Chicago newspapers—which were reprinted around the country; and 2) descriptions of the spectacle of each debate, including the effect on the towns or cities where they were held.

The transcripts of the debates would get a second life when Lincoln published them in 1860. Since there was no book publisher in Springfield, Lincoln turned to Follett, Foster, and Company of Columbus, Ohio. This book was based on Lincoln's scrapbook collection of newspaper clippings during the debate season. The Ohio publisher printed fifty thousand copies of the debate transcripts. Douglas would object to the publication, especially what he called the "alterations and mutilations" of his spoken words in the seven speeches.[46]

In the end, Douglas reveled in his Illinois election—the "glorious triumph achieved by the Democracy" over "the Enemies of the Constitution and Union."[47] Even if he would never have the fire-eaters in the South on his side, Douglas was the champion of moderates around the country, not just in Illinois. Even some Southern newspapers were content with his win over Lincoln. The *Richmond Dispatch* reported on a mass meeting of eleven thousand in Chicago after Douglas's election and ran the *Chicago Times*'s verbatim transcript of the speech in which Douglas hailed the Illinois Democratic Party's "noble victory" over "the combined forces of abolitionism and its allies."[48] Later, the *Dispatch* would cover the annual Virginia Democratic conference in which a resolution was made congratulating Douglas and the party for the "defeat of a dangerous sectional organization."[49] The *Richmond* (VA) *South* also saw the Illinois senator's position on slavery as far more reasonable than that of the Republicans.[50]

Lincoln, on the other hand, looked to the future, writing: "The cause of civil liberty must not be surrendered at the end of one, or even, one hundred defeats."[51]

The American press covered the debates as spectacle, reporting what was said, how the men appeared, and the environment in which they spoke. Ultimately, the debates would prove to be a dress rehearsal for the 1860 presidential election.

Notes

1 Harold Holzer, ed., *The Lincoln-Douglas Debates: The First Complete, Unexpurgated Text* (New York: Fordham University Press, 2004), 3.
2 "The Canvass in Illinois," *Randolph County Journal* (Winchester, IN), October 28, 1858, 2.
3 David W. Bulla and Gregory A. Borchard, *Journalism in the Civil War Era* (New York: Peter Lang, 2010), 194.
4 David Herbert Donald, *Lincoln* (New York: Simon & Schuster, 1995), 217.
5 "The Campaign—Douglas among the People," *Chicago Times*, August 22, 1858.
6 Holzer, *The Lincoln-Douglas Debates*, 88.
7 "The Campaign—The Discussion at Freeport," *Chicago Times*, August 30, 1858.
8 "Douglas-Lincoln," *Missouri Republican* (St. Louis), September 2, 1858, 2.
9 Edwin Erle Sparks, ed., *The Lincoln-Douglas Debates* (Chicago: Hall & McCreary, 1918).
10 *Chicago Press and Tribune*, August 20, 1858.
11 "The Campaign—Douglas at Jonesboro," *Chicago Times*, September 17, 1858.
12 "The Jonesboro Debate," *Chicago Press and Tribune*, September 17, 1858.
13 "The Audience at Charleston," *Chicago Times*, September 21, 1858.
14 "Galesburg Debate," *Galesburg* (IL) *Democrat*, October 9, 1858.
15 "The Galesburg Debate—A Word to the Committee of Arrangements," *Chicago Press and Tribune*, October 5, 1858.
16 Quoted in Allen Guelzo, *Lincoln and Douglas: The Debates That Defined America* (New York: Simon & Schuster, 2008), 224.
17 "Lincoln Gets Douglas Down," *Quincy* (IL) *Whig*, October 15, 1858, 2.
18 David W. Bulla and Gregory A. Borchard, *Journalism in the Civil War Era* (New York: Peter Lang, 2010), 196.
19 "The Canvass in Illinois," *New York Tribune*, October 8, 1858, 6.
20 "Conclusion of the Whole Matter," *New York Tribune*, August 26, 1858, 7.
21 "Exhausted to the Dregs," *New York Herald*, October 13, 1858, 4.
22 "The Next Presidency—Seward and Douglas—Joint Stock Combination," *New York Herald*, November 20, 1858, 4.
23 "The Political Contest in Illinois," *New York Times*, August 23, 1858, 5.
24 "The 'Little Giant' and Mr. Lincoln," *Millersburg* (OH) *Holmes County Republican*, October 28, 1858, 1.
25 Ibid.

26 "Douglas and Lincoln: Their Respective Platforms," *Weekly Portage Sentinel* (Ravenna, OH), September 23, 1858, 1.

27 "Lincoln Trotted down to Egypt," *Ottawa* (IL) *Free Trader*, September 25, 1858, 2.

28 Quoted in "Illinois Politics—Douglas and Lincoln," *Memphis Daily Appeal*, October 8, 1858, 2.

29 "Vice President Breckinridge's letter," *Alexandria* (VA) *Gazette*, October 25, 1858, 2.

30 "Speech of Hon. James L. Orr," *Keowee Courier* (Pickens Court House, SC), August 28, 1858, 2.

31 "Douglas and Lincoln," *Nashville* (TN) *Patriot*, September 6, 1858, 2.

32 "Stephen A. Douglas," *Nashville* (TN) *Union and Patriot*, October 27, 1858, 2.

33 "The Douglas and Lincoln fight in Illinois," *Clarksville* (TN) *Chronicle*, September 24, 1858, 2.

34 "Shall Douglas Be Defeated," *Memphis Daily Appeal*, October 2, 1858, 1.

35 "Mr. Lincoln of Illinois," *New Orleans Daily Crescent*, September 22, 1858, 2.

36 "What will be the next of the Opposition to Democracy," *North-Carolina Standard* (Raleigh), September 15, 1858, 1.

37 "The Illinois Canvass," *Yazoo* (MS) *Democrat*, September 11, 1858, 1.

38 "The Illinois Contest—Douglas and Lincoln," *The Prairie News* (Okolona, MS), September 16, 1858, 2.

39 "The Douglas Campaign in Illinois," *Dallas* (TX) *Herald*, October 24, 1858, 1.

40 "Revulsion of Parties in the United States," *The Spectator*, November 6, 1858, 1,166.

41 "Revelations of the Honorable A.G. Brown," *The Spectator*, October 23, 1858, 1,119.

42 "The United States," *Morning Post* (London), September 29, 1858, 6.

43 "Scraps from the Americans Papers," *Kentish Gazette* (Canterbury, Kent), November 23, 1858, 8.

44 "American Politics," *Jersey Independent and Daily Telegraph* (Saint Helier, Jersey), December 13, 1858, 4.

45 "News from the United States," *The Daily Colonist* (Victoria, BC), December 18, 1858, 4.

46 David H. Leroy, *Mr. Lincoln's Book: Publishing the Lincoln-Douglas Debates* (New Castle, DE: Oak Knoll Press, 2009), 83.

47 Robert W. Johannsen, *The Letters of Stephen A. Douglas* (Urbana, IL: University of Illinois Press, 1961), 430.

48 "The Douglas Celebration in Illinois—Speech of the Senator," *Richmond* (VA) *Dispatch*, November 23, 1858, 1.

49 "The Virginia State Democratic Convention," *Richmond* (VA) *Dispatch*, December 6, 1858, 1.

50 *Richmond* (VA) *South*, October 20, 1858.

51 Abraham Lincoln to Henry Asbury, November 19, 1858, in Roy P. Basler, ed., *Collected Works of Abraham Lincoln*, vol. 3 (New Brunswick, NJ: Rutgers University Press, 1953), 339.

PART III

The Election of 1860 and the Crisis of Secession

17

THE DEMOCRATS DIVIDE

Newspaper Coverage of the 1860 Presidential Conventions

Brian Gabrial

[handwritten annotation: Northern Democrates wanted a Proslavery candidate?]

As the nation headed into the 1860 election season, it would not be business as usual. For one thing, the aftershocks of John Brown's raid on Harpers Ferry continued to haunt the world of politics despite it being months since his neck met the noose. Indeed, his very name became an agitprop for Southern Democrats and their newspaper allies to use against the upstart Republicans. "The success of such a party," a *Richmond Enquirer* editorial said shortly after Republicans met for their May convention, would encourage "all the John Browns that now secretly await an opportunity to avenge the death of the pioneer in the cause of which is to be perfecter and finisher."[1] But while the Republicans were settling on their antislavery platform and candidate, the Democratic Party was falling apart. The first Democratic convention failed to select a presidential candidate, and the second convention would fail as well. All the while, a significant portion of the Democratic delegates had already decided party unity did not matter unless they secured a proslavery platform and the right candidate to support it.

On some level, the irony of the 1860 election was that the political party with the most to lose over slavery was the party most in favor of it. Democrats, North and South, wrestled with a single question: How much should slavery be supported? The *Richmond Enquirer* articulated the Southern view perfectly in its almost daily reminder of what the party should stand for on the "Slavery Question": "Absolute non-interference," its editor wrote, "by the General Government, to introduce or to exclude slaves, in any of the territories of the Union."[2] No less insistent on that point was the *Charleston Mercury*'s Robert Barnwell Rhett, who warned ten days before the Democrats met for their first convention in Charleston, that the South "should take her rights and destiny into her own hands" if Northern voters voted against a proslavery candidate.[3] The fire-eater demanded that the election results should give Southerners "peace with our rights and institutions." For Rhett and others like him, this meant the victor in the next presidential election had to profess non-interference when it came to slavery or, as Rhett wrote, Southerners would "take a separation of our Union with them."[4] By the time the Democratic delegates first convened in April, the Southern wing of the party had adopted a take-no-prisoners strategy. By the end of June, the results of that strategy would rend the party in half, leaving two presidential tickets—a Southern half and a mostly Northern half—to campaign against the Republicans and the Constitutional Union Party (the new fourth party composed mostly of former Whigs united in support of the Union).

One observer watching the Democrats' internecine war was Murat Halstead, a reporter for the *Cincinnati Commercial* who covered all the conventions that year.[5] Halstead likened it all to a family feud, with each faction accusing the other in the harshest terms "of being factionists, bolters, traitors, incendiaries, etc., etc.— epithets conveying imputations offensive, in a political sense, being exhausted in vain efforts on both sides to do justice to the subject."[6]

The Democrats had chosen Charleston, "the queen of the South," as their convention city, perhaps as a way to appease the Southern Democrats who were becoming increasingly agitated.[7] The convention began on April 23 in Institute Hall with 303 delegates from thirty-two states.[8] Halstead described the hall as a place with a "good deal of gaudy and uncouth ornamentation" that could "contain about three thousand people."[9] It was hot, too, with the thermometer often topping 100 degrees. Such temperatures did not encourage a collegial spirit.

From New York, James Gordon Bennett made his position clear on the extraordinary importance this convention was to have. On the day it convened, his *New York Herald* observed, "the action of this Convention bears intimately upon the fate of the Union, and upon the existence of the last of the party organizations which once comprised delegates from every section of the confederacy."[10] Bennett asked if the delegates would "save the country from revolution and destruction by the fanaticism of the abolitionist and black republican demagogues."[11]

Many delegates from the North and West hoped the party would adopt the 1856 Democratic platform which advocated popular sovereignty as its proslavery plank, and support its grand architect, Stephen Douglas of Illinois, as its presidential nominee. But many of the Southern delegates viewed Douglas as a villain for his non-committal support of slavery.[12] Fights over the platform began from the start. Historian Stan M. Haynes writes, "It was widely known before the convention that the Democratic parties of the states of the Deep South had instructed their delegates to walk out of the convention if it did not adopt a platform favoring full federal protection of slavery in the territories."[13] Fifteen Southern states, as well as California and Oregon, voted for just such a majority platform, while the Douglas supporters created a minority platform that essentially reaffirmed the

[handwritten margin note: Douglas was radical to the majority of Southern delegates]

1856 popular sovereignty stand. As reporter Halstead put it, "There is an impression prevalent this morning that the Convention is destined to explode in a grand row."[14]

When the vote came, the minority platform won approval. *following Douglas* Right on cue, delegates from Alabama, Arkansas, Florida, Louisiana, Mississippi, Texas, South Carolina, and Georgia walked out. For Alabama's Senator William Yancey, this was exactly what he wanted. According to the historian Douglas Egerton, it was premeditated because it would provide the justification Yancey and other secessionists were seeking to separate the South from the North.[15] "There was a Fourth of July feeling in Charleston last night," observed Halstead after the walkout. "There was no mistaking the public sentiment of the city. It was overwhelmingly and enthusiastically in favor of the seceders [*sic*]."[16] Yancey spoke to a supportive crowd about forming a "Constitutional Democratic Convention," and, as Halstead wrote, the senator said that "even now, the pen of the historian was nibbed to write the story of a new Revolution." That spurred someone in the crowd to yell, "Three cheers for the Independent Southern Republic."[17] An *Enquirer* editorial later described the event as follows:

> All honor, then, to the retiring delegates! Whatever credit may be attributable to other soldiers in the battle of constitutional right, these cavaliers who *par excellence* have assumed the post of danger, maintain a claim to national gratitude, second to that of none of their peers.[18]

This separatist sentiment did not spread northward. A *New York Times* editorial expressed exasperation at the Southern delegates, having little sympathy for their point of view: "The Slave States have substantially controlled the policy of the Federal Government for the last fifty years.... And for the last few years they have held it by coercion, —by menaces, by appeals to the fears of the timid, the hopes of the ambitious, and the avarice of the corrupt, in the Northern States."[19] The editorial continued: *↖ this shit dates back to Jefferson*

> One thing is very certain: —the South must make up its mind to lose the sway it has exercised so long. The scepter is passing from its hands. Its own imprudencies have hastened the departure of its power, but it has always been merely a question of time. The South can either accept it as inevitable and make the best of it, —or plunge the whole country into turmoil, and bring down swift ruin upon its own borders, in the vain contest against national growth and development.[20]

Meanwhile, the convention continued. Halstead observed about the remaining convention delegates, "Our North-western friends use language about the South, her institutions, and particularly her politicians, that is not fit for publication, and my scruples in that respect are not remarkably tender.... They say they do not care

reaction of minority Southern reporter

a d—n where the South goes, or what becomes of her."[21] But Douglas supporters could not get the required two-thirds of the 303 total to win the nomination. The convention adjourned, agreeing to meet again in Baltimore on June 18 with a hope that cooler heads would prevail.

The *New York Herald* understood the ramifications: "The old democratic party has breathed its last, and to-day we enter upon a new era in national politics."[22] The *Richmond Dispatch* described the events differently: "The disruption of the Charleston Convention is looked upon by many as the parting of another strand in the cable which holds the ship of state to its anchorage and has thus far saved it from the sea of uncertainties which must follow a dissolution of the Union."[23]

Douglas certainly had his supporters, but he was also seen by others, from within and from without his party, as someone who would do anything to achieve political success. "If to be vilified is to be recommended to the protection of the higher powers, the Senator from Illinois is decidedly in the way of victory," a *New York Times'* editorial said. "A better-abused man has not recently been known in our public affairs. The fanatics of New-England and the secessionists of the Gulf States, agreeing in nothing else, agree in denouncing the champion of Popular Sovereignty."[24]

In the intervening weeks between the first Democratic convention in Charleston and the second in Baltimore, the Republicans met for their convention in Chicago, which had its share of machinations and maneuverings. Like the intense antagonism that faced the front-runner Democrat Stephen Douglas, Republican front-runner Senator William H. Seward of New York found himself losing out to dark horse Abraham Lincoln. The *New York Herald's* James Gordon Bennett took special delight in Seward's embarrassment, writing, "... Wm. H. Seward, the real organizer of the party and its principles, is thrown aside as a worn-out and useless politician."[25] Likewise, an editorial in the *Richmond Dispatch* accused Seward of using all means at his disposal for increasing his political power: "His talent was for undermining that which others had effected at great pains and cost[.] [H]e elevated himself by pulling down that which others had built up. He fomented political strife, sectional convulsions and civil war, for the purpose of becoming President."[26]

The editor at the *Richmond Enquirer* seemed less gleeful, but it was not because Seward lost but because Lincoln won: "The latter [Seward] has talents and ambition which might have operated to restrain his prejudices; but the former [Lincoln], an illiterate partizan [*sic*], is without talents, without education, possessed only of inveterate hatred of slavery and his openly avowed predilections for negro equality to recommend him to his party." As the editorial continued, it identified the Republican contest as one between two detestable options: "The selection between the traitors was decided by the intensity of the treason, and Lincoln, surpassing Seward in the bitterness of his prejudices and in the insanity of his fanaticism, won the prize."[27]

FIGURE 17.1 "The Great Exhibition of 1860," lithograph published by Currier & Ives, 1860. This cartoon satirizes the Republicans' attempts to deemphasize their antislavery position in the 1860 election. Abraham Lincoln is drawn riding a rail labeled "Republican Platform." His lips are padlocked shut and he is tied by a string to Horace Greeley (left), who is playing an organ labeled "New York Tribune." Greeley says, "Now caper about on your rail Abraham...." Lincoln replies, "Mum." William H. Seward (dressed in both male and female attire) stands in the background with a wailing black child: "It is no use, trying to keep me and the 'Irrepressible' infant in the background; for we are really the head and front of this party." On the right are New York editors Henry J. Raymond of the *New York Times* and James Watson Webb of the *Courier and Enquirer*, both of whom supported the Republican Party. (From the Library of Congress Prints and Photographs Division Washington, DC. <www. loc.gov/item/2003674593/> Reproduction Number: LC-USZ62-14226 (b&w film copy neg.). Currier & Ives, "The Great Exhibition of 1860.")

Lincoln had his press supporters, especially the *New York Times.* "Not only do his opponents feel justified in overwhelming him with abuse," an editorial said, "but his friends err in the opposite extreme by fairly smothering with ill-judged praise. The early incidents of his boy-life are dug from oblivion." The newspaper editor warned not to underestimate Lincoln, despite his working-class background: "... it is not worth while to make the mistake of supposing that the fact of splitting rails, or of living in a log cabin in early life, constitute reasons for electing any man President. Mr. Lincoln had brains, energy, perseverance and virtue enough to escape from the necessity of splitting rails and devote himself to higher objects and more widely useful pursuits.... If he were *still* a mere splitter-of-rails no one would dream of making him President."[28]

The *New York Herald*'s Bennett, true to his racist form, asserted that the public is "tired of the fanatical outcry about the nigger, and do not at all sympathize in the branding of a great social institution at the south as an evil and a crime."[29] Bennett added, "Beside this, there is an inherent antagonism on the part of every white man to the admission of the negroes to that social equality which must follow the conferring of political equality upon four or five millions of them."[30]

A *Richmond Enquirer* editorial, on May 15, reminded Democrats that would be heading to Baltimore that they must consider the only possible party platform, which was the majority platform put forth at Charleston. In this way, "the true doctrine of 'non-intervention' [can] be carried out by restraining the territorial legislatures, and every other organization, from interfering, with the equal rights of persons and property, constitutionally pertaining to the citizens of each and every State."[31] Three days later, another editorial asserted forcefully, "No argument, however plausible, no appeals however urgent, and no considerations of party policy however serious, will induce the abandonment of the right of protection by the Southern States.... If the property of a Southern man is less sacred than that of our Northern brethren, of what use or value has the Union become?"[32] The editors begged the party leaders to get their act together: "Continued discord and division of the party will be the defeat of both sections, the triumph of their common enemy, the destruction of the rights of the South, and the dissolution of the Union."[33] Two weeks later the newspaper would express the problem among Democrats this way: "The 'nigger question' has stirred up the virulence of a fanaticism as intolerant as the religion of the Koran, and enlisted an army of fanatics mad with the lust for landed acquisition and the greed of political predominance."[34]

As for the seceding Democrats, they met at Richmond to play a waiting game. The *New York Times* would publish this about the delegates and their mini-convention: "It was intended to intimidate the Northern Democrats by imposing spectacle of a united South, solemnly and sullenly resolved to have no further concert with the North, except upon their own terms and conditions."[35] These rebel delegates, however, had a plan. They wanted to see what would happen at the Baltimore convention.

The Democrats' second national convention would meet at Baltimore's Front Street Theater on June 18. Halstead described the convening delegates as proceeding to the city "in a state of stimulated enthusiasm, and partial blindness."[36] The partial blindness, according to the reporter, pertained to their inability to conceive of the Southern delegates truly breaking away. But according to Halstead, they "did not know the power and desperation of the South, and were foolish enough to believe the opposition to them in that quarter would quietly subside." Bennett, who clearly hated Republicans and abolitionists, blamed them for the impending crisis facing the Democrats and country. "The crisis is a great one," his newspaper read on the day the Baltimore convention met. "The destructive tendencies of the ideas they have proclaimed as the rule of government are universally acknowledged," the paper noted, "and all hope to be saved from the terrible ordeal. Safety lies in the union of the national sentiment against the revolutionary assault, and this union is rendered to all appearance impracticable by the personal contests of ambitious aspirants."[37]

Things did not go smoothly in Baltimore. Halstead would call Baltimore a "political St. Domingo,"[38] an allusion to the violent Haitian slave revolt that expelled the French and a common, Antebellum metaphor for disaster. Dissension immediately erupted over sitting the delegates who had walked out at Charleston. Delegates eventually rejected an initial vote for a loyalty oath requiring that all delegates uphold the outcome of the convention, regardless of the nominee.[39] Not all seceding delegates had returned, and so a controversy swirled about those vacant seats. One Pennsylvania delegate upset over the apparent desire to reseat the delegates from the seceding states said, "They put themselves from us, not we from them.... We adjourned for what? For the purpose of enabling those states in the South, whose delegates had seceded to fill up the places of those who had left us."[40]

The fight over credentials continued into the third day of the convention, and rival delegates fought with words and fists. As historian Haynes noted, "Delegates beating each other up right and left did not bode well for a peaceful resolution of the issues facing the convention."[41] The controversy would not be easily resolved. A majority report from the committee wanted to exclude the seceding delegates, replacing them with alternative delegates. A minority report wanted the opposite, claiming the departing delegates had never resigned. Isaac Stevens from Oregon, a member supporting the minority view, warned that to exclude the seceding delegates would irrevocably harm the party: "It is a question not simply of the integrity, but the existence of the Democratic party in several States of this unity. It is a question whether the Democratic party in said State shall be ostracized and branded as unworthy of affiliation with the national organization."[42] It was too much for Bennett at the *New York Herald*, who wrote on June 21, "The work of the dismemberment of the democratic party, so promisingly begun at Charleston, has been renewed at Baltimore, with every indication of a Mississippi steamboat explosion."[43] Indeed, by the fifth day of the convention, a compromise over the seating of the delegates would fail. Haynes observed, "On Friday, June 22, 1860, the American Civil War began.... On this day, the great Democratic Party of the United States irreconcilably split...."[44] Delegate after delegate began leaving the convention, including Virginia delegates who had not left in Charleston.

Much of the ire focused on Douglas supporters who were controlling the proceedings. Even more problematic, there were not enough delegates left to give any candidate the necessary two-thirds majority needed by party rules to win the nomination. That did not stop things from proceeding to a vote. Eventually, Douglas and his running mate, Herschel Johnson, would get the nomination with 181 out of 194 votes left.

Down the street, the seceding delegates soon held their own convention, with none other than the former president of the Charleston and Baltimore

conventions, Caleb Cushing, at the helm. This convention went to work immediately and quickly nominated Vice President John C. Breckinridge of Kentucky
for president and Joseph Lane as his running mate. This became the ticket for
the Southern Democratic Party. Their nomination gained the quick endorsement
of the *Richmond Enquirer*, which noted on June 29, "There is, we believe, one
reason, sufficient of itself to justify this action. It was the only means left in their
power by which to vindicate entirely the constitutional rights of the people of the
whole Union, and especially of the Southern States."[45] The newspaper applauded
Breckinridge's support for the Dred Scott decision (which allowed for the expansion of slavery throughout the Western territories), calling him the "very man for
the times" and saying that American voters would support such a candidate with
"so many elements of power and success."[46]

It had been a sad state of affairs in Charleston and Baltimore, one that left
the Democratic Party split with little hope to win the presidential election.
Interestingly, the *Richmond Dispatch*, following the collapse at Charleston, blamed
the Northern press and the (now technologically possible) rapid spread of news
for promulgating the most "fabulous reports of men and things at the South,"

FIGURE 17.2 "National," lithograph by unknown author and publisher, 1860. This
political cartoon, also called "Dividing the national map," shows clear understanding
of the sectional nature of the 1860 political election. The illustration depicts Lincoln
and Douglas fighting for the western part of the country, Breckinridge ripping apart
the South, and John Bell standing on a chair with a pot of glue in his hand trying to
repair the map of the nation. (From the Library of Congress Prints and Photographs
Division Washington, DC. <www.loc.gov/resource/cph.3a12915/> Reproduction
Number: LC-USZ62-10493 (b&w film copy neg.). "National.")

noting that these stories promoted "falsehood and ignorance" more than "truth and knowledge." "Probably, a hundred years ago, before railroads, steamboats, telegraphs, or daily papers," the editorial said, "the people of the North and South had much more enlightened and accurate views of each other's character and habits than they have at this hour."[47]

The *New York Times*, however, lay the responsibility for the collapse of the Democratic Party squarely at the feet of the party itself: "The party has become so thoroughly demoralized, all the great principles which formerly gave it power and popularity have so completely disappeared, that its leading members have no faith in its future, and no heart for its expiring struggles."[48]

In the 1860 presidential election, the North and West voted for Lincoln, giving him 180 Electoral College votes and 1,865,908 popular votes (39.9%). Eleven Southern states voted for Breckinridge, totaling seventy-two electoral votes and 848,019 popular votes (18.1%). Douglas carried the state of Missouri (nine electoral votes) and three electoral votes from New Jersey, but received 1,380,202 popular votes (29.5%), making him the second highest vote getter. John Bell, the candidate of the Constitutional Union Party, won three slave states—Virginia, Kentucky, and his home state of Tennessee—receiving thirty-nine electoral votes and 590,901 popular votes (12.6%).[49] The North and West had won the election for Lincoln, who had received majority votes from the states that combined to give him the Electoral College vote. It was a sectional contest, and by 1860, the North and the West held a majority of the votes in the Electoral College.

The division of the Democratic Party was the first step toward the secession of the Southern states. The fire-eaters would get their way, turning political defeat into revolution.

Notes

1 "The Prospect before us," *Richmond* (VA) *Enquirer*, May 22, 1860, 2.
2 "Cardinal Democratic Principles on the Slavery Question," *Richmond* (VA) *Enquirer*, May 8, 1860, 2.
3 "The Democratic Party," *Charleston* (SC) *Mercury*, April 13, 1860, 2. Three days later, the newspaper's editorial said, "[The South] must make it known to her confederates in the Union that at the next Presidential, the question is not only a Democratic or Black Republican President, but union or disunion." ("The Democratic Party," *Charleston* (SC) *Mercury*, April 16, 1860, 2.)
4 Ibid.
5 See Murat Halstead, *Caucuses of 1860, A History of the National Political Conventions of the Current Presidential Campaign* (Columbus, OH: Follett, Foster and Company, 1860).
6 Halstead, *Caucuses of 1860*, 229.
7 Stan M. Haynes, *The First American Political Conventions: Transforming Presidential Nominations, 1832–1872* (Jefferson, NC: McFarland and Company, 2012), 152.
8 Ibid., 153.
9 Halstead, *Caucuses of 1860*, 4–5.

10 "The Charleston Convention—Candidates and Complications," *New York Herald*, April 23, 1860, 6.
11 "The Lifting of the Veil at Charleston—Signs of the True Influence There," *New York Herald*, April 24, 1860, 6.
12 Haynes, *The First American Political Conventions*, 152.
13 Ibid., 153.
14 Halstead, *Caucuses of 1860*, 23.
15 Douglas R. Egerton, *Year of Meteors: Stephen Douglas, Abraham Lincoln, and the Election that brought on the Civil War* (New York: Bloomsbury Press, 2010), 11.
16 Halstead, *Caucuses of 1860*, 76.
17 Ibid., 75.
18 "Where we stand," *Richmond (VA) Enquirer*, May 8, 1860, 2.
19 "The Disruption of the Democratic Party," *New York Times*, May 4, 1860, 4.
20 Ibid.
21 Halstead, *Caucuses of 1860*, 87.
22 "Split in the Charleston Convention—Cycles of Political Parties," *New York Herald*, May 1, 1860, 6.
23 "The Charleston Convention," *Richmond (VA) Dispatch*, May 7, 1860.
24 "Douglas and the South," *New York Times*, May 16, 1860, 4.
25 "The Chicago Nominations and Platform—Their True Reading," *New York Herald*, May 19, 1869, 6.
26 "The Late W. H. Seward," *Richmond (VA) Dispatch*, May 24, 1860, 2.
27 "The Prospect before us," *Richmond (VA) Enquirer*, May 22, 1860, 2.
28 "The Sorrows of a Candidate," *New York Times*, June 9, 1860, 4.
29 "The Nominations at Baltimore—Attitude of the Country on the Question of the Day," *New York Herald*, May 11, 1860, 6.
   Editor's Note: This is a work of history, a major focus of which was the racial hatred that existed in the United States. This racial hatred was openly reflected in the language of the newspapers of the time. That language, when provided in direct quotations, is generally retained in this book in order to accurately reflect the historical context.
30 Ibid.
31 "The Baltimore Convention," *Richmond (VA) Enquirer*, June 15, 1860, 2.
32 "The Platform at Baltimore," *Richmond (VA) Enquirer*, May 18, 1860, 2.
33 "The Prospect before us," *Richmond (VA) Enquirer*, May 22, 1860, 2.
34 "The Signs of the Times," *Richmond (VA) Enquirer*, June 5, 1860, 2.
   Editor's Note: This is a work of history, a major focus of which was the racial hatred that existed in the United States. This racial hatred was openly reflected in the language of the newspapers of the time. That language, when provided in direct quotations, is generally retained in this book in order to accurately reflect the historical context.
35 "The Richmond Convention," *New York Times*, June 13, 1860, 4.
36 Halstead, *Caucuses of 1860*, 228.
37 "The Baltimore Convention—Its Composition and Attendant Dangers," *New York Herald*, June 18, 1860, 4.
38 Halstead, *Caucuses of 1860*, 229.
39 Haynes, *The First American Political Conventions*, 157.
40 Ibid., 157.
41 Ibid., 160.
42 *Official proceedings of the Democratic national convention, held in 1860, at Charleston and Baltimore* (Cleveland, OH: Nevin's Print, 1860), 177.

43 "The Baltimore Convention—The Political Parties of the Day and the Spoils," *New York Herald*, June 21, 1860, 4.

44 Haynes, *The First American Political Conventions*, 162.

45 "The Virginia Democracy," *Richmond* (VA) *Enquirer*, June 29, 1860, 2.

46 "Our Noble Presidential Candidates," *Richmond* (VA) *Enquirer*, July 6, 1860, 1.

47 "A Humbugged World," *Richmond* (VA) *Dispatch*, May 5, 1860, 2.

48 "The Presidential Election," *New York Times*, July 4, 1860, 4.

49 "Presidential Election of 1860: A Resource Guide," *Library of Congress*, Digital Reference Section, accessed March 25, 2018, www.loc.gov/rr/program/bib/elections/election1860.html.

18

FANNING THE FLAMES

Extremist Rhetoric in the Antebellum Press

Phillip Lingle

In the decade preceding secession and the Civil War, the perception each side had of the other was largely informed and shaped by the increasingly divisive and activist press. While two events stand out as watershed moments—Representative Preston Brooks's caning of Senator Charles Sumner in 1856, and John Brown's raid at Harpers Ferry in 1859—earlier incidents foreshadowed and shaped press reactions. Given the already tense and increasingly partisan attitude of the time, it is no wonder that the North and South understood the events of the day very differently. In the wake of each occurrence, the press and leadership on each side depicted these actions as being representative of the radical agendas of the majority on the other side. This extremist rhetoric contributed to division, and ultimately, secession and war.

Opinion-makers created narratives in which their opponents threatened the very existence of the fabric of their society. Northerners portrayed the attack on Senator Sumner as representing the violence of an autocratic ruling class of slaveholders. This ran counter to what they regarded as a "free society." Southern politicians and editors depicted the actions of John Brown as representing a call to slave revolt and violence against Southern whites. Servile insurrection was the worst nightmare of the slaveholding South. Public reactions to these events were pointed out as signifying radical sentiments on both sides. These fears were not without some basis in fact. If one looked hard enough, and these men did, there were precedents for both concerns. A Southerner, particularly a member of the group known as the fire-eaters, could discern a pattern of increased slave insurrection plots during the latter part of the 1850s, especially surrounding the presidential election of 1856. A Northerner, especially one opposed to slavery, could see a pattern of violent overreaction by Southerners to criticism of their peculiar institution as representative of the irrationality of Southern people.

The best-known American slave uprising was Nat Turner's Revolt, which took place in August 1831. Turner and his followers killed about sixty whites before it was over.[1] This timing coincided with the rise of antislavery sentiment in the North, especially in New England. Abolitionist William Lloyd Garrison had begun publishing his antislavery paper, *The Liberator*, several months before Turner's Revolt, and to some Southerners, that was no coincidence. The September 15 Washington, DC, *National Intelligencer* reprinted a letter that declared, "An incendiary paper, 'The Liberator,' is circulated openly among the free blacks of this city; and if you will search, it is very probably [*sic*] you will find it among the slaves of your county." On September 3, 1831, Garrison addressed Turner's Revolt in an article entitled the "The Insurrection": "For ourselves, we are horror-struck at the late tidings." He claimed he had always stressed the importance of following a non-violent, Christian approach to persuading the world of the sinfulness of slavery. However, in the same article, he warned, "Wo to this guilty land, unless she speedily repent of her evil doings! The blood of millions of her sons cries aloud for redress! *IMMEDIATE EMANCIPATION* alone can save her from the vengeance of Heaven, and cancel the debt of ages!" To Garrison, the carnage of the Turner insurrection was but a taste of things to come, unless slavery was everywhere abolished.

On November 30, 1831, the Georgia General Assembly reacted to *The Liberator* by voting to offer $5,000 for the capture of Garrison.[2] Under a law passed by the State of Georgia, anyone convicted of distributing publications designed to incite servile insurrection was subject to the death penalty.[3]

A marked increase in reports of slave insurrection conspiracies occurred in 1856, an election year featuring the first presidential candidate of the avowedly antislavery Republican Party, John C. Frémont. The editor of *The New York Herald* on December 11, 1856, wrote, "The idea, no doubt, was that with Fremont's election all the negroes of the South would be instantly emancipated or supported from the North in a bloody revolt." In Clark County, Virginia, a plot involving both slaves and free blacks to attack Harpers Ferry was discovered, three years before John Brown's Raid.[4] According to a *Baltimore Sun* article on December 27, 1856, some of the captured conspirators testified in court: "they had heard white men and negroes talking [that] if Fremont was elected they would be free, and as they knew he was not, they were prepared to fight for it." In Dover, Tennessee, a plot was uncovered that involved not just slaves and free blacks, but white accomplices. Three of these white men were beaten and driven out of the state. A fourth, according to the Maysville, Kentucky, *Eagle* on January 8, 1857, "was captured in Memphis, taken across to Arkansas, and given a thousand lashes." At least nineteen black conspirators were hanged.[5]

Public reaction in the North to these slave plots depended on the position of the speaker on the issue of slavery. Horace Greeley's antislavery *New York Tribune* proclaimed, "Let the South with her growing insurrections look to it.... The manacles of the slave must be stricken off,"[6] while the Democratic-leaning

New York Herald, referring to Greeley, wrote, "It is painful to see the apparent gusto with which our nigger-worshipping contemporary of the *Tribune* gloats over the news of projected Southern servile insurrections."[7] The editors of the Ohio *Western Reserve Chronicle*, on December 24, 1856, chided Southerners who during the past election were "singing paeans to that benign institution" of slavery, but now, "They cannot sleep peacefully in their beds for fear of an uprising among their slaves." And *The Indiana True American*, on December 26, 1856, stated, "As the South have insisted the North has no right to interfere with its domestic institutions, of course they will not ask us to help keep their property from shooting them."

The Southern press had a pronounced opinion on the role of the North and the emergence of the Republican Party as a cause for the recent slave disturbances. *The Weekly North Carolina Standard* editorialized that the revolts throughout the South were "natural products of the sectional war being waged against the South, and particularly of the recent organization of a sectional party running a sectional candidate."[8] The *Richmond Enquirer* also attributed recent slave unrest to the emergence of the "sectional" Republican Party, and reflected on how "fatally the success of the Black Republicans would have affected the social security of the slave-holding States."[9] To Southerners, the rise of the Republican Party and the growing voice of abolitionism in the North were part and parcel of an increasing hostility towards the South. While it is important to recognize that there were no actual uprisings in 1856, the fears in the South were very real. The number of blacks, both slave and free, hanged, shot, or whipped to death reflects how seriously Southerners took the threat of insurrection.

Some Southerners reacted violently to political attacks as well. In May 1856, South Carolina Representative Preston Brooks had beaten Massachusetts Senator Charles Sumner with a cane in retaliation to a Sumner speech. Northerners considered this the ultimate display of Southern violence and irrationality. This was not, however, the first instance of a Southern member of Congress resorting to violence that year.

On January 29, 1856, Representative Albert Rust of Arkansas had assaulted Horace Greeley of *The New York Tribune* over remarks made by Greeley about a resolution proposed by Rust. The day before, Greeley had written, "I have had some acquaintance with human degradation, yet it did seem to me today, that Rust's resolution in the House was a most discreditable proposition." Rust had approached Greeley on a Washington, DC, street and struck him with a cane. Greeley was not seriously injured. A typical Southern reaction to the incident may be found in the *New Orleans Daily Crescent* of February 9, which emphasized that Rust, previous to the beating, had asked Greeley whether he was a "combatant," meaning a fighting man. The paper reported that Greeley had apparently answered in a tone which Rust resented, and after Greeley had feebly defended himself, Rust finished the job with his cane, giving Greeley a "good walloping." Because Greeley did not affirm that he was a fighting man and therefore an equal to Rust, Rust felt justified in the beating.

On May 8, 1856, another member of Congress with Southern ties resorted to force—in this case deadly force. California Congressman Philemon Herbert, a native of Alabama, entered Willard's Hotel in Washington, DC, and demanded breakfast. A member of the restaurant staff told him that it was too late, but Herbert demanded to be served. Herbert ordered Irish waiter Thomas Keating to get him his breakfast "damn quick." According to *The Evening Star* of Washington, DC, Keating refused, and Herbert reportedly told him, "Clear out you damn Irish son of a bitch!"[10] and struck Keating across the face. A scuffle broke out between Herbert and Keating, as well as Keating's brother and Herbert's companion, William Gardiner. Herbert drew a gun and fatally shot Thomas Keating. Within days, Horace Greeley's *Tribune* attributed Herbert's actions to Southern traits, such as easily taking offense and no control over "agitations of passion." On May 10, Greeley's paper stated prominently, "The killing of a menial is not considered a grave offense south of Mason-Dixon Line." On May 20, a *Chicago Tribune* editorial claimed that Herbert would not be held accountable because, "the murdered man was only a laborer, a servant, an Irishman....The murderer belongs to the upper crust Democracy; he goes for Slavery extension—for oligarchy." The Northern press characterized the incident as typical of the overwhelming violence and aggression inherent in the ruling classes of the South—those irrational representatives of the Slave Power.

Southern accounts attributed the incident to lower-class Northerners not knowing their place and thereby putting themselves in danger by acting inappropriately towards their social superiors. An editorial in a small Alabama paper, the *Mail*, warned Northern white laborers, "we hope the Herbert affair will teach them prudence."[11] The editors of the Charlotte, North Carolina, *Western Democrat*, on May 13, blamed "the insolence" of the waiter for the incident and went so far as to state that Herbert only shot to prevent himself being killed.

Some two weeks after the Herbert incident, Massachusetts Senator Charles Sumner began a two-day oration entitled "The Crime Against Kansas." In it he railed at length against slavery, slaveholders, and Southern politicians, all related to the ongoing disturbances in Kansas. He used language highly charged with violent and sexual implications. Of the expansion of slavery into Kansas Territory, he declared, "It is the rape of a virgin Territory, compelling it to the hateful embrace of Slavery; and it may be clearly traced to a depraved longing for a new slave State, the hideous offspring of such a crime, in the hope of adding to the power of slavery in the National Government."[12] He also made numerous personal attacks on South Carolina Senator Andrew Butler, who was not present in the Senate chambers. Of the elderly Senator, Sumner said, "The Senator from South Carolina has read many books of chivalry, and believes himself a chivalrous knight with sentiments of honor and courage. Of course, he has chosen a mistress to whom he has made his vows, and who, though ugly to others, is always lovely to him; though polluted in the sight of the world, is chaste in his sight—I mean the harlot, slavery."[13] He warned against "the Slave Power, which, with loathsome folds, now coiled about the whole land."[14] He responded to charges of Republican section-alism by again mentioning the Senator from South Carolina: "I affirm that the Republican party of the Union is in no just sense sectional, but, more than any other party, national; and that it now goes forth to dislodge from the high places of the government the tyrannical sectionalism of which the Senator from South Carolina is one of the maddest zealots."[15] He mentioned Butler dozens of times, but never by name, and implied that he was a liar: "This senator, in his labored address, vindicating his labored report—piling one mass of elaborate error upon another mass—constrained himself, as you will remember, to unfamiliar, decencies of speech."[16]

Illinois Senator Stephen A. Douglas, who was also attacked by Sumner, remarked that the speech would get the Massachusetts senator shot. South Carolina Congressman Preston Brooks, a relative of Butler, could not let the insults to his family and State go unpunished. After consulting with three friends, congressmen Lawrence Keitt and James Orr of South Carolina and Henry Edmundson of Virginia, Brooks concluded that Sumner was not a gentleman, and therefore could not be challenged to a duel. Keitt later noted that, according to the Southern code of honor, a gentleman like Brooks could not challenge a "cur" such as Sumner to a duel, but had to "beat him like a cur."[17] As historians Lorman A. Ratner and Dwight L. Teeter Jr., in their 2003 book, *Fanatics and Fire-eaters*, explained, "Brooks chose to punish Sumner in a way consistent with the way in which a southern planter and a gentleman would treat someone beneath him in rank and station."[18]

On May 22, Brooks, accompanied by fellow congressmen Keitt and Edmunson, entered the Senate chambers. They found Sumner at his desk and, after waiting for the gallery to clear, Brooks confronted Sumner and beat him senseless with his cane while Keitt brandished his own cane and kept others from coming to Sumner's aid.[19]

The sensationalism and sectionalism of the coverage of the earlier Herbert incident set the stage for the even more lurid and partisan coverage of the attack on Sumner.[20] In support of Brooks, the June 3 edition of the *Federal Union* of Milledgeville, Georgia, declared, "We believe there are some kinds of slander and abuse, for the perpetration of which, no office or station should protect a man from deserved punishment. Massachusetts has no right to complain, for she has for a long time been without the pale of the constitution and the laws of the Union by virtue of an act of her own legislature." (This was a reference to Massachusetts state laws that sought to circumvent the Fugitive Slave Law.) *The State Register* of Springfield, Illinois, blamed the speech, claiming it "un-surpassed in blackguardism anything heard on the floor of the Senate." The editors went on to describe Sumner: "He is a base, lying, blackguard, a bully without courage, a peace man and a blusterer, a provoker of fights, and a non-resistant—in short a heterogeneous conglomeration of everything knavish, mean and cowardly."[21] The May 31 edition of the *Richmond Whig* speculated on the severity of Sumner's injuries. It suggested this was a ploy to "strengthen the sympathy awakened for him among his confederates at the North, [those] Nigger-worshipping fanatics of the male gender, and weak-minded women and silly children, are horribly affected at the thought of blood oozing out from a pin-scratch."[22] Not only were they suggesting that Sumner was, in their words, "playing possum," they equated Northern men with women and children.

Among early Northern accounts, there is a hint of agreement with Brooks's assessment of Sumner as less than a real man. Horace Greeley's *New York Tribune*, on May 24, bemoaned the fact that Northern Senators "had always lacked manly self-assertion." Also on May 24, the *Pittsburg Gazette* suggested, "if our Representatives won't fight when attacked, let us find those that will." This sentiment faded as the Northern narrative evolved until many in the North came to view the attack on Sumner as part of a pattern of violence by Southerners against Northerners caused by an innate violence in Southern men accompanied by the expectation of deference. The June 11 *Pittsburg Gazette* viewed the attacks by Rust, Herbert, and Brooks as being part of a pattern of behavior. The *Portland Advertiser* said the attack on Sumner was a "representative act": "It is the outgrowth and deed of a system, and Brooks himself is as much the instrument of slavery as his cane is his own instrument. Given the slave plantation and all its appliances, and a Herbert, a Rust and a Brooks are the natural results."[23] As Ratner and Teeter note, the attack on Sumner was to Northerners another, "clear example of barbaric, unrepublican, and arrogant behavior of Southern slave-holders in Congress."[24]

John Brown's 1859 raid on the Federal armory at Harpers Ferry was another touchstone event in the run-up to secession in 1860. Southern opinion-makers universally charged Brown with seizing the armory as a first step to initiating a slave insurrection. In the North, leaders had to be careful not to be thought of as approving of the lawlessness of Brown's actions, no matter how much they might agree with his antislavery thinking. This was especially true of the new Republican Party. The 1860 presidential election was only their second entry into national politics, and the South had already blamed them for the slave plots of 1856. In his Cooper Union Speech, Abraham Lincoln distanced himself and his party from the whole affair: "You charge that we stir up insurrections among your slaves. We deny it; and what is your proof? Harper's Ferry! John Brown!! John Brown was no Republican; and you have failed to implicate a single Republican in his Harper's Ferry enterprise."[25]

The press, North and South, were not as clearly divided over John Brown as they were over the Sumner caning. Like Northern politicians, the press of the North could not be seen as supporting Brown's actions. In addition, from the first day of the raid until long after Brown's execution, rumors of new insurrections, rescue attempts, and other wild stories filled the papers. The earliest *New York Tribune* article, on October 18, 1856, was confusing and speculative: "A most extraordinary telegraphic bulletin startled the whole country yesterday—one importing that *an Insurrection had just broken out at Harper's Ferry, Virginia, and that it was the work of negroes and Abolitionists!*" *The Charleston Mercury*, four days later, was full of exciting

Act of Congress in the year 1859, by FRANK LESLIE, in the Clerk's Office of the District Court for the Southern

NEW YORK, SATURDAY, NOVEMBER 5, 1859.

HARPER'S FERRY INSURRECTION—BURYING THE DEAD INSURGENTS.

FIGURE 18.1 "Harper's Ferry Insurrection—Burying the Dead Insurgents," *Frank Leslie's Illustrated Newspaper*, November 5, 1859. John Brown's Raid and its aftermath filled the front pages of America's periodicals in fall 1859. This front-page illustration from *Frank Leslie's Illustrated Newspaper* introduces its November 5, 1859 follow-up story about the raid. (From the Library of Congress Prints and Photographs Division Washington, DC. <www.loc.gov/item/94504951/> Reproduction Number: LC-USZ62-109750 (b&w film copy neg.). *Frank Leslie's Illustrated Newspaper*, "Harper's Ferry Insurrection—Burying the Dead Insurgents.")

details: "From the accounts given of the Harper's Ferry business, it would seem that it was concocted two months since at the Ohio State Fair, by Brown and other confederates, and that its object was to raise the slaves in that country, kill all persons interfering or in the way, and carry them off to freedom north of the Mason and Dixon's line." The *Chicago Press and Tribune*, on October 22, offered a different theory: The Democratic Party, especially Southern Democrats, caused Brown's Raid. By continually supporting slavery, and by insisting that fugitives be returned, Democrats "put to work bands of reckless and bloody men, like Brown and his confederates, who will shrink at nothing, stop at nothing in the gratification of their instincts of fanaticism and revenge." In contrast, the *Memphis Appeal*, on October 25, noted that while many Northern papers claimed that Brown was a madman, the cause of his madness was the teaching of the abolitionists—"anarchy, bloodshed, and confusion."

Following a trial for treason, Brown was swiftly found guilty and sentenced to hang. In the days leading up to his execution, and following, some Northern papers changed their tone slightly. While still condemning Brown's actions, they began to extol the virtues of his devotion to the abolition cause. Following Brown's sentencing, the *New York Tribune* stated, "the champion of the slaveholding class will put to death the champion of the slave." Northern papers began referring to Brown as a martyr. A December 3 *Pittsburg Gazette* article about Brown's execution stated, "While millions of prayers went up for the old martyr yesterday, so millions of curses were uttered against the hellish system which so mercilessly and ferociously cried out for his blood." Even before the hanging, some Southern newspapers warned that Virginia may be creating a martyr. The New Orleans *Times-Picayune* cautioned on November 16, "If Brown is executed, we are warned that a great sympathy will grow up in the North, that he will be made a hero and a martyr. His name will be the watchword with which to kindle a fire of indignation and hatred against the South that will utterly consume away all hope of resisting the popular triumph of Abolitionism at the polls." The *Chicago Press and Tribune* on December 2, the day of Brown's execution, wrote, "We have firm belief that this execution of Brown will hasten the downfall of that accursed system against which he waged war. Throughout all this land, men will not fail to see that there is a conflict between the principles of humanity that have obtained a lodgment in every human heart, and obedience to laws which all have tacitly agreed to support."

By 1860, the rise of the Republican Party had become the central political issue in the nation. Senator Robert Toombs of Georgia, in a January 24, 1860, address to the Senate, said, "These public enemies are Abolitionists, ...this coalition has but one living, breathing principle and that is hatred of the people and institutions of the slaveholding states."[26] Congressman Lawrence Keitt of South Carolina, Preston Brooks's friend and advisor for the Sumner caning, made very clear the perceived connection between the Republicans, abolitionists, and the Harpers Ferry incident. He claimed, "There is an indissoluble connection between the principles of the Republican Party...and their ultimate consummation in blood and rapine on the soil of Virginia."[27] (Keitt's brother had been murdered in his bed by four of his own slaves. Three of the four were recent purchases from Virginia, and Keitt believed that exposure to Northern abolitionists, not the institution of slavery, caused the level of hatred needed to cut a man's throat while he lie in bed.)[28]

Newspapers, North and South, created narratives of the events of this tumultuous time to fit the visions of the political party they supported. Politicians of the day incorporated these narratives to fit their own visions for the future of the country. The South increasingly believed that the entire Northern people were hostile to them. The generation of leaders in power in the South convinced themselves that slavery was positive and natural. Slavery, white supremacy, and

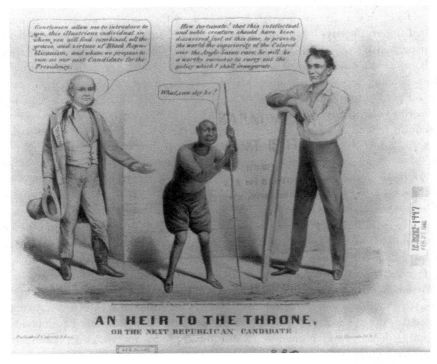

FIGURE 18.2 "An Heir to the Throne, or the Next Republican Candidate," lithograph published by Currier & Ives, 1860. During and after the 1860 presidential election, extremist rhetoric intensified. This openly racist political cartoon, published and distributed by Currier & Ives, argues that the Republicans' next presidential candidate will be a black man. On the left, Horace Greeley proposes the candidate, while on the right, Abraham Lincoln says, "How fortunate! that this intellectual and noble creature should have been discovered just at this time, to prove to the world the superiority of the Colored over the Anglo Saxon race, he will be a worthy successor to carry out the policy which I shall inaugurate." The candidate, pictured in the center, is a deformed African man, who had recently been featured as the "WHAT IS IT" in P. T. Barnum's Museum on Broadway (see poster in background). (From the Library of Congress Prints and Photographs Division Washington, DC. <www.loc.gov/item/2003674574/ > Reproduction Number: LC-USZ62-1997 (b&w film copy neg.). Currier & Ives, "An Heir to the Throne, or the Next Republican Candidate.")

Southern society were so intertwined that one could not separate them; nor did many Southerners wish to. Any perceived threat to slavery threatened not just the institution of slavery, but the identity, security, and honor of Southerners. Southern papers exacerbated this crisis mentality by spinning the discourse about events like the caning of Sumner and the attack at Harpers Ferry in such a way that each new incident of friction between the two sections caused greater irritation. The South saw insults and threats in any action by the North. The Sumner incident and the panic caused by Brown's raid convinced Northerners that the South was irrational and violent. With each new incident, the North became less interested in accommodating the South. The election of Lincoln and the ascendancy of the Republican Party fed Southern fears about the potential loss of both slavery and their white supremacy. Despite every Republican protestation to the contrary, the South believed the Black Republicans would destroy the institution of slavery. For Southerners, to threaten slavery was to threaten not only the Southern way of life, but Southern lives as well.

Notes

1 Eric Foner, ed., introduction to *Nat Turner* (Englewood Cliffs, NJ: Prentice-Hall, 1971), 9.

2 *Acts of the General Assembly Passed in Milledgeville Georgia in November and December, 1831* (Milledgeville, GA: Prince & Ragland Printers, 1832), 255.

3 *Acts of the General Assembly Passed in Milledgeville Georgia in November and December, 1829* (Milledgeville, GA: Camak & Ragland, 1830), 171.

4 Harvey Wish, "The Slave Insurrection Panic of 1856," *The Journal of Southern History* 5, no 2. (May 1939): 221, www.jstor.org/stable/2191583.

5 Ibid., 211.

6 *New York Tribune*, December 13, 1856.

7 *New York Herald*, December 12, 1856.
    Editor's Note: This is a work of history, a major focus of which was the racial hatred that existed in the United States. This racial hatred was openly reflected in the language of the newspapers of the time. That language, when provided in direct quotations, is generally retained in this book in order to accurately reflect the historical context.

8 *The Weekly North Carolina Standard* (Raleigh), December 17, 1856.

9 *Richmond Enquirer*, December 16, 1856.

10 *Evening Star* (Washington, DC), May 9, 1856.
    Editor's Note: This is a work of history, a major focus of which was the racial hatred that existed in the United States. This racial hatred was openly reflected in the language of the newspapers of the time. That language, when provided in direct quotations, is generally retained in this book in order to accurately reflect the historical context.

11 Katherine A. Pierce, "Murder and Mayhem" in *Words at War: The Civil War and American Journalism*, ed. David B. Sachsman et al. (West Lafayette, IN: Purdue University Press, 2008), 92.

12 Charles Sumner, "Crime Against Kansas" (speech, Washington, DC, May 19–20, 1856), https://archive.org/details/crimeagainstkans00sumn.

13 Ibid.

14 Ibid.

15 Ibid.

16 Ibid.

17 Eric H. Walther, *The Fire-Eaters* (Baton Rouge: Louisiana State University Press, 1992), 175.

18 Lorman A. Ratner and Dwight Teeter, Jr., *Fanatics and Fire-eaters* (Urbana, IL: University of Illinois Press, 2003), 41.

19 Eric H. Walther, *The Fire-eaters* (Baton Rouge: Louisiana State University Press, 1992), 172.

20 Pierce, "Murder and Mayhem," 85–86.

21 "Assault in the United States Senate Chamber," *Illinois State Register* (Springfield, IL), May 26, 1856.

22 Editor's Note: This is a work of history, a major focus of which was the racial hatred that existed in the United States. This racial hatred was openly reflected in the language of the newspapers of the time. That language, when provided in direct quotations, is generally retained in this book in order to accurately reflect the historical context.

23 "A Sermon in Brooks," *Weekly Advertiser* (Portland, ME), June 10, 1856.

24 Lorman and Teeter, *Fanatics and Fire-eaters*, 37.

25 Abraham Lincoln, "Cooper Union Address" (speech, New York, February 27, 1860), www.abrahamlincolnonline.org/lincoln/speeches/cooper.htm.

26 *The Congressional Globe: Containing the Debates and Proceedings of the First Session of the Thirty-Sixth Congress: Also of the Special Session of the Senate* (Washington, DC: John C. Rives, 1860), 88, http://digital.library.unt.edu/ark:/67531/metadc30806/.

27 Ibid., 220.

28 Walther, *The Fire-Eaters*, 184–85.

19

THE FIRE-EATING *CHARLESTON MERCURY*

Stoking the Flames of Secession and Civil War

Debra Reddin van Tuyll

Antebellum Southerners were of three minds when it came to the question of secession. Some Southerners—Andrew Jackson, for example—were vehemently devoted to the Union. They believed nothing would ever justify splitting the nation. Appropriately enough, they were referred to as Unionists. Others believed the Constitution allowed for secession but believed that breaking up the Union required some sort of extreme provocation from the North, and if it came to that point, the South should secede as a unit rather than as individual states. These were called cooperationists.[1] Finally, some believed that the only possible response to the actions of the North was secession, which would secure "liberty, equality, and self-government."[2] These were the fire-eaters.

Each group had its official organs, newspapers that raised support for political candidates and enforced party discipline. Unionist and cooperationist supporters and newspapers far outnumbered the fire-eaters, but the fire-eaters were so vehement, so caustic, that their influence was far greater than their numbers.[3]

No one really knows where the term fire-eaters came from, but it was in common use in the 1850s by Northerners and Southerners alike to describe someone who was far outside the political mainstream.[4] The most dangerous fire-eaters were a small group of Southern men—perhaps a dozen in total—whom others respected and looked to for leadership and guidance.[5] Their ranks included prominent politicians, jurists, and writers: William Lowndes Yancey of Alabama, Joseph Brown of Georgia, Louis T. Wigfall of Texas, Laurence M. Keitt of South Carolina, William Barksdale of Mississippi, Robert Barnwell Rhett of South Carolina, Nathan Beverley Tucker of Virginia, and James DeBow of Louisiana— just to name a handful of the best known. Several, including DeBow and Rhett, were journalists whose publications served as the main organs for the extreme secessionists.[6]

By the late 1850s, when the secession debate reached its apex, South Carolinian Robert Barnwell Rhett was the undisputed ringleader, the dean, one might say, of the fire-eaters. He earned that title for several reasons. First, early in his career, Rhett, with John C. Calhoun's help, gained power and influence in South Carolina, the epicenter of the secessionist movement. He, along with Franklin Elmore, a former nullifier and editor of the Columbia *South Carolinian*, ran a powerful political clique in South Carolina known as the Regency. Second, in the late 1850s, as the "irrepressible conflict" hulked nearer and nearer, Rhett's sons would affiliate and eventually purchase the most important fire-eating organ in the South, the *Charleston Mercury*.[7]

Both Southern newspapers and Southern newspaper readers recognized the influence of the *Mercury*. In a letter to the paper, a correspondent who signed his letter "A Southerner," wrote that he considered the Charleston paper to be "the staunchest, ablest and most sagacious exponent of southern opinion and State Right principles in the South."[8] Late in the Antebellum period, the New York *Journal* wrote that its most valued Southern exchange was with the *Mercury*, in large measure because of its excellent editor, Robert Barnwell Rhett, Jr., and because of its outstanding business reporting—an interesting perspective since the paper was best known as a political journal. The *Journal* referred to Rhett, Jr. as "a scholar as well as an able editor."[9] The *Mercury* agreed, boasting that it was the most "able and veteran leader of southern journalism in the cause of southern rights."[10] The *New York Times* agreed as well, writing that the *Mercury* was "the recognized leading newspaper organ of extreme southern opinion."[11]

Robert Barnwell Rhett, Sr., had long been an important financial backer of the *Mercury*, a Calhoun political organ, which from the 1840s was owned by his brother-in-law. When Rhett and Calhoun parted ways, the *Mercury* cut its ties with Calhoun and followed Rhett. The paper's readers, accustomed to the *Mercury* being a Calhoun newspaper, dropped their subscriptions in droves.[12] The paper would pass through a variety of hands until the Rhetts took complete ownership in the late 1850s and built it into the leading Southern rights journal in the country. The paper's small subscription list belied its influence in national and sectional affairs.[13]

This chapter examines a sample of issues of the *Charleston Mercury* from July 1, 1857, when Rhett's name first appeared on the masthead, through December 21, 1860, when the paper published its story about the South Carolina secession convention's decision to act on its own and secede from the United States. The sample included issues from the first and the fifteenth day of each month, or the closest day to those dates if they were not available. On occasions when a major issue was brewing, such as John Brown's trial, the sample was expanded to include full coverage. This study found that the *Mercury* was a perpetual supporter of Southern rights, states' rights, and secession, essentially in that order, and that other newspapers, fire-eating sympathizers as well as Unionists, considered the paper to be America's leading voice of disunion.[14]

Fire-eaters and the Partisan Press

The *Charleston Mercury*'s support for Southern rights, states' rights, and secession surprised no one. Newspapers in the Antebellum period were avowedly partisan and had been since the early Republic. Newspapers commonly aligned with political parties and spoke as organs for platforms and candidates. Some newspapers were devoted to a cause or an issue. For example, temperance, abolitionist, even suffragist papers were commonly available. Only a handful of newspapers fell into the fire-eating, ultra-secessionist camp. They wanted Southern rights and property protected in the territories as well as in the Southern states. Some even supported reopening the African slave trade. They were skeptical that the federal government would keep its promises, would not try some sort of slick legal or other maneuvering, and so they advocated for Southern secession as the only real means of protecting the region's peculiar institution.[15]

The *Mercury*, as John C. Calhoun's chief organ, had led the Southern Rights camp—politicians and organs—essentially since the Nullification Crisis of 1832. Little changed with regard to editorial position when the Rhetts became partial owners in 1857, and took over full ownership in 1858. The paper was "a pioneer of southern interests and sentinel of plantation slavery."[16]

The *Mercury*'s approach was an interesting one. While the paper editorialized about issues, these stories appeared far less frequently than stories reprinted from other newspapers with, perhaps, an editorial note added as an introduction. Or, equally common, the *Mercury* would "editorialize" on an issue by engaging in a back-and-forth conversation with another paper. That is, the *Mercury* would report that a particular paper had said something about a certain issue or had interpreted a *Mercury* column in a particular way, and then the Charleston journal would offer its response—or, often, critique—of the other newspaper's position.

The Lecompton Constitution

The first sectional issue that arose after the Rhetts attained editorial control over the *Mercury* was the Lecompton Constitution, one of four constitutions Kansas considered for presentation to Congress in its attempt to obtain statehood.[17] The territorial legislature, which was charged with framing the constitution, was proslavery, but the settlers were mostly free-soilers. Kansas Governor Robert J. Walker was a proponent of slavery, but he insisted that the proposed constitution, along with the antislavery Topeka Constitution, be put to a popular vote.[18] Each side boycotted the vote on the opposing constitution. Despite some waffling, President James Buchanan supported the Lecompton Constitution, but the faction of the Democratic Party led by Stephen A. Douglas sided with the Republicans in opposing it.

The Rhetts' *Mercury* saw the issue as the straw that broke the Democratic Party into two factions and believed the Union itself would be the next to fall.[19]

The *Mercury* was inclined to lay the responsibility for the debacle at the feet of two people: President Buchanan and his Kansas territorial governor, Walker. The paper reported on the issue with very little original copy, even editorials. Instead, it republished and responded to stories from other Southern journals, sometimes with an editorial note at the top. The paper was particularly fond of publishing rundowns of where other papers stood on issues. On September 1, 1857, it published the stances on Kansas of some seven newspapers.[20]

In one of its rare in-house-written editorial pieces, the paper observed that many believed the issue at hand was whether Kansas would come into the Union as a free state or a slave state. The bigger problem, according to the *Mercury*, was the affect the admission of Kansas would have on the political power structure. If Kansas came in as a slave state, the South would gain two senators, which would help keep the balance of power between the sections, at least for a while. The paper ominously reported that the Richmond *Examiner* had predicted Kansas would come in free, that Missouri would do away with slavery within a decade, and that Virginia would soon follow suit. This eventuality, the *Mercury* argued, was the reason why the South had to rule herself "or surrender all hope of peace or security."[21]

Articles such as this one led other papers to question the *Mercury*'s loyalty to the Democratic Party. The Lynchburg, Virginia, *Republican*, as well as the Boston *Traveler*, accused both the paper and South Carolina congressmen of not being Democrats. The *Mercury* responded that the Democratic Party had waxed and waned for its entire thirty-year existence, and it wondered if the party would last past the next presidential election. The Lecompton issue was just the latest to work toward tearing the party to shreds, but regardless, the paper pledged to stand by the principles of the Democratic Party.[22]

When Congress failed to approve the admission of Kansas under the Lecompton Constitution, the *Mercury* turned its scrutiny on Stephen A. Douglas, the leader of the Democratic Party whom the paper viewed as a traitor for siding with the Republicans in opposing the constitution.[23]

Secession, Slavery, and John Brown

Long before the 1860 presidential election pushed the secession question to the tipping point, the *Mercury* and the Rhetts had agitated for an immediate breakup of the Union as the only means available to protect slavery. They also were not entirely opposed to the idea of reopening the slave trade. The availability of more slaves meant their cost would go down and farmers' profits would go up, the paper explained in one edition. In another, the paper reprinted a letter from Alabama fire-eater William Lowndes Yancey arguing that federal laws against reopening the slave trade should be repealed and the issue left to the individual states.[24] Publishing the letter was an indication that the *Mercury* was at least passively in agreement with Yancey. The looming 1860 election, however, would launch the

Mercury into overdrive as it strove to convince Southerners that secession was the only alternative.

That is not to say the *Mercury* was ever unwavering on the question of secession. Almost as soon as he gained control of the paper, the younger Rhett wrote frequently on the issue. In one of his early pieces, he wrote that he believed there must be some way to keep the Union together without South Carolina being forced to submit to the "voluntary vassalage under which we have groaned for these forty years."[25]

The great fear was the rise of the Republican Party. In May 1859, the *Mercury* reported, "The six New England States send twelve Senators and twenty-nine members of the House of Representatives to Congress, everyone one of whom at the next session will be Republicans in politics."[26] In other words, the South was about to be outnumbered in Congress. On April 15, 1859, the *Mercury* reprinted a piece from the Washington, DC, *Union* that argued the Republican success in elections and policies was due to "the triumph of sectionalism and faction over the Constitution."[27] In the *Mercury*'s view, any Southerner who took any position other than immediate secession was weak on the question of Southern rights. This was the heart of its objections to Stephen A. Douglas as a presidential candidate.

Events at the end of 1859 heightened the call for secession. News of John Brown's raid on the arsenal at Harpers Ferry, Virginia, raised the bar for Southern hysteria beyond anything it had ever contemplated before.

The *Mercury* received its first dispatch about the Harpers Ferry raid on October 18, two days after the event, and even then it only received partial news because the telegraph office closed early. The report was sketchy and inaccurate. It maintained that government workers at the arsenal in Harpers Ferry had started a protest along with some abolitionists and some slaves who "have been forced to co-operate." The entire town was in the possession of the insurgents, the paper reported, and all the telegraph lines and train tracks had been cut. The paper expected the insurrection to have been put down by the time readers received their newspapers.[28] In another story, it reported that one hundred Marines with canon arrived from Washington to deal with the insurrectionists.[29]

The *Mercury* reported that it saw no reason for alarm, but it seized the opportunity to push even harder for a Southern nation. The paper declared, "The great source of evil is that we are under one government with these people." The paper took the raid to be an indication of a widespread plan in the North for "insurrections throughout the South." The purpose of these insurrections, the paper speculated, was to get slavery overthrown with the aid of the US military.[30] The paper anticipated that insurrections such as the one in Harpers Ferry would become more common.[31] The paper claimed the South had many enemies in the North and that unless a change came soon to Northern society, a "great conflict must come."[32]

The *Mercury* picked up a good bit of coverage from Northern newspapers. It reprinted an article from the New York *Herald* that reminded its readers that, although some Northern journals were referring to John Brown as "Mad Brown," Republican newspapers were applauding the insurrection and Brown's involvement in the bloody fighting in Kansas earlier in the decade. The story alleged that abolitionist leaders were fully aware that Brown's raid was in the works, even though they denied any knowledge.[33] The *Mercury* also reprinted an article from the New York *Express* that questioned who had paid for Brown's raid. The implication was that he had had financial assistance from abolitionists.[34]

By the time of Brown's trial, the *Mercury* scoffed at the idea that the raid had been an insurrection, since no slaves were involved, at least not voluntarily. It was a trifling affair, the paper concluded. The significance of the raid grew out of the twenty-five years Northerners had spent agitating about slavery, breaking up churches, running off slaves, and excluding Southerners and their property from territories. The recent events in Virginia, the paper stated, were but a preview of what awaited the South should it elect to stay in the Union. "Our connection with the North, is a standing instigation of insurrection in the South."[35]

The 1860 Democratic National Convention

The big-picture issues going into the convention were the expansion of slavery into the territories and popular sovereignty.[36] The convention could not settle these questions; that resolution would come only after America's bloodiest war. The convention would, however, provide a stage upon which Southerners could contest Stephen A. Douglas's Freeport Doctrine, which maintained federal law could not protect slavery in territories where the people did not want it. Southern Democrats were appalled by the notion that popular sovereignty could overturn slavery.

In the run-up to the convention, Southern state conventions adopted platforms that were uncompromising on the question of the extension of slavery into the territories and popular sovereignty. The Richmond *Enquirer* correspondent who covered the 1859 Mississippi state Democratic Convention commended the platform they adopted and predicted Virginia would follow suit in its own convention.[37]

The Griffin, Georgia, *Independent South* published—and the *Mercury* reprinted—an article asserting that some believed that the Democratic Party was irretrievably broken and recommended creating a new party devoted to Southern rights. The article stated that the work of creating a new Southern Rights party would be working in "the interest, the honor, and the equality of the South."[38] South Carolinians took the call to heart, according to the *Mercury*, and called for a Southern convention in Richmond in October 1859 to organize a party "to act upon national politics."[39]

By the end of 1859, with the Democratic National Convention only half a year away, the *Mercury* was becoming frantic about the fate of the South if it did not secede. The paper published one article calling for the Charleston convention to be limited to only Southern states or to be canceled.[40] In another, it claimed none of the Democratic candidates, including immediate secessionist John Breckinridge himself, was sound on Southern rights. This was one of the few articles in which the *Mercury* even mentioned Breckinridge, who would have logically been that paper's candidate of choice. Instead of promoting Breckinridge, however, the *Mercury* spent most of its editorial capital on denigrating Douglas. In that way, its coverage of the election was both skewed and odd. Most nineteenth-century newspapers would name a candidate it endorsed, even run his name above the masthead in every issue for months before the election. The *Mercury* did not do this in the 1860 election. It recommended no candidate, only advocated against Douglas for what the editors saw as his antislavery stance.[41]

As the Democratic National Convention approached, the *Mercury* declared the Democratic Party, as an effective, national organization, to be dead. It was no more than a faction. "The Northern and Southern members share no common principles," the editors wrote. "On every issue vital to the South, the Northern members agree with the Black Republicans, and this result the South herself has produced by her weakness and timidity." Rhett believed it was time for the South to end its association with those who would deny her rights. It was time for Southerners to prepare to act. The election, he told his readers, was not a question of Democrat or Republican. It was a question of union or disunion.[42]

And he was right. On April 30, delegates from Alabama, Mississippi, Louisiana, South Carolina, Arkansas, Texas, Florida, and part of Delaware walked out of the Charleston convention after the party declined to support a platform plank to protect slavery in the territories and another to protect slavery in general. The convention adjourned while the Georgia and Virginia delegations caucused over what to do. According to the *Mercury*, the withdrawal of the states was calm, quiet, and firm, and grounded in differences of principles.[43]

The net result of the Charleston convention was that the national Democratic Party was rent into two factions. Lincoln's election had been virtually guaranteed before the convention, but the walkout absolutely sealed the outcome. The convention would hold fifty-seven ballots, and while Douglas would win on each of them, he never got enough votes to secure the nomination. As a result, the convention adjourned and continued six weeks later in Baltimore. A fight over credentials in Baltimore would lead to a second walkout and end in the nomination of Douglas as the Northern Democratic candidate. Those who walked out would assemble and nominate Kentuckian John Breckinridge as the Southern Democratic candidate.[44]

The 1860 Presidential Campaign and Election

Even before the Northern and Southern Democrats had their candidates nominated, the *Mercury* was campaigning against Lincoln. Like so many other Southern publications, one technique the *Mercury* used was to make fun of

Lincoln's looks and height.[45] The paper also reported on opposition to Lincoln and warned its readers not to believe what they might be reading about Lincoln in Northern journals.[46] Nevertheless, the *Mercury*'s New York correspondent fully expected Lincoln would end up victorious. He wrote in early October that the Republicans were raising thousands of dollars for the campaign and were able to do so because this was the first time they had a viable presidential candidate.[47]

Meanwhile, Southern militias were drilling in public squares in preparation for the outcome of the election.[48] South Carolina was organizing units of Minute Men, arming, drilling, and equipping them for whatever might happen in November.[49] In the North, equivalent groups called Wide Awakes were parading in Washington, DC, and Wheeling, Virginia, and talks were already going on regarding how the government would be reshuffled once Lincoln won.[50] In North Carolina, the state bank had suspended discounting until the election crisis passed.[51]

By November 1, the *Mercury* was reporting on many Southerners' acceptance that Lincoln would win the election. It published a speech in which the Hon. John Townsend asked his fellow South Carolinians whether they planned to remain passive and "be handed over to the butcher like sheep," or whether they were ready to act.[52]

Following the election, the *Mercury* published a letter from "J.L." that no doubt reiterated the editors' sentiments. The writer maintained that he was revolted by the idea that an unfit man could be made president by a single section of the country and would rule over all of it. The paper recommended immediate secession and urged unity among Southerners. It was even willing to accept a fusion of cooperationists and secessionists as delegates to a Southern convention.[53]

The debate in South Carolina through November and early December was over whether to take immediate, independent action or to wait for the South to secede as a unit.[54] The answer to that question was evident. Citizens throughout South Carolina were putting up palmetto tree bunting and dragging palmettos from the swamps into their homes and businesses. The *Mercury* was particularly impressed with Mr. Joseph F. Church, "a popular plumber," who brought a tree to his office from Long Island, which had been important in the American Revolution.[55]

In Columbia, the state capital, some six hundred Minute Men, along with volunteers and local fire companies, staged a "magnificent torchlight demonstration." James L. Orr, a leading state politician, spoke to the procession and proclaimed himself a convert to the idea of secession.[56] Every day, the *Mercury* was hearing from more and more men who were volunteering for service in the South Carolina military—men from as far away as Baltimore, which had offered a company of one hundred men, and New Jersey, which had offered one of five hundred.[57]

Finally, on December 20, 1860, the *Mercury* got its fondest wish. South Carolinians met in convention in Charleston's Institute Hall and passed an Ordinance of Secession. The paper called it a day that would mark an "epoch in the history of the human race." Arrogant tyrants had been overthrown. South Carolina had regained its sovereign powers and "become one of the nations of the earth."[58] The South Carolina national flag would fly, briefly, over the state until other Southern states in early 1861 would follow the Palmetto state out of the Union and into a Southern Confederacy.

As the 1850s progressed, Southerners increasingly viewed the North, and particularly the rise of the Republican Party, as "an existential threat not only to their livelihood, but also to their lives."[59] When Southern Democrats walked out of the 1859 Democratic National Convention in Charleston, secession was all but guaranteed. The fire-eaters, it would appear at the time, had won the day. Their cheerleader and mouthpiece, the *Charleston Mercury*, played no small role in what seemed at the time to be a long-sought-after victory.

The paper achieved its goal by unusual journalistic means. It never named a candidate it endorsed, a most uncommon action for an Antebellum paper. Instead, it spent its ink on denigrating both Lincoln and Douglas, and in advocating for secession and Southern rights. It espoused an idea, not a party, not a party's candidate. The *Mercury* stood for Southern rights, and its editors saw secession as the only way to protect those rights.

The *Mercury*'s editors wrote little original copy. Instead, they relied on other voices—newspapers and correspondents. By republishing or publishing those pieces, the *Mercury* signaled its approbation to their position, and set the agenda for other fire-eating newspapers to follow. In a sense, it served as a vast clearinghouse for the exchange of ideas among fire-eating politicians and journalists—a public forum in which the fire-eaters proposed and debated strategies to achieve their cherished goal of a separate Southern nation.

They achieved their goal. They won their battle. In the end, however, they would lose the war. The *Mercury* itself would shutter its windows and fold its last edition in February 1865 as Union General William T. Sherman's troops swept a path through South Carolina.[60]

Notes

1 Steven A. Channing, *Crisis of Fear: Secession in South Carolina* (New York: W.W. Norton & Co., 1974), 83.

2 Michael F. Holt, *The Political Crises of the 1850s* (New York: W. W. Norton & Co., 1978), 253.

3 Lorman A. Ratner and Dwight L. Teeter, Jr., *Fanatics and Fire-eaters: Newspapers and the Coming of the Civil War* (Urbana, IL: University of Illinois Press, 2003), 33; Holt, *The Political Crises of the 1850s*, 219–20.

4 Eric H. Walther, *The Fire-Eaters* (Baton Rouge, LA: Louisiana State University Press, 1992), 2.

5 William C. Davis, *Rhett: The Turbulent Life and Times of a Fire-Eater* (Columbia, SC: University of South Carolina Press, 2001), 583.

6 Walther, *The Fire-Eaters*, xi.

7 H. Hardy Perritt, "Robert Barnwell Rhett: Disunionist Heir to John C. Calhoun, 1850–1852," *Southern Speech Journal* 24:1 (1958): 38–55; Nancy McKenzie Dupont,

"Mississippi's Fire-Eating Editor Ethelbert Barksdale and the Election of 1860" in David B. Sachsman, S. Kittrell Rushing, and Debra Reddin van Tuyll, eds., *The Civil War and the Press* (New Brunswick, NJ: Transaction Publishers, 2000), 137; Walther, *The Fire-Eaters*, 128.

8   "From a Correspondent of the Mercury," *Charleston Mercury*, May 16, 1859, 1.

9   "The Charleston Mercury," *Charleston Mercury*, July 1, 1859, 1.

10   "The Charleston Mercury," *Charleston Mercury*, July 15, 1859, 2.

11   "The Charleston Mercury," *Charleston Mercury*, July 15, 1859, 1.

12   John S. Coussons, "Thirty Years with Calhoun, Rhett, and the Charleston Mercury: A Chapter in South Carolina History" (PhD diss., Louisiana State University and Agricultural and Mechanical College, 1971), v; Walther, *The Fire-Eaters*, 125.

13   Granville T. Prior, "The Charleston Mercury" (PhD diss., Harvard University, 1946), 457.

14   Robert E. Bonner, *Mastering America: Southern Slaveholders and the Crisis of American Nationhood* (Cambridge: Cambridge University Press, 2009), 236; Peter Bridges, *Pen of Fire: John Moncure Daniel* (Kent, OH: Kent State University Press, 2002), 168.

15   Jim A. Kuypers, *Partisan Journalism: A History of Media Bias in the United States* (Lanham, MD: Rowman and Littlefield, 2014), 15, 30; Richard D. Brown, *Knowledge is Power: The Diffusion of Information in Early America, 1700–1865* (New York: Oxford University Press, 1989), 37.

16   Coussons, "Thirty Years with Calhoun, Rhett, and the Charleston Mercury," 8; Steven G. Saltzgiver, "Consensus and Public Debate: The Washington Press During the Nullification Crisis, 1832–1833" in David B. Sachsman, S. Kittrell Rushing, and Debra Reddin van Tuyll, eds., *The Civil War and the Press* (New Brunswick, NJ: Transaction Publishers, 2000), 33.

17   Francis Howard Heller, *The Kansas State Constitution: A Reference Guide* (Santa Barbara, CA: Greenwood Press, 1992), 1–4.

18   Kenneth M. Stampp, *America in 1857: A Nation on the Brink* (New York: Oxford University Press, 1992), 167–80.

19   "The Delta as an Independent Southern Journal—What It is For and What It Is Not For," *Charleston Mercury*, July 1, 1857, 2.

20   "Kansas and the Administration," *Charleston Mercury*, September 1, 1857, 2.

21   "The Importance of Kansas to the South," *Charleston Mercury*, September 1, 1857, 2.

22   "Correspondence of the Mercury," *Charleston Mercury*, August 15, 1857, 2; "Northern Views of Southern Parties," *Charleston Mercury*, August 15, 1857, 2.

23   Robert W. Johannsen, "Stephen A. Douglas, 'Harper's Magazine,' and Popular Sovereignty," *Mississippi Valley Historian Review* 45:4 (1959): 606; "Governor Wise's Letter," *Charleston Mercury*, January 15, 1858, 2; "The Democratic Caucus in Washington," *Charleston Mercury*, April 1, 1858, 3.

24   Edmund Ruffin, "Slavery and Free Labor Defined and Compared," *Charleston Mercury*, July 1, 1959, 1; William Lowndes Yancey, "Mr. Yancey and the Slave Trade," *Charleston Mercury*, July 15, 1950, 2.

25   "Non-Intercourse," *Charleston Mercury*, July 3, 1857, 2.

26   *Charleston Mercury*, May 2, 1859, 1.

27   "The Character of Parties," *Charleston Mercury*, April 15, 1859, 1.

28   *Charleston Mercury*, October 18, 1859, 2; *Charleston Mercury*, October 19, 1859, 3.

29   Ibid., October 18, 1859, 3.

30   "The Plan of Insurrection," *Charleston Mercury*, November 1, 1859, 1.

31   "The Harper's Ferry Insurrection," *Charleston Mercury*, October 19, 1859, 1.

32 "The Harper's Ferry Conspiracy," *Charleston Mercury*, October 25, 1859, 1.

33 "Mad Brown's Insurrection," *Charleston Mercury*, October 25, 1859, 1.

34 "Who Paid the Expenses," *Charleston Mercury*, October 25, 1859, 1

35 "The Insurrection," *Charleston Mercury*, October 31, 1859, 1.

36 Owen. M. Peterson, "The South in the Democratic National Convention of 1860," *Southern Speech Journal* 20:3 (1955): 212.

37 "Virginia and the Mississippi Platform," *Charleston Mercury*, August 1, 1859, 1.

38 "A Third Party Movement," *Charleston Mercury*, August 1, 1859, 1.

39 *Charleston Mercury*, August 15, 1859, 2.

40 "The Charleston Convention," *Charleston Mercury*, December 1, 1859, 1.

41 "Speech of Hon. Jabez L. M. Curry," *Charleston Mercury*, December 15, 1959, 4; "Correspondence of the Mercury," *Charleston Mercury*, January 1, 1860, 1; "Mr. Douglas's Article," *Charleston Mercury*, September 16, 1859, 1.

42 "The Democratic Party," *Charleston Mercury*, April 16, 1860, 1.

43 *Proceedings of the Democratic National Convention Held in 1860 at Charleston and Baltimore* (Cleveland, OH: Nevins First, Plain Dealer Job Office, 1860), 58–65; "The Convention Breaking Up," *Charleston Mercury*, May 1, 1860, 1.

44 James Leonidas Murphy, "Alabama and the Charleston Convention of 1860" in *Studies in Southern and Alabama History*, George Petrie, ed. (Montgomery, AL: Alabama Historical Society, 1905), 259.

45 *Charleston Mercury*, June 1, 1860, 1.

46 "Lincoln's Radicalism Proved—His Obedience to the Abolition Idea," *Charleston Mercury*, September 15, 1860, 1.

47 "Our New York Correspondence," *Charleston Mercury*, October 1, 1860, 1.

48 "Meetings," *Charleston Mercury*, October 1, 1860, 2.

49 "The Minute Men," *Charleston Mercury*, October 15, 1860, 1.

50 *Charleston Mercury*, October 15, 1860, 2; "The Presidential Distractions," *Charleston Mercury*, October 15, 1860, 1.

51 "Southern Banks Preparing for Political Crisis," *Charleston Mercury*, October 15, 1860, 1.

52 "An Address of Hon. John Townsend," *Charleston Mercury*, November 1, 1860, 4.

53 "The Election to the Convention" and "State Rights Feeling in New York," *Charleston Mercury*, November 15, 1860, 1.

54 "The Meeting in Edgefield," *Charleston Mercury*, November 15, 1860, 1.

55 "A Palmetto Tree," *Charleston Mercury*, November 15, 1860, 2.

56 "Doings at the State Capital," *Charleston Mercury*, November 15, 1860, 3.

57 "More Volunteers," *Charleston Mercury*, November 15, 1860, 2.

58 "The 20th Day of December, in the Year of Our Lord, 1850," *Charleston Mercury*, December 21, 1861.

59 David Goldfield, "How the Charleston Elite Brought on the Civil War," *New York Times*, April 21, 2017.

60 Charles Beales, *War Within a War: The Confederacy Against Itself* (Philadelphia: Childon, 1965), 152; George Rable, *The Confederate Republic: A Revolution Against Politics* (Chapel Hill, NC: University of North Carolina Press, 1994), 175.

20

"OUR ALL IS AT STAKE"

The Anti-Secession Newspapers of Mississippi

Nancy McKenzie Dupont

During the secession crisis of 1860–1861, thirty newspapers were regularly published in Mississippi, the second state to leave the Union. Copies from fifteen of these newspapers are still available. Three of these fifteen newspapers took an editorial stance opposed to immediate secession. Though the powerful fire-eating newspapers that called for immediate secession may have overshadowed them, these anti-secessionist newspapers are worthy of note for their courage, their conviction, and the strength of the arguments they presented to the people of Mississippi.

Two of these anti-secessionist newspapers were located in the state's larger Mississippi River towns; they were the Natchez *Courier* and the Vicksburg *Whig*. The third anti-secessionist newspaper was the Brandon *Republican*, located in a small town near Jackson, the state capital. Each had a distinct style, but the editors had some similarities: they were all small businessmen, they were all born in states that would side with the Union during the Civil War, and they all wrote passionately against immediate secession, which they considered the worst mistake Mississippi could make.

The *Courier*

The Natchez *Courier* has a long and confusing history with several incarnations and mergers with other newspapers, but in 1860 it had been stable for a few years. Giles M. Hillyer became editor of the *Courier* in 1850 and began issuing a daily paper in 1852.[1] Hillyer, a Connecticut native, was forty-one at the time of the secession crisis and owned $14,000 in assets. He had a wife and four children, but he owned no slaves.[2]

Hillyer was active in Mississippi politics in the 1850s. In 1855, he won a seat in Congress as a Know-Nothing candidate and was reelected in 1857, despite a decline in Know-Nothingism in Natchez.[3] For the most part, Hillyer and his newspaper appealed to the well-to-do class of Natchez, such as planters, bankers, and large merchants, but the *Courier*'s circulation was far below that of its competitor, the Natchez *Free Trader*.

The *Courier* was considered to be well written, but it was weak in local coverage and concentrated more on national and international issues than Mississippi news.[4] This trend was evident in other newspapers during the secession period, though as the crisis of 1860 progressed, the amount of Mississippi news, and accounts from other Southern states, increased.

The *Whig*

For much of its existence, the Vicksburg *Whig* was led by journalist and historian William H. McCardle. But in 1858, Marmaduke Shannon began editing the newspaper after being the sole owner for sixteen years.[5] Born in New Jersey, Shannon was thirty-three years old during the secession crisis. He owned eleven slaves, but there is no information about what real property he might have owned other than the newspaper.[6]

Between 1840 and early 1860, the *Whig*, which was the only Mississippi newspaper that published daily, was considered one of the leading publications in the state.[7] But by July of 1860, the *Whig* was only publishing once a week. On May 9, 1863, during the siege of Vicksburg, the *Whig*'s printing plant burned and the newspaper stopped publication.[8] The fire was begun, probably, by an artillery shell from a Federal sniper. Shannon left the newspaper business and became Sheriff of Vicksburg after the war.[9]

The *Republican*

Brandon, Mississippi's *Republican* was edited by A. J. Frantz, a newspaperman with a colorful style who left historians with an excellent record of his later life. Frantz started the *Republican* in 1850 and was publisher until at least 1876.[10] Journalist R. H. Henry worked for Frantz in his early years and wrote about him in *Editors I Have Known Since the Civil War*.

According to Henry, Frantz was born in Maryland and came to Mississippi as a printer. He was a strong writer who used simple language and slang in his columns. He had a philosophy about newspapering that made his knowledge of the outside world somewhat limited: Frantz rarely read a paper printed outside the state, holding to the idea that the way to make a Mississippi paper was to print Mississippi news, which he did almost exclusively. The only exceptions were on election nights, when Frantz made special efforts to get the news by wire.[11] As a boss, Frantz was unfriendly except for the one night a week ("mailing night") when he treated his employees to beer at a local bar.[12]

A fugitive shot Frantz in the shoulder and groin during a sheriff's posse ride in the winter of 1860. Frantz was on crutches for the first two years of the Civil War, but the *Republican* never missed publishing a single issue. During the height of the Civil War, when Sherman's army was in Mississippi, Frantz put up a small printing press in the woods so his paper would come out on time.[13]

The Union Organs Speak Out

News of Lincoln's election trickled in rather than striking suddenly. Hand-counting paper ballots is a slow way of tallying votes, but it was the only method available to nineteenth-century Americans. After local votes were tallied, newspapers disseminated the information by telegraph, rail, and the post, with the telegraph the only method providing instantaneous communication from one point to another. Six days after the election, the Natchez *Free Trader* complained about the slowness of the incoming election returns, but was able to provide its readers with one indisputable fact: Lincoln, a sectional candidate from a sectional party, had won the election. There was then a more urgent question to answer: How would Mississippi and the rest of the slave states respond to the election of a Northern antislavery president?

On November 14, 1860, the Vicksburg *Whig* decried the secession sentiment in South Carolina: "It appears fanaticism is about to do its worst. Our heart sickens at the rashness of a misguided and demagogue-ridden commonwealth."[14] On page three, the *Whig* made its strongest statement against secession in a position usually reserved for its political endorsements. The symbol of an American Flag was followed with what the *Whig* called its motto: "We join ourselves to no party that does not carry the flag and keep step to the music of the Union."[15] Though the motto had the appearance of a permanent addition to the newspaper, it did not appear again. However, the flag symbol was a frequent feature of the secession crisis editions of the *Whig*. The November 21 edition carried these words above the flag: "The Union must and shall be preserved."[16]

A special session of the Mississippi legislature in November called for a special convention to meet on January 7. An election for delegates to the convention would be held on December 20.[17] The South Carolina legislature had called for a special convention to begin meeting on December 17.[18] The Georgia legislature had done the same, with its convention scheduled to meet on January 16.[19] For an information distribution system depending on boat, rail, and telegraph, Mississippi newspapers made a valiant effort to keep their readers informed about reactions to Lincoln's election. There was another task for the newspapers as well: promoting their own version of the proper response to the Republican victory. The December 20 election day would be the first test of the strength of their arguments.

The last days of November were marked by an argument between the pro-secession Jackson *Mississippian* and the Unionist *Whig* over the right of a state to secede. On November 21, the *Mississippian* wrote that secession was the only acceptable response to Lincoln's election, and that secession was a legal remedy. At the end of the Revolutionary War, the *Mississippian* contended, "Each of the States mutually acknowledged the separate and independent sovereignty of the other."[20] The argument continued: "As separate States they joined this confederacy, and some of them to make assurance doubly sure, in so many words joined upon the condition of their right to withdraw. The other recognizing this right as clearly deducible from the nature of the compact, received them upon that condition."[21] The *Mississippian* concluded, "Let us take our own rights and institutions in our own keeping. There is safety in no other course."[22]

The *Whig* accused *Mississippian* editor Ethelbert Barksdale of lying. It argued that no state had joined the Union under the condition of leaving when they pleased:

> We say it, without fear of contradiction, and it is indeed what every attentive reader of history must know to be the truth, that not a single State joined the Union upon any such condition, and indeed upon any condition of any kind. In advocating the right of secession, let the State organ state the facts. We hope it will correct its statement on this point, or else point out the page of history that records the event which we say never occurred.[23]

The *Mississippian*'s next edition counter-attacked, writing that the *Whig* desired to "conceal the enormity of Black Republican crimes against the Constitution." The *Mississippian* accused the *Whig* of trying to convince Mississippians that they "live under a consolidated government of unlimited power." It called the *Whig*'s rebuff "quite a lecture in the Sir Oracle style." The *Mississippian* then, in two different columns, defended its position by quoting from the Constitution ratification statements made by Virginia and New York. The Virginia document stated that the state would resume powers delegated to the federal government "whenever the same shall be perverted to her injury or oppression."[24] The New York document was quoted in this way: "That the powers of the Government may be re-assumed by the people whenever it may become necessary to their happiness."[25]

There were only twenty-four days from the time of the calling of the special state convention to the election of delegates, and Mississippi newspapers redoubled their efforts to make their points before the convention convened. Though candidates for the election did not have to state their position on secession, many did. But there was an air of confusion, since so little time had passed since the presidential election. Mississippi newspapers could not invent party affiliations that would apply to large groups, so inexact terms such as secessionist, separate-state action, immediate secessionist, Southern rights, and Unionist were applied. Delegates to the secession convention represented a range of political affiliation and parties, with one delegate, Edward F. McGehee of Panola County, listing his political affiliation as "Extremely and intensely Southern."[26]

In the days leading up to the state convention election, Mississippi newspapers clarified their positions on the secession issue. Rabid fire-eaters, such as the

Mississippian, called for separate state action, meaning that each Southern state would secede from the Union individually to demonstrate its outrage over Lincoln's election. Newspapers endorsing Unionism called for delay and ultimately cooperation, meaning the Southern states would meet in convention and take action together, possibly seceding as a group. Even a staunchly Union newspaper like the *Whig* tried to help readers understand the two positions: "Every man styled a Union man in the State is a co-operationist. They believe that the South ought to assemble and unitedly demand an acknowledgement of her rights, and a guarantee of their concession. There is another party which favors the doctrine of separate State secession."[27] The *Whig* wanted the legislature to find out what the people of Mississippi wanted, and it lamented that the state convention had been called so hastily, fearing the convention's members might act out of excitement.

The pro-Union Brandon *Republican* outlined the position of the state cooperationists. "Secession is not the remedy,"[28] the *Republican* wrote as it called for a Southern convention. "The united voice of the people of the south shall be a supreme command to us, and we are willing now, as heretofore, to pledge ourselves to whatever it may be,"[29] the *Republican* promised. The benefits, in the newspaper's view, were numerous: "The whole South would make a respectable government, able and willing to protect itself against any adversary. In extent, it would be an empire; in population, it would be three times as large as the colonies were when our fathers struck for liberty and struck successfully."[30]

The *Republican* believed separate secession by Mississippi was the road to ruin for the South:

> Secession can give neither effectual resistance nor strength, nor safety, nor union, nor stability, nor respectability. We would present the miserable spectacle of a weak and imbecile power, without an army or a navy, without population or territory, large enough to command respect. We would be kicked and cuffed like the little government of Genoa, and hectored over by every strong power on the face of the globe.[31]

The *Republican* then turned its attention to an insult being used against cooperationists—that they were "submissionists." The insult, the newspaper stated, was coming from "a few hot-headed Disunionists in our midst who are in the habit of denouncing every man who differs with them."[32] The hot-heads were frequently "men who never did own a negro, and who never expect to own one, and sometimes from men who were born and raised in the North."[33] It is they who were submissionists, the *Republican* contended, because of all they were willing to turn over to the North:

> The territories, all the public treasure, the navy, the Capitol, all the public buildings, which cost us millions upon millions of dollars, all our interest

in the Patent Office, the Smithsonian Institute, the Observatory, the Washington Monument, the Star Spangled Banner, all our national airs, the glorious Fourth of July, and the Constitution itself.[34]

The *Republican*'s advice to the hotheads was to let Mississippi stay in the Union and fight for its rights as well as its claims to the list of American property and privileges.

For the *Whig*, secession meant not only the loss of an American birthright, it would be the end of peace in Mississippi. "Are the people prepared for revolution and civil war? We think not,"[35] the *Whig* told its readers. It would also be expensive: "At the lowest possible estimate, it will cost twenty-five million dollars to maintain the State of Mississippi out of the Union. All of this will have to be raised by direct taxation on the people."[36] The *Whig*'s appeal to the people was to stay in the Union, at least until the fall of 1861 when the Southern states should decide together what should be done. The *Whig* believed it could summarize its position simply, and it tried to do so in a single, though lengthy, sentence: "We desire to see the existing causes of division between the two sections settled in the Union, upon a permanent and honorable basis that will insure tranquility and harmony to the entire nation; failing in that, we desire to see the whole body of Southern States in united phalanx, moving in a solid column to the establishment of their independence out of the Union."[37]

One of the last appeals from the *Whig* was a dire warning of what secession would mean. "Our all is at stake," the *Whig* wrote. "Our liberties, our lives, our fortunes, and a false step may ruin all…. Mississippi has but few arms, less munitions of war, no credit, and her treasury has been empty since July."[38] Loyalty to the South was one thing, but a complete trust in the secession newspapers was foolish, according to the *Whig*. "The secession papers say if Mississippi goes out of the Union, she will force the other Southern States out. Indeed! We rather guess, though, if they have the same Southern spirit we claim, they will not be forced to do anything."[39]

Mississippi voters proved they could not be forced to do anything as well. Despite the panic over Lincoln's election and the thousands of words the newspapers wrote, only 60 percent of the voters who cast ballots in the November presidential election went to the polls on December 20.[40] Unionists or cooperationists ran for election in thirty-four of Mississippi's sixty counties.[41] The number of Unionists and cooperationists in Mississippi was in stark contrast to the unanimous vote of the South Carolina convention to leave the Union, a decision made on the same day as Mississippi's state convention election.

As in the presidential election, returns from the state convention election were slow to be reported in the newspapers. The secession newspapers quickly claimed a sweeping victory, and they were right. Candidates who claimed they were in favor of immediate secession dominated the number who would assemble in Jackson on January 7.[42] But there was no legal reason why all of the candidates

FIGURE 20.1 "The 'Secession Movement,'" lithograph published by Currier & Ives, 1861. This illustration shows the feeling of anti-secessionists. Namely, that the secessionist states were being led over a cliff to their doom. Florida, Alabama, Mississippi, and Louisiana, depicted as men riding donkeys, are led by South Carolina to the precipice. South Carolina, in pursuit of a "Secession Humbug," is riding a pig. The four trailing states follow blindly, denouncing the Union. In the foreground, Georgia rides separately down an incline: "We have some doubts about the end of that road and think it expedient to deviate a little." (From the Library of Congress Prints and Photographs Division Washington, DC. <www.loc.gov/item/2003674576/> Reproduction Number: LC-USZ62-32995 (b&w film copy neg.). Currier & Ives, "The 'Secession Movement.'")

could not change their minds several times in the eighteen days between the election and the convention, and the newspapers made use of the opportunity, at times appealing to the delegates directly as the full effects of South Carolina's secession became known.

To the anti-secession newspapers, the fight was not over but the odds were against them. The Natchez *Courier* admitted that the state convention election returns were a disappointment to its cause: "The co-operationists in the Convention will not be as numerous as we wished; but they are too many in numbers, too respectable in character, and represent too large a body of their fellow citizens, to be either brow-beaten or cheated out of their rights."[43] The most passionate last-minute, pre-convention appeal came from the Vicksburg *Whig*, which warned Mississippians that they would lose in a pro-secession decision. "Our fathers gave us a goodly heritage, rich in present blessings, and all glorious with the golden promise of yet richer fruition in the years to come,"[44] the *Whig* told its readers. "And now, when some of the heirs attempt to mismanage this grand plantation, shall we, who are equal heir, 'secede' and leave the whole inheritance?"[45] There was no reason to give up on the United States now, the *Whig* wrote: "The Federal Union has proved a wonderful success; it has made us mighty as a people—invincible against the world, and prosperous beyond example. All admit the perfection of our system of Union; and that would be perfect if the hands of the North were tied on the subject of slavery."[46] The *Whig* added: "Separate or partial secession abandons all, and leaves us weak and powerless for all time to come."[47]

Mississippi's weak position, should she secede, would leave the state with little protection if war broke out, according to the *Whig*. And the newspaper believed that war was inevitable. In one of its last editions issued before the beginning of the convention, the Whig attacked those who advocated immediate secession: "The leaders pretend to believe that secession will be peaceable. If so, why is the Governor of Georgia providing to tax her people a million of dollars to purchase arms and to provide for war? Why are the leaders so active to organize military companies all over Mississippi?"[48]

There was still hope, in the *Whig*'s view. Two days into the convention, on January 9, the *Whig* reported optimistically that there were sixty-eight delegates in favor of immediate secession and twenty-nine supporting cooperation. Though it was publishing its newspaper in Vicksburg, the *Whig* had a message for the delegates meeting in Jackson: "Let each member feel that he is acting for the whole state, let his first thought be for undivided sentiment, and the assembling of the convention will not be in vain."[49]

By January 9, the Natchez *Courier* had lost its patience with the convention. It proudly announced that Mississippi was "still in the union," but that that might be a temporary situation. The *Courier* had a warning for the assembled delegates: "The people are patient, but they will kick against any high-handed measures of revolution done without their consent."[50]

The Day of Decision

The convention's final vote on secession, taken January 9, 1861, was eighty-four to fifteen in favor. The newspapers opposed to immediate secession had, in all probability, a tough decision to make on how to present the news from Jackson. The Vicksburg *Whig* was a model of graciousness and concession as it began its remarks: "The act has been done, and it is not becoming for one single son of the State to refuse to yield." The *Whig* did not hide its disappointment, but urged unity in a time of crisis: "We did not approve it. But it is not for us, nor for any other citizen of the State, to set ourselves or himself against the Act of the State. ... as a citizen of the State, we feel bound to yield, not a sullen and forced submission, but a willing and ready obedience to its solemn mandate."[51]

The *Whig* proved its new support for the cause by reporting meticulously on the establishment of the new Confederate government and by printing the inaugural address of President Jefferson Davis in its entirety.[52] The most chilling report in the post-secession *Whig* was an announcement from the Catholic Bishop of Mississippi informing his clergy that the Sunday liturgy had to be changed: "In the 'Prayer for the President of the United States' use the words Governor of the state instead of 'President of the United States'; —also in the 'Prayer for Congress' substitute the words This State for 'These United States,' —and the word Legislature in place of 'Congress.'"[53]

The Natchez *Courier* was less magnanimous in defeat. The newspaper called the work of the convention, which passed secession without putting the question to a vote of the people, an act of "dictatorial oligarchy."[54] The *Courier* called for the people to reject secession, and if they refused, it would be "another proof of how easily the people can be deprived of one of their 'inalienable rights.'" In February, the *Courier* had harsher words still for former Senator Jefferson Davis, who had been appointed provisional president of the Confederacy: "Neither in character nor in politics, has he any hold upon the confidence of the people. He has at last attained that for which he has been struggling for ten years past—a Presidency. We may admire the ingenuity with which he has attained his object, and yet despair the tricks by which he has crawled to it."[55]

The *Courier* felt comfortable in its tirade because Adams County's delegation had voted solidly against secession. Other votes against secession came from nine other counties, with the largest concentration in the extreme northeastern section of the state. All four delegates from Tishomingo County and two of the four from Itawamba County had voted against the secession ordinance. Other no votes came from Washington, Warren, Attala, Rankin, Franklin, Amite, and Perry counties. In Adams and Warren counties, competing newspapers had promoted vastly different solutions to the sectional crisis. In all other counties, there were no newspapers or the newspapers are no longer extant.

The writings of the anti-secession newspapers provide the thinking of the opposition during the secession crisis in Mississippi, and they provide support for the assertion that secession was debated in Mississippi rather than universally accepted in some romantic notion of Southern rights and dignity. As Mississippi novelist James Street wrote in 1942, "The idea that the South rose to a man to defend Dixie is a stirring legend…and nothing more."[56]

Notes

1 *Mississippiana: Union List of Newspapers*, vol. 2 (Science Press, 1971), 52.
2 US Census Bureau, Eighth Census of the United States, 1860.
3 D. Clayton James, *Antebellum Natchez* (Baton Rouge, LA: Louisiana State University Press, 1968), 126–27.

4 Ibid., 230.

5 *Mississippiana*, 72.

6 US Census Bureau, Eighth Census of the United States, 1860.

7 *Biographical and Historical Memoirs of Mississippi* (Chicago, IL: Goodspeed Publishers, 1891), 244.

8 Samuel Carter III, *The Final Fortress: The Campaign for Vicksburg 1862–1863* (New York: St. Martin's Press, 1980), 203.

9 Ibid., 310.

10 *Mississippiana*, 6.

11 R. H. Henry, *Editors I Have Known Since the Civil War* (New Orleans: Press of E. S. Upton Company, 1922), 35.

12 Ibid.,37.

13 Ibid., 39.

14 *Whig* (Vicksburg, MS), November 14, 1860, 3.

15 Ibid., 3.

16 *Whig* (Vicksburg, MS), November 21, 1860, 3.

17 Ralph A. Wooster, *The Secession Conventions of the South* (Princeton, NJ: Princeton University Press, 1962), 28.

18 Ibid., 15.

19 Ibid., 82.

20 *Mississippian* (Jackson, MS), November 21, 1860, 1.

21 Ibid.

22 Ibid.

23 *Whig* (Vicksburg, MS), November 28, 1860, 3.

24 *Mississippian* (Jackson, MS), December 5, 1860, 1.

25 Ibid.

26 The list is found in *Journal of the State convention and Ordinances and Resolutions Adopted in January 1861* with an Appendix (Jackson, MS: E. Barksdale, state printer, 1861).

27 *Whig* (Vicksburg, MS), December 5, 1860, 3.

28 *Republican* (Brandon, MS), December 6, 1860, 2.

29 Ibid.

30 Ibid.

31 Ibid.

32 Ibid.

33 Ibid.

34 Ibid.

35 *Whig* (Vicksburg, MS), December 12, 1860, 1.

36 Ibid.

37 Ibid.

38 Ibid.

39 *Whig* (Vicksburg, MS), December 19, 1860, 1.

40 Avery O. Craven, *The Growth of Southern Nationalism, 1848–1861* (Baton Rouge, LA: Louisiana State University Press, 1951), 364.

41 Ralph A. Wooster, *The Secession Conventions of the South* (Princeton: Princeton University Press, 1961), 56.

42 Ibid., 58.

43 *Courier* (Natchez, MS), January 4, 1861, 4.

44 *Whig* (Vicksburg, MS), December 26, 1860, 1.

45 Ibid.
46 *Whig* (Vicksburg, MS), January 2, 1861.
47 Ibid.
48 Ibid.
49 *Whig* (Vicksburg, MS), January 9, 1861, 1.
50 *Courier* (Natchez, MS), January 9, 1861, 4.
51 *Whig* (Vicksburg, MS), January 23, 1861, 1.
52 *Whig* (Vicksburg, MS), February 20, 1861, 1; February 28, 1861, 1.
53 *Whig* (Vicksburg, MS), January 23, 1861, 1.
54 *Courier* (Natchez, MS), January 10, 1861, quoted in Percy L. Rainwater, *Mississippi, Storm Center of Secession 1856–1861* (Baton Rouge, LA: Otto Claitor, 1938), 215.
55 *Courier* (Natchez, MS), February 16, 1861, quoted in Rainwater, *Mississippi*, 215. .
56 James Street, *Tap Roots* (New York: The Dial Press, 1942), 10.

21

EXCHANGE ARTICLES CARRIED BY THE NEW YORK *EVENING POST*, DECEMBER 13–31, 1860

Erika Thrubis

The New York *Evening Post* was a Unionist, anti-slavery daily paper with a circulation of about 7,300.[1] William Cullen Bryant, the *Evening Post's* editor-in-chief between 1829 and 1878, was a known abolitionist and Free-Soiler who thought of the *Evening Post* as "both an educational tool and moral force with which to diffuse to the common man practical information and encourage prudence."[2] John Bigelow, a pro-Union Republican supporter, edited the paper until January 1861. During Bigelow's twelve years of proprietorship, the *Evening Post* prospered; its net income during the time more than quadrupled.[3]

The *Evening Post* regularly featured correspondents from both Washington and Charleston, the latter of which the *Evening Post* referred to as "our fire-eating correspondent." The *Evening Post* and other papers commonly shared— or exchanged—content with each other. These shared articles, called exchanges, were a common form of publication in the *Evening Post*. Almost 40 percent of the content reviewed for this chapter originated as exchanges.[4] The *Evening Post* was well known and respected. Its content was shared via the exchange network in many other papers, including famed fire-eater Robert Barnwell Rhett's *Charleston Mercury*, as well as the Missouri *Democrat*, *Providence Evening Press*, and the Boston *Journal*.[5]

On December 20, 1860, Robert Barnwell Rhett's thirty-year campaign for disunion finally succeeded, allowing his *Charleston Mercury* to print a broadside boasting "THE UNION IS DISSOLVED!" while the bells from Institute Hall rang in celebration.[6] Rhett and 168 fellow fire-eaters unanimously passed their Ordinance of Secession, which was followed by celebrations across the Southern states, according to the New York *Evening Post*, including the firing of one hundred guns in the air, military parades, cheering in the streets, impromptu speeches, and the cheery ringing of bells.[7]

But not all Southerners felt secession or the events leading up to it were cause for celebration. An exploratory analysis of the *Evening Post* revealed an interesting pattern. Exchange articles from Southern papers often represented pro-Union rather than pro-Secession sentiments. It was this purposeful publication of Union-supporting exchanges that sparked the idea for this study. This chapter evaluates patterns found in exchange articles published in the New York *Evening Post* between the dates of December 13 and December 31, 1860.

Newspapers were instrumental in shaping the nation's beliefs leading up to and during the Civil War. The power of the press was well recognized during the period. An exchange article from the Richmond *Enquirer* published on December 17, 1860, in the *Evening Post* blamed "the presses and leaders" of New York for producing "the present state of public affairs," specifically pointing to the *Tribune*, *Times, Commercial, Evening Post*, and *Courier*.[8]

Between December 13 and December 31, 1860, exchange articles represented 103 of 299 secession-related articles in the New York *Evening Post*, about a third of its secession-related content. In the *Evening Post*, exchanges from more than one newspaper were commonly grouped into one article. After separating the *Evening Post's* grouped exchanges into unique entries, exchanges represented a total of 38 percent (117 of 310) of secession-related articles.[9]

In related research conducted at Wayne State University, Professor Michael Fuhlhage studied the content of the Missouri *Democrat* from December 13 to December 31, 1860, and found that exchange articles represented 135 of 278 secession-related articles (49 percent).[10] Likewise, he found that almost half of the secession-related content of the Boston *Journal* consisted of exchange articles (164 of 343).[11] On the other hand, Fuhlhage found that the Macon *Telegraph* included twenty-nine exchanges of its eighty-six secession-related articles (34 percent) between December 13 and December 28, 1860.[12] Fuhlhage, Sarah Walker, Jade Metzger, and Nicholas Prephan found that the *Charleston Mercury*, owned by famous fire-eater Robert Barnwell Rhett, featured 113 exchange articles between December 13 and December 27, 1860, representing about 29 percent of the 391 secession-related articles reviewed.[13] They found that the New Orleans *Picayune* included exchange content in only 33 of 205 articles (16 percent).[14]

Members of the press during the time found value in what other papers and other members of the press had to say. John Bigelow, the editor of the *Evening Post*, later noted that the *Charleston Mercury* was such an influential paper that "it was rare for the *Evening Post* to go to press in those days without an extract, longer or shorter, from the editorial columns of the *Mercury*."[15] In fact, exchange articles from the *Mercury* were featured ten times in the thirteen dates analyzed.

Exchanging with the New York *Evening Post*

The *Evening Post's* exchange network in 1860 was vast, including papers from the North, South, border states, Washington, DC, as well as across the Atlantic.

The *Evening Post* published thirty-eight exchange articles from twenty Northern papers, including three papers from Illinois, one from Massachusetts, one from New Jersey, one from Ohio, ten from New York (including the *Associated Press*), and four newspapers from Pennsylvania. The *Evening Post* published fifty-one exchange articles from twenty-five Southern papers: four newspapers from Alabama, one from Florida, five from Georgia, two from Louisiana, two from Mississippi, two from North Carolina, two from Tennessee, two from Virginia, and five papers from South Carolina. A total of eight exchange articles from five different papers were published from border states, including three newspapers from Maryland and two from Missouri. The *Evening Post* also published twelve articles from five Washington, DC newspapers, five articles from the *London Times*, one from the Montreal *Gazette*, and one citing only "Canadian papers." Finally, one exchange article was only cited as from "a morning paper."[16]

To gather data for this chapter, thirteen editions of the New York *Evening Post* between the dates of December 13 and December 31, 1860, were assessed for secession-related content. These dates represent six papers printed before, one printed day-of, and six papers printed after South Carolina's secession declaration. Editions including December 16, 23, 25, 26, 28, and 30 were not published or not available for analysis and therefore excluded. A total of 299 secession-related articles of all types were recorded from the New York *Evening Post*. Exchange articles (98) were then filtered from the data set. Because the *Evening Post* often combined exchanges from more than one newspaper into one article, entries featuring more than one publication were separated into one unique entry to properly categorize each contributing paper. After this data cleaning, exchanges represented 117 of 310 entries.

Of the 117 exchange articles in the data, the majority (69 percent or 81 articles) represented some type of sentiment, anti- or pro-, regarding secession. Thirty-six articles were neither pro- nor anti-secession. Not surprisingly, in the Northern-based Republican *Evening Post*, most exchanges represented anti-secession sentiment, a total of fifty-six of the eighty-one sentiment-coded articles. While Northern papers represented half of the anti-secession sentiment, international, border-state, and the DC papers also contributed anti-secession sentiment. A total of nineteen anti-secession sentiment articles were published in exchanges from papers originating in the South, or were classified anti-secession because of strong commentary from the *Evening Post* accompanying the exchange.

In comparison to the anti-secession exchanges, a much smaller number, twenty-five, were coded as pro-secession. Of these pro-secession exchanges, as one might expect, exchanges originating from Southern papers represented the clear majority—80 percent, or a total of twenty.

Some name-calling occurred in the exchanges of the *Evening Post* between December 13 and 31, 1860, including positive and negative terms such as *traitors, patriot, Black Republican, fire-eater,* and *loyalist,* among others. In the 117 exchange articles analyzed, twenty-eight included some type of name-calling. Name-calling occurred in both the paper of origin and accompanying comments from the New York *Evening Post.* While both negative and positive name-calling occurred, negative name-calling was far more apparent. Pejorative name-calling occurred in twenty-five exchanges compared to only three positive name-calling instances.

Violence was also a topic of the exchanges in the *Evening Post.* Talk of violence occurred a dozen times during the analyzed date range.

Slavery was discussed eighteen times in the articles the *Evening Post* published through its exchange network. These articles were characterized as antislavery, proslavery, talking about slave insurgence, or Personal Liberty Bills. Ten of the eighteen were antislavery articles.

With the nation in distress, it is no wonder that many articles included speculation about the future. In fact, more than two-thirds, seventy-nine of the 117 exchange articles, included some sort of speculation. Twenty-five exchange articles from Northern newspapers engaged in speculation, including nine that discussed the protection of the Union or the Constitution, and six that discussed future military movement or the possibility of war. Thirty-three exchange articles from Southern newspapers talked about war or military action, calls to slow down secession, or saving or protecting the Union or Constitution.

The *Evening Post* also contained exchange articles that talked of military movements as well as political movements or political talk. Twenty-three of the sixty-nine exchange articles coded for political and military talks came from Northern newspapers. Seven of these articles vilified Buchanan and five discussed the forts in Charleston. The *Evening Post* published thirty-two articles referring to political or military movements from Southern papers, including thirteen articles relating to the forts in Charleston, four concerning economics, four vilifying Lincoln, and three relating to Lincoln's coming inauguration. General military action was discussed twice, and one article argued that secession or war was inevitable.

Commentary from the *Evening Post*

The New York *Evening Post* often added its own commentary to the exchange articles. Sarcasm, defined as quips or comments that could be perceived sarcastically, occurred in ten of the 117 exchange articles. Five were in pro-secession articles originating from Southern newspapers.

In seven articles, the *Evening Post* employed a communicative tactic in which it would refer to an exchange and make deliberate counterpoints, or statements of disagreement, to individual arguments within the exchange. Five occurred with exchanges originating from the South and two from the North.

The New York *Evening Post* commonly added commentary to exchange articles advocating pro-Union views, supporting Unionist actions, expressing antisecession sentiment, and regarding the South in general. In an article referencing a call to prevent Lincoln's inauguration by force, the *Evening Post* prefaces the exchange with the following statement: "We quote most of the article to show our readers for what diabolical work the mind of the South is being prepared by government agency."[17] Both pro-Union support and anti-secession commentary occurred eleven times.

The *Evening Post* often added short quips to set the scene or provide background context for an exchange article the reader was about to read. For example: "The Columbus (Georgia) *Sun* publishes the following letter from [General] Wise. It gives the reasons for the ex-governor's proposed fight 'in the Union.'"[18] This scene setting occurred nineteen times across papers from the North, South, Washington, DC, and border states.

This study provides a record of the extensive use of articles written by other newspapers in the *Evening Post* during a critical moment in the history of the Republic, and finds patterns in these articles, including secession sentiment, name-calling, violence, slavery, speculation, political talk and military movements, and finally, commentary by the *Evening Post*. It shows that readers in New York were receiving news and opinion from newspapers throughout the nation, including states that were in the midst of seceding from the Union. Thus, the words of even the Southern fire-eaters were reaching far beyond their local circulation. And indeed, many newspapers, though local in origin, were national in reach.

Notes

1 John Bigelow, Statement, November 15, 1860, "Vol. 76: Financial Statements, Accounts, Personal Statements, *New York Evening Post*," *John Bigelow Papers, 1839–1912*, Manuscripts and Archives Division, New York Public Library.

2 Gregg MacDonald, "William Cullen Bryant's 30-Year Crusade Against Slavery," in *Words at War: The Civil War and American Journalism*, eds. David B. Sachsman, S. Kittrell Rushing, and Roy Morris Jr. (West Lafayette, IN: Purdue University Press, 2008), 61.

3 John Bigelow, *Retrospections of an Active Life*, vol. I (New York: The Baker & Taylor Co., 1909), 319–24. During the time of analysis, Bigelow was editor and proprietor of the New York *Evening Post*, but retired from his position a mere fifteen days following the review period of December 13 through 31, 1860. Bigelow sold his interests to Parke Goodwin, son-in-law of William C. Bryant. His last day at the *Evening Post* was January 15, 1861.

4 *Evening Post* (New York), December 13–15, 1860; December 17–22, 1860; December 24, 1860; December 27, 1860; December 29, 1860; and December 31, 1860.

5 Michael Fuhlhage, Sarah Walker, Jade Metzger, Nicholas Prephan, "Charleston *Mercury*, 13–27 December 1860 Secession Coverage Database," unpublished data (2016); Michael Fuhlhage, "Missouri *Democrat* (St. Louis, MO), 13–22 December 1860 Secession Coverage Database," unpublished data (2017); Michael Fuhlhage, "Boston *Journal*, 13–22 December 1860 Secession Coverage Database," unpublished data (2017); Tabitha Cassidy, "Providence *Evening Press*, 13–22 December 1860 Secession Coverage Database," unpublished data (2017).

6 William C. Davis, *Rhett: The Turbulent Life and Times of a Fire-Eater* (Columbia: University of South Carolina Press, 2001).

7 "The Secession of South Carolina. Rejoicing in Southern States," *Evening Post* (New York), December 21, 1860, 3.

8 "An Invitation Declined," *Evening Post* (New York), December 17, 1860, 2.

9 *Evening Post* (New York), December 13–15, 1860; December 17–22, 1860; December 24, 1860; December 27, 1860; December 29, 1860; and December 31, 1860.

10 Fuhlhage, "Missouri *Democrat* (St. Louis, MO)," unpublished data (2017).

11 Fuhlhage, "Boston *Journal*," unpublished data (2017).

12 Fuhlhage, "Macon (GA) *Telegraph*, 13–27 December 1860 Secession Coverage Database," unpublished data (2016).

13 Fuhlhage, Walker, Metzger, and Prephan, "Charleston *Mercury*," unpublished data (2016).

14 Fuhlhage, Walker, Metzger, and Prephan, "New Orleans *Picayune*, 14–27 December 1860 Secession Coverage Database," unpublished data (2016).

15 John Bigelow, *Retrospections of an Active Life*, 304.

16 *Evening Post* (New York), December 13–15, 1860; December 17–22, 1860; December 24, 1860; December 27, 1860; December 29, 1860; and December 31, 1860.

17 "The President's Dogs Barking at General Scott," *Evening Post* (New York), December 31, 1860, 2.

18 "Why Governor Wise Intends to 'Fight in the Union,'" *Evening Post* (New York), December 15, 1860, 3.

22

WAR OF WORDS

Border State Editorials During the Secession Period

Melony Shemberger

Antebellum editors were not expected to be independent or objective. Rather, their role was to lead their readers through a "war of words" that helped to shape America. This "war of words" was intense in the border states of Kentucky, Delaware, Maryland, and Missouri during the secession crisis.

Border states shared ties to both the Union and the Confederacy in varying degrees. The commonwealth of Kentucky adopted a policy of neutrality until September 1861, when those favoring the Union gained control of the legislature. Nevertheless, a sizable pro-Confederate presence remained in parts of the state, and soldiers from Kentucky served in both the Union and Confederate armies. Furthermore, although Kentucky had adopted a policy of neutrality, Louisville was officially a Confederate city from April through mid-September 1861. Not long after Union forces occupied the city, *The Louisville Daily Journal*, in its October 31, 1861 issue, printed a front-page article, "Declaration of Southern Independence," which dominated an entire column.[1]

In the 1860s, Americans' interest in the looming Civil War, and its collateral effects, caused a spike in demand for news reports. Secession created days of heightened concern for the border states, and newspapers were one of the few resources that people relied upon for war news.[2]

Editorials explain the news, fill in background, forecast the future, and pass moral judgment.[3] Border state editorials published during the secession crisis and in the early months of the Civil War tell the story of the conflicts of the time. This chapter explores the positions of border states through the lens of various newspaper editorials from the time prior to and after the 1860 election through the secession crisis to the early months of war. Since Kentucky and Missouri were severely divided, with Kentucky remaining neutral, this study heavily involved newspaper editorials published in these two states. Insights gleaned from the newspaper editorials of Delaware and Maryland, both of which remained loyal to the North based on their substantial support to the war effort with men and material, are applied throughout this chapter.

Newspapers in the four border states did not fit neatly into the Northern or Southern press. In fact, a division of the press was more apparent within these states' borders, with editors mostly choosing sides based on their region and the impact on it from issues such as slavery.

Before the war, the *Lexington Observer & Reporter* published an editorial from the *Nashville Banner* that provided early indications of the coming conflict. The editorial criticized the Democratic Party in the South for agitating the issue of slavery by exciting "the prejudices of the Southern people against their Northern brethren," and for hoping "to lead them on step by step until they are eventually ready to go to any extreme to which it may suit the purposes of political tricksters to advise them."[4]

In Kentucky, the proponents of slavery came from the southern and western sections of the state, where the lifestyle most resembled that of the Deep South, especially in terms of crop distribution. Newspapers in west Kentucky supported the Confederacy, and editors' choice of words in editorials demonstrated the raw passion this region had for the South. Throughout 1861, most issues of the *Columbus Crescent, Hickman Courier,* and *Paducah Herald*—all in the Jackson Purchase area of west Kentucky—contained content praising the Confederacy and cursing the Union.[5] George Warren, editor of the *Courier,* dismissed any resolutions favoring the Union as "fustian, bombast and nonsense."[6] Len G. Faxon, a veteran Democratic politician and editor of the *Crescent,* also called for disunion. In an editorial just as the state legislature convened to consider the national crisis, he wrote, "Kentucky may as well prepare immediately to go with the Southern States into a separate Confederacy—the sooner she does so the better it will be for her."[7] John C. Noble, editor of the widely read *Paducah Herald,* praised Confederate President Jefferson Davis and Vice President Alexander H. Stephens as "two of the wisest, calmest, firmest and bravest men in the South."[8]

Despite this sentiment, Kentucky stayed loyal to the Union. Newspapers in the northern and eastern regions of the state urged a moderate Unionist course for all border states. But while most Kentuckians approved the legislature's Unionism, those in the far western part of the state did not.[9] The *Hickman Courier* wrote: "We know not what are the feelings of the citizens of the upper portion of Kentucky, but we can not believe their representatives are reflecting their feelings in the present Legislature."[10]

Between December 1860 and June 1861, eleven Southern states seceded, leaving the border states torn between loyalty to their sister slave states and the nation that had elected Abraham Lincoln as its president. Delaware rejected an invitation to join the Confederacy early in 1861, and through the war remained loyal to the North, mobilizing its industries to provide supplies for the Union Army. Despite some Southern sentiments, the state never seriously threatened to leave the Union. The Wilmington *Delaware State Journal and Statesman* wrote that the election was a "well-contested, perfectly constitutional and fairly fought battle, in which all kinds of combination were formed by the common enemy to defeat the people." Nevertheless, the editorial, with the headline "Stand Firm," advocated Unionism for Delaware.[11]

In Maryland, newspaper editorials generally reflected Union sentiments. An editorial in the April 4, 1861, issue of the *Port Tobacco Times and Charles County Advertiser*, just days before the start of the war, recounted resolutions adopted by a mass convention of Frederick County declaring "undying attachment to the Union":[12]

> The Convention was of opinion that no cause has yet arisen sufficient to justify a dissolution of the Union, and while they would preserve the country and the government from destruction at the hands of the Secessionist, they would with equal force and determination oppose the aggressions of the fanatic of the North. They further believe that under existing circumstances secession is not the proper remedy for our complaints; but that a strong fight, a firm fight and a fight together—at the ballot-box—*in* the Union—is more likely to prove effectual than all the efforts which can be made *out* of the Union.... Now, so far as standing firm by the Union is concerned, we certainly endorse the idea....[13]

Newspaper editorials often reflected the identities and perspectives of their communities. C. P. Anderson, editor of the *Weekly California News* in California, Missouri, penned an editorial, "Misrepresentations of the People of Our Town," that explained the character of the community. The editorial painted the town as a place where visitors from free states "will find a kind and hospitable reception from our people..." as long as they are not "negro-worshipers, negro thieves, horse thieves, nor persons who are disposed to tamper with or place themselves on an equality with our negro population."[14]

In the days before the war started, newspapers discussed whether the Lincoln administration would opt for peace or war. The *Weekly California News*, on April 13, 1861, featured an editorial titled "Peace or War" from the *St. Louis Herald*. In it, a climate of uncertainty is painted, giving the reader a pessimistic dose of realism:

> We confess that things have a greatly less pacific look than they had. Coercion is perhaps not *determined on*. But there are signs that the administration are strongly *inclined* to use it. The preparations for war are unmistakable. The clouds are evidently gathering. Whether they are to discharge themselves in fire and blood, or will finally disperse without such a catastrophe, remains to be seen. Meanwhile all is suspense, uncertainty, anxiety. Nothing is certain. Emphatically—in a peculiar and most extraordinary sense may it be said now of our painful condition—no man knoweth what a day may bring forth.[15]

Throughout 1861, many Southerners in Missouri and Kentucky wanted to stay in the Union but out of the war.[16] This aspiration echoed through various newspaper editorials. *The Randolph Citizen* in Huntsville, Missouri, straddled between support of the North and the South. However, between May and August 1861, the newspaper stopped publishing after the area came under siege from the Union. Even when the paper resumed on August 29, 1861, the editor and publisher, Francis M. Taylor, wrote—rather solemnly yet optimistically—that the paper would not participate in the war's discourse:

> In regards to the position we deem best to be taken in regard to our troubles, we have no other policy to recommend than one of peace. If everyone would endeavor to cultivate friendly feelings towards his neighbor, and willingly concede to him the same liberty of opinion and expression that he himself claims, it would greatly mitigate the misfortunes by which we are surrounded. ...
>
> We intend, temporarily, to lay the pen "on the shelf." Results are now reached more by the sword than reason. The latter term is becoming somewhat obsolete, and rapidly losing ground. But we look anxiously forward to a brighter day, when peaceful discussion and the quiet ballot, rather than the bullet and reckless passion, shall decide the destinies of the people.[17]

As the war escalated, newspaper editorials sometimes pleaded for the war to end. In one of its editorials, *The Daily Commonwealth* of Kentucky reflected on a series of events that, it argued, prolonged the war, beginning with how the April 1860 Democratic Convention, the largest assembled at that time, could have lessened the need for war had Stephen Douglas won the nomination. "No one doubts that a compromise in that convention would have saved the Union," according to the editorial, which also argued that the war could have ended on other occasions within a year.[18] *The Louisville Daily Democrat*, a Union supporter, preferred the days before the war when the Union was undivided, writing that Kentuckians should side with "the Union unconditionally" and "stand by the Union with all its risks":[19]

> Here, in this Commonwealth, we are specially opposed to division. We shall not abide enemies for centuries to come North or South of us. We don't intend to have the soil of Kentucky as a battlefield, where rival nations shall fight for empire.[20]

The Louisville Daily Express took its coverage a step further by publishing statistics on the costs of the war. This paper also documented an estimate of the value of the various kinds of property, such as cotton, bridges, railroads, sugar, buildings and more, all of "which has been ruthlessly destroyed by the rebels since the commencement of the present war."[21] The paper also printed an editorial titled

"War and Its Results": "War cannot be carried on without expense. Every bullet that is projected, every cap that is exploded, and every grain of powder ignited, costs something."[22]

The bold voice of newspaper editorials during this period shaped the "war of words." Many editors indulged in "fine writing." Editors appeared to vie with one another to see who could use the most flowery language in presenting content to their readers.[23] Simple, direct language was not sufficient. Some editors were fierce in their publications. Len G. Faxon of the *Columbus Crescent*, for instance, called the Yankees: "bow-legged, wooden shoed, sour craut stinking, Bologna sausage eating, hen roost robbing Dutch sons of ----."[24]

Throughout American history, newspaper editorials at times have been known to ignite reactions from readers. Newspaper reports related to the country's political tensions, and the war became the primary material readers wanted. Border state editors were aware of the need their papers filled at this critical time, and they wrote passionately and persuasively, either in support of the Union or the Confederacy. Their editorials expressed views that both reflected and helped shape public opinion, providing context for their readers and perhaps even educating them.

In their "war of words," newspaper editorials of the secession period sought to make sense of the times, offering their readers a framed perspective. Through their words, readers received a greater comprehension and appreciation of the hardships plaguing the border states during this great crisis.

Notes

1 *Louisville* (KY) *Daily Journal*, October 31, 1861, 1.
2 R. J. Brown, "How the South Gathered News During the Civil War," *HistoryBuff.com*, accessed June 2, 2014.
3 William Pinkerton, in W. L. Rivers, B. McIntyre, and A. Work, *Writing Opinion: Editorials* (Ames, IA: Iowa State University, 1988), 25.
4 *Lexington Observer & Reporter*, February 22, 1860.
5 Berry F. Craig, "Kentucky's Rebel Press: The Jackson Purchase Newspapers in 1861," *Register of the Kentucky Historical Society* (1977): 20–27.
6 Ibid.
7 Excerpt from the *Hickman Courier* in the *Louisville Courier*, January 17, 1861.
8 *Paducah Herald*, February 22, 1861, in Craig, "Kentucky's Rebel Press," 27.
9 Craig, "Kentucky's Rebel Press," 22.
10 Excerpt from the *Hickman Courier* in the *Louisville Courier*, February 6, 1861.
11 *Delaware State Journal and Statesman* (Wilmington, DE), January 1, 1861.
12 *Port Tobacco* (MD) *Times and Charles County Advertiser*, April 4, 1861, 2.
13 Ibid.
14 *Weekly California News*, April 13, 1861, 2.

Editor's Note: This is a work of history, a major focus of which was the racial hatred that existed in the United States. This racial hatred was openly reflected in the language of the newspapers of the time. That language, when provided in direct quotations, is generally retained in this book in order to accurately reflect the historical context.

15 *Weekly California News,* April 13, 1861.

16 Craig, "Kentucky's Rebel Press," 26.

17 *The Randolph Citizen* (Huntsville, MO), August 29, 1861, 1.

18 *Daily Commonwealth,* August 19, 1862, 3.

19 *Louisville Daily Democrat,* April 9, 1862, p. 1.

20 Ibid.

21 *Louisville Daily Express,* May 19, 1862, 2.

22 Ibid.

23 Herndon J. Evans, *The Newspaper Press in Kentucky* (Lexington, KY: The University of Kentucky Press, 1976), 42.

24 Craig, *Hidden History of Kentucky in the Civil War* (Charleston, SC: The History Press, 2010), 14.

Editor's Note: This is a work of history, a major focus of which was the racial hatred that existed in the United States. This racial hatred was openly reflected in the language of the newspapers of the time. That language, when provided in direct quotations, is generally retained in this book in order to accurately reflect the historical context.

ABOUT THE EDITORS

David B. Sachsman holds the George R. West, Jr. Chair of Excellence in Communication and Public Affairs at the rank of professor. He came to the University of Tennessee at Chattanooga in August 1991 from California State University, Fullerton, where he had served as dean and professor of the School of Communications. Previously, he was chair of the Department of Journalism and Mass Media at Rutgers University.

Dr. Sachsman is the director of the annual Symposium on the 19th Century Press, the Civil War, and Free Expression, which he and S. Kittrell Rushing founded in 1993. Dr. Sachsman is an editor of *The Civil War and the Press*, a book of readings drawn from the first five conferences, published by Transaction Publishers in 2000, and he is an editor of a three-book series based on conference papers published by Purdue University Press. The first book in the series, *Memory and Myth: The Civil War in Fiction and Film*, which was published in 2007, was nominated for the Lincoln Prize and exhibited at the Frankfurt Book Fair. The second book in the series, *Words at War: The Civil War and American Journalism*, was published in 2008, and the third book, *Seeking a Voice: Images of Race and Gender in the 19th Century Press*, was published in 2009. Transaction published another book in 2013, *Sensationalism: Murder, Mayhem, Mudslinging, Scandals, and Disasters in 19th-Century Reporting*, and in 2014, it published *A Press Divided: Newspaper Coverage of the Civil War*. In 2017, Transaction (now Routledge/Taylor & Francis) published *After the War: The Press in a Changing America, 1865–1900*.

Dr. Sachsman also is known for his research and scholarly activities in environmental communication and environmental risk reporting and for the three editions of *Media: An Introductory Analysis of American Mass Communications* (which he wrote with Peter M. Sandman and David M. Rubin, and for which he wrote the history chapter). A journalist by trade, Dr. Sachsman also has written about the suburban press. In 2005, Dr. Sachsman headed the team appointed to evaluate the US Agency for International Development's environmental education and communication efforts in more than thirty countries across twelve years. From 2000 to 2010, Dr. Sachsman was engaged in a research project on environmental reporters with Dr. JoAnn Valenti, now retired from Brigham Young University, and Dr. James Simon of Fairfield University. This project resulted in the publication of five refereed journal articles and a book, *Environment Reporters in the 21st Century*, which was published by Transaction in 2010. Dr. Sachsman and Dr. Valenti are currently editing the forthcoming *Routledge Handbook of Environmental Journalism*.

Dr. Sachsman received his BA in English from the University of Pennsylvania and his MA and PhD in communication from Stanford University. He has been teaching journalism to students and professionals since 1969. In 1998, he received the *Yale Daily News*'s Braestrup Fellowship and gave two presentations on journalism ethics at Yale University. In 2003, he delivered the Medart Lecture (on "Mass Media and War") at Maryville University in St. Louis. Dr. Sachsman served as a Senior Fulbright-Hays Scholar in 1978–1979 in Nigeria, where he helped plan for the development of one of the first mass communication graduate degrees in West Africa.

The Hazel Dicken-Garcia Award for Distinguished Scholarship in Journalism History was awarded to Dr. Sachsman at the twenty-fifth annual Symposium on the 19th Century Press, the Civil War, and Free Expression in 2017. Dr. Dicken-Garcia headed the selection committee.

Gregory A. Borchard is a professor in the Hank Greenspun School of Journalism and Media Studies at the University of Nevada, Las Vegas. He has written numerous books, book chapters, and articles on journalism history, including his 2019 Routledge book, *A Narrative History of the American Press*, which provides an overview of the events, institutions, and people who have shaped the press. Together with David W. Bulla, he is the author of *Lincoln Mediated: The President and the Press through Nineteenth-Century Media*, which was published by Transaction (now Routledge) in 2015, and *Journalism in the Civil War Era*, which was published by Peter Lang in 2010. He is also the author of *Abraham Lincoln and Horace Greeley*, published by Southern Illinois University Press, now in its second edition. He is the editor of *Journalism History*, a quarterly journal published by the History Division of the Association for Education in Journalism and Mass Communication.

Borchard has attended the Symposium on the 19th Century Press, the Civil War, and Free Expression at the University of Tennessee at Chattanooga since 1999. His Symposium paper on "Publishing Violence as Art and News: Sensational Prints and Pictures in the 19th-Century Press" was published as a chapter in *Sensationalism: Murder, Mayhem, Mudslinging, Scandals, and Disasters in 19th-Century Reporting*. His paper on "Taking No Right for Granted: The Southern Press and the 15th Amendment" was published in *Words at War: The Civil War and American Journalism*. In 2012, he received the Symposium's Hazel Dicken-Garcia Award for Distinguished Scholarship in Journalism History. Two of his papers received the Symposium's Award of Excellence. He earned a PhD in mass communication from the University of Florida in 2003, and a master's and bachelor's from the University of Minnesota.

CONTRIBUTORS

Brie Swenson Arnold is an associate professor of history at Coe College in Cedar Rapids, Iowa. She has an MA and a PhD in history from the University of Minnesota. Her research interests center on nineteenth-century race, gender, and political culture. Her recent publications include "'To Inflame the Mind of the North': Slavery Politics and the Sexualized Violence of Bleeding Kansas" (*Kansas History: A Journal of the Central Plains*, 2015) and "An Opportunity to Challenge the 'Color Line': Gender, Race, Ethnicity, and Women's Labor Activism in Late Nineteenth-Century Cedar Rapids, Iowa" (*Annals of Iowa*, 2015).

Dianne M. Bragg is an associate professor in the Journalism and Creative Media Department at The University of Alabama. She has a PhD in mass communication, with a concentration in history, and an MA in journalism. She holds leadership positions with several academic organizations, including serving as the 2017–2018 president of the American Journalism Historians Association. Her research interests lie in Antebellum newspapers, the politics of slavery, and late nineteenth- and early twentieth-century women journalists.

David W. Bulla is an associate professor of communication at Augusta University. Bulla focuses his research on the history of press performance in times of conflict. His first book, *Lincoln's Censor*, was published by Purdue University Press in 2008. Bulla and Gregory A. Borchard have published two books together, *Journalism in the Civil War Era*, in 2010, and *Lincoln Mediated: The President and the Press through Nineteenth Century Media*, in 2015. Bulla earned a PhD in mass communication from the University of Florida in 2004 and an MA in journalism from Indiana University in 2001.

Mary M. Cronin is a professor in the Department of Journalism and Media Studies at New Mexico State University. Her research interests center on nineteenth- and early twentieth-century press performance and press issues during the US Civil War era. She is the author (or co-author) of three books and numerous scholarly articles. Prior to her academic career, Cronin worked as a reporter, copy editor, and assistant news editor at newspapers in Massachusetts, New Jersey, and Florida.

Nancy McKenzie Dupont is a professor of journalism at the University of Mississippi School of Journalism and New Media. Though she teaches broadcast journalism, her research has focused on the nineteenth-century press of Mississippi and Louisiana. She is a co-author of *Journalism in the Fallen Confederacy* and has written numerous book chapters and journalism articles about the Antebellum press of Mississippi and New Orleans. She is a frequent contributor to the Symposium on the 19th Century Press, the Civil War, and Free Expression. She holds a PhD from the University of Southern Mississippi.

Brian Gabrial is the Wise Endowed Chair in Journalism at Northwestern State University in Louisiana and emeritus professor in the Department of Journalism at Concordia University in Montréal. His current research explores intersections between national identity and the press at critical junctures in nineteenth-century North America. He received his MA and PhD from the University of Minnesota's School of Journalism and Mass Communication. His book *The Press and Slavery in America* is available from the University of South Carolina Press.

William E. Huntzicker taught journalism and media history at Minnesota and Wisconsin universities for thirty-five years. He has written *The Popular Press, 1833–1865* (1999) about Antebellum and Civil War newspapers and *Dinkytown: Four Blocks of History* (2016) about a small Minneapolis business district, and numerous academic articles and book chapters, mostly on nineteenth-century journalism history. Bill is proud of his involvement in the Symposium on 19th Century Press, the Civil War, and Free Expression and its publications. He grew up on ranches in southeastern Montana, studied history at Montana State, earned a PhD in American studies from the University of Minnesota, and worked as a reporter in Minneapolis and Miles City, Montana.

Phillip Lingle is a master's student in the Public History program at New Mexico State University. He is also pursuing graduate certificates in Museum Studies and Cultural Resources Management. He holds a BA in history from South Dakota State University. His area of interest is conflict in American history and its portrayal in both conventional and unconventional media. He has written on sensationalism in the Antebellum press on the eve of secession, as well as on Native American use of ledger art to depict clashes on the Great Plains in the nineteenth century.

Dea Lisica received her MA in English, with an emphasis in rhetoric and writing, from the University of Tennessee at Chattanooga. She serves as assistant editor for the West Chair of Excellence at UTC, assistant director of the Symposium on the 19th Century Press, the Civil War, and Free Expression, and the curator of the Symposium Collection. She has received a "with" on this book, as well as for *After the War: The Press in a Changing America, 1865–1900* (edited by David B. Sachsman), for her role in the editing and publishing process.

Nicole C. Livengood is an associate professor of English and director of the First Year Seminar at Marietta College, a small liberal arts college in southeastern Ohio. Her research focuses on Antebellum-era periodicals; nineteenth-century American women writers; and 1840s anti-abortion rhetoric in texts varying from trial pamphlets to medical journals.

Milad Minooie earned his PhD in mass communication from the University of North Carolina at Chapel Hill and his MA in communication from the University of Texas at Arlington. He is the co-author of *Agendamelding: Newspapers, Television, Twitter, and Civic Community*, a book on how audiences process media messages. Dr. Minooie has spent years as a journalist in Tehran. He is interested in the relationships among government, media systems, and press freedom.

Abigail G. Mullen earned her PhD in history from Northeastern University in 2017. Her dissertation focuses on international relations in the Fist Barbary War (1801–1805). While at Northeastern University, she worked on the Viral Texts project, a digital project that investigates the spread of reprinted texts throughout networks of print. She currently works at the Roy Rosenzweig Center for History and New Media at George Mason University, where she manages Tropy software to help researchers organize and describe their research photos.

Erika Pribanic-Smith is an associate professor of journalism at the University of Texas at Arlington. A past president of the American Journalism Historians Association, she specializes in research examining political communication in the print media of the nineteenth- and twentieth-century American South. Specifically, she focuses on the use of editorials and letters to the editor to disseminate political ideology. In addition to various book chapters, Pribanic-Smith has published her work in *American Journalism*, *American Periodicals*, *Journalism History*, *Kansas History*, and *Media History Monographs*.

Katrina J. Quinn, PhD, is Communication Department chair and professor at Slippery Rock University in Pennsylvania, where she teaches writing and strategic communication courses. Dr. Quinn's methodology is interdisciplinary, bringing critical literary perspectives to journalistic texts. She has published a number of articles and chapters on nineteenth-century journalism, including works on epistolary journalism, sensationalism, and first-person travel writing from the American frontier. Among her projects are chapters in *Sensationalism: Murder, Mayhem, Mudslinging, Scandals, and Disasters in 19th-Century Reporting* (2013) and *After the War: The Press in a Changing America, 1865–1900* (2017), both edited by David Sachsman.

James Scythes earned a BA in history at Rowan University and has a MA in history from Villanova University. He is an assistant professor of history at West Chester University of Pennsylvania. His research interests focus on Antebellum America, the American Civil War, and nineteenth-century military history. His first book, *"This Will Make a Man of Me": The Life and Letters of a Teenage Officer in the Civil War*, was published by Lehigh University Press in 2016.

Donald L. Shaw is Kenan Professor Emeritus in the School of Media and Journalism of the University of North Carolina at Chapel Hill. Along with Maxwell McCombs, he formulated the agenda-setting theory of the media, launched with their 1972 *Public Opinion Quarterly* article. His work as a historian in newspaper content analysis has been pivotal in establishing newspapers as historical primary sources. He is a retired US Army colonel and former director of the North Carolina Selective Service.

Melony Shemberger, EdD, is associate professor of journalism and mass communication at Murray State University in Murray, Kentucky. Her primary research interests are journalism history, sunshine laws, pedagogy, and public relations topics. Shemberger was an award-winning news reporter, specializing in the education and court beats. Shemberger has a BA from Western Kentucky University, with a double major in mass communication, and history and government; an MA in mass communication from Murray State; an MA in management from Austin Peay State University, Clarksville, Tennessee; and an EdD in administration and supervision, with a concentration in higher education, from Tennessee State University.

Thomas C. Terry is a professor in the Department of Journalism and Communication at Utah State University, Logan. He earned his MA and PhD at the University of North Carolina at Chapel Hill. He is a past president of the Illinois Press Association, launched a Shakespearean Festival, founded two newspapers, and owned a group of community newspapers in western Illinois. His research is in the African American press, eighteenth- and nineteenth-century media history, and agenda setting.

Erika Thrubis earned her MA in communication studies from Wayne State University and BA in advertising from Michigan State University. She currently works full-time in health care and resides in the Detroit-metro area with her husband and two dogs. Her chapter was originally presented at the Symposium on the 19th Century Press, the Civil War, and Free Expression in 2017.

Debra Reddin van Tuyll is a professor of communication at Augusta University. She is the author and/or editor of six books, including *The Confederate Press in the Crucible of the American Civil War*. She is also one of the two founders of the annual conference on transnational journalism history. Her interest there is on the early Irish-American press. But the Civil War press will always be her first love.

INDEX

Note: Page numbers in *italics* refer to figures.

36°30' parallel 4, 27, 77

Abolition of the Slave Trade Act 126, 131
abolitionism: and colonization 131; growth of 39, 46, 78, 214; and John Brown 228; and Kansas-Nebraska Act 114–122; and Nat Turner's revolt 43–44; press and 5, 42, 46, 61–67, 71, 105–106, 180–181, 213, 245; and Republican Party 27, 28, 219
ACS (American Colonization Society) 131
Adams, J. 3, 52
Adams, J. F. 146
Adams, W. 194
Adams Sentinel (Gettysburg, PA) 130
Africa 44, 45, 125–127, 130–134, 258
African Americans 3, 94, 131
agenda setting theory 16–21
agrarianism 2
Alabama: Nullification Crisis 56–57; number of slaves 3; presidential convention delegation 203, 229; secession 7
Alexandria Gazette 32, 193
Alien and Sedition Acts 3, 52, 58n27
Allen, B. F. 26
Amelia Island 128
American Colonization Society *see* ACS
American Flag 236
American Party 121, 157, 170

"American System" 4, 50, 56
Anderson, C. P. 253
Anderson, R. 24
Angelica (ship) 133
Anthony, D. R. 74
Anthony, S. B. 74
anti-secessionist press 234–242
AP (Associated Press) 5
"Appomattox Syndrome" 15
Argus (Albany, NY) 43
Arizona 73
Arkansas 14, 203, 213, 229
Associated Press *see* AP
Athenian (Athens, GA) 55
Athens (TN) *Post* 73, 106, 108, 194
Atlas (Albany, NY) 103
Attucks, C. 94, 95
Auburn American 172
Augusta Chronicle 116
Augusta Chronicle and Sentinel 29, 33
Austin 83, 84, 85, 87

backbone 155–162
Bailey, G.: abolitionism 61, 62–63, 67, 93; on Compromise of 1850 64; on *Dred Scott v Sandford* 66; on Kansas-Nebraska Act 65; and Lecompton Constitution 65
Bake-pan for the Dough-faces, A (pamphlet) 160
Baker, J. H. 164n32

Baltimore: building of slave ships 127, 132;
Democratic Convention 102, 206–208,
229; military volunteers 230
Baltimore Daily Exchange 28
Baltimore Republican 130
Baltimore Sun 125, 134, 182, 213
Bangor Daily Whig & Courier 118, 119, 150
Bangor Democrat 71
Barksdale, E. 223, 237
Bartlett, D. W. 111n4
Battle of Horseshoe Bend 42–43
Beard, C. 15
Beard, M. 15
Beecher, H. W. *145*, 156
Bell, J. 17, 30, 81, 185, *186*, *208*, 209
Bell, P. H. 88
Benedict, G. W. 73
Bennett, J. G.: on abolitionism 115; on
Brooks and Sumner 146; cartoon *173*;
on Democratic conventions 202, 207;
on Lincoln-Douglas debates 192; on
Republican convention 204, 206;
sensationalism 182
Benton, J. 158, 159, 168
Benton, T. H. 81, 158, 168
Bermuda 131
Bigelow, J. 245, 246
Bill of Rights 2
Bingham, K. S. 169
Binmore, H. 190
Birch, J. H. 71, 72
Birch, W. F. 72
Black Republicanism 119, 149, 158, 193,
214, 229, 237
"Bleeding Kansas" 7, 27, 114, *173*
Blow, T. 185
Borchard, G. 1–12, 167–176
Boston 40, 115, 116, 128, 130, 142
Boston Daily Advertiser 120–121
Boston Daily Atlas 116, 142–143, 144–145,
150
Boston Daily Commonwealth 91–98
Boston Massacre 93–94
Boston Patriot 128, 128n36, 128n37, 129n38
Boston Pilot 172
Boston Post 103
Boston Tea Party 92, 93
Boston Transcript 103
Boyce, W. 31
Bragg, D. M. 9, 10, 114–122, 141–152
Brazil 125, 128
Breckinridge, J.: cartoons 185, *186*,
208; *Charleston Mercury* and 229;

Lincoln-Douglas debates 193;
presidential election 30, 31, 208, 209;
Southern Democratic Party 17, 28, 30,
208, 229; vice president 175
Britain: abolition of slavery 61; Abolition
of the Slave Trade Act 126; coverage
of Lincoln-Douglas debates 195–196;
Dana on 93; House of Commons report
135; Royal Navy 126, 128, 129; Slavery
Abolition Act 126; slave-trading 126;
West Africa Squadron 126
Brodnax, W. 45
Brooklyn Daily Eagle 157, 172
Brooks, P. *10*, 141, 144–149, 152, 212, 214,
216–217
Brown, G. W. 71
Brown, Jesse 19
Brown, John 28–29, 98, 119–122, 201, 212,
217–221, 227–228
Brown, W. W., *Experience, or How to Give a
Northern Man a Backbone* (play) 156–157
Brownlow, W. 17, 71, 174
Brownlow's Knoxville Whig 71, 174
Brownson, O. A. 172
Brownson's Quarterly Review 172
Bryant, W. C. 180, 245
Buchanan, J.: cartoon 185, *186*; on Civil
War 14; as doughface 155, 157–160; and
Douglas 192–193, 196; and Lecompton
Constitution 225–226; presidential
election 7, 102, 141, 148–152, 171, 174,
175; Republican smears 172; on slave
ships 132; support of Pierce and Douglas
170; on Supreme Court 179–180
Budget (Troy, NY) 103
Bulla, D. W. 8–9, 11, 61–67, 188–197
Burlington Free Press 73, 108
Burns, A. 92, 95–98, 142
Butler, A. 142–143, 144, *145*, 147, 215–216

Calhoun, J. C. 19, 24, 25, 45, 50–51, *80*, 83,
224–255
California: and Compromise of 1850 26,
64; Democratic Convention 202; as free
state 25, 26, 64, 79, 81, 170; Frémont
and 168, 170, 171; gold rush 25, 81;
presidential election 174
Camden and Lancaster Beacon 52, 53
Canfield, A. W. 83
Cape Verde Islands 128
capital, relocation of 3
capital punishment 134
Carey, J. 6

Caribbean 125, 126, 128
Carolina Times (Durham, NC) 122
Cass, L. *80*, 102, 107, 143–144, *149*, 155
Catholicism 167, 170–174, 175, *186*
Caudill, E. 17
Cecil Whig (Elkton, MD) 104
Chamberlain, R. 73
Charleston (NC) 131–132, 146, 189, 191, 245, 248; Democratic Convention 30, 201–204, 206–208, 228–229, 254
Charleston Courier 31, 33, 51, 52–53, 56
Charleston Mercury 223–231; closure of 231; on the Constitution 33, 56; and Democratic Convention 201, 228–229; on Douglas 193, 229; exchange network 231, 245; on John Brown 29, 120–122, 217–218, 227–228; Lecompton Constitution 225–226; and Lincoln 229–230; Ordinance of Secession 245; on Pierce 106; presidential election 150, 151, 152, 229–231; as proslavery 5, 182; on secession 31, 53, 227, 246; on Sumner and Brooks 143, 146, 149; on Turner 43
Chattanooga Rebel 26
Cherokees 55
Chicago 109, 196; Republican Convention 204
Chicago Press and Tribune 188, 189, 190, 191–192, 218–219
Chicago Times 116, 188, 190–191, 196–197
Chicago Tribune 65, 109, 131–133, 194, 215
Christian Inquirer 106
Church, J. F. 230
Cincinnati 62, 107; Democratic Convention 148
Cincinnati Commercial 30, 194, 202
Cincinnati Daily Press 75n12
Civilian (TX) 87
Claremont Eagle 105
Clarksville Chronicle 194
Clay, C. 71
Clay, H. 4, 56, 64, *80*, 81, 85, 131
Clotilde (ship) 137n78
coastal slave trade 130–131
Cohen, B. 16
Cole, A. N. 169
colonial newspapers 17–18
colonization movement 131
Columbus Crescent 252, 255
Compiler (Richmond, VA) 40, 44
Compromise of 1790 3

Compromise of 1850 6; Democratic Party and 102; Pierce and 168–169; political divisions and 25; press coverage of 26, 64; Texan press coverage 77–88
Compromise Tariff 19, 49
Confederacy: border states and 251–252; establishment of 14; press and 242; South Carolina and 29, 56–57, 231
Connecticut, slavery abolition 126
Connecticut Courant 129
Conservative (Leavenworth, KS) 74
Constitution: economic system 66; and Fugitive Slave Law 115; Lincoln-Douglas debates and 193, 194; Minkins and 92; Nullification Crisis and 52, 53, 55, 56; Pierce and 103; protection of slavery 61, 62–63, 179, 183; Republican Party and 227, 237; Three-Fifths Compromise 1
Constitutional Convention 23, 126
Constitutional Union Party 17, 30, 185, 202, 209
Constitutional Whig (Richmond, VA) 40
Converse, J. O. 70
cooperationists 223, 230, 238–239
Cordova, P. 85
cotton 127, 130
county elections 17
Courier (Natchez, MS) 234–235, 240–242
Craft, W. and E. 92
Craven, A. 14–15
Crawford, M. J. 29
Cronican, M. 83
Cronin, M. M. 9, 77–88
Cross, E. E. 74
Cuba 127, 128, 131, 134
Curtis, G. T. 93
Cushing, C. 155, 166n64, 208
Cushney, W. H. 84–85
Cutler, C. 127

Dahomey, King of 127
Daily Alabama Journal (Montgomery) 115
Daily Cleveland Herald 117, 151–152
Daily Colonist (Victoria, BC) 196
Daily Commonwealth (Frankfort, KY) 254
Daily Delta (New Orleans) 182–183
Daily Evening Transcript (Boston) 127
Daily Morning News (Savannah) 117, 141–142, 144
Daily National Intelligencer (Washington) 28, *118*, *120*, *142*, 213
Daily News (Nashville) 72

Daily Picayune (New Orleans) 121
Daily Scioto Gazette (Chillicothe, OH) 118, 145
Daily Sentinel (New York) 46
Daily Southern Guardian (Columbia, SC) 33
Dallas Herald 195
Dana, R. H. 93
Davis, C. 93
Davis, J. 148, 242, 252
Davis, M. K. 40
Dayton Daily Empire 71
Dayton, W. L. 169
Deal, B. F. 194
Declaration of Independence 1, 2, 9, 181
Delaware 229, 251, 252–253
Delaware State Journal and Statesman (Wilmington) 253
Democrat (St Louis) 182, 245, 246
Democratic Banner (Bowling Green, MO) 71, 75n12
Democratic Party: and abolitionism 114–115; Baltimore Convention 102–105, 206–208; border states 252; Calhoun and 58n12; cartoons *10, 149, 161*; Charleston convention 30, 201–204, 206–208, 228–229, 254; Cincinnati Convention 148; divisions in 17, 25, 27–28, 30, 102, 141, 201, 208–209, 225–226, 229, 231; doughfaces 156; Douglass and 62, 66; Era of Good Feelings 4; Henderson and 75n12; and John Brown 219; midterm elections 107–110; Northern Party 82, 111, 155, 206; press and 71–72, 83, 106, 182, 190–192, 206; radical Copperhead faction 71; Southern Party 43, 103, 108, 157, 185, *186*, 192–194, 201–202, 208, 219, 231; Texas 78, 83
Democratic Telegraph and Texas Register (Houston) 84, 85, 86, 87
DeMorse, C. 84, 85, 87
District of Columbia, relocation of capital to 3
Dobbin, J. C. 102
Dooley, P. 70
doughfaces 105, 155–162
Douglas, S.: cartoons *149, 186; Charleston Mercury* on 227, 229; Compromise of 1850 64, 81; as doughface 155; on *Dred Scott v Sandford* 181; Freeport Doctrine 228; and Kansas-Nebraska Acts 27, 65, 106–107, 170; and Lecompton Constitution 225, 226; and Lincoln 181,

188–197; and popular sovereignty 179; presidential nomination 28, 30, 202, 207, 209; Sumner and 142–144, 216
Douglass, F. 5, 20, 61–62, 64–67, 191
Douglass' Monthly 61
Doyle, J. 119–120
Dred Scott v. Sandford 7, 66–67, 175, 179–186, 190–191, 194, 208
duels 73–74, 147, 216
Dupont, N. M. 12, 234–242

Eagle (Maysville, KY) 213
Echo (ship) 131–132
editorials 50
editors: changes in 70; multiple professions 73; physical pressures 72–73; political pressures 70–72; social pressures 73–74; violence towards 71–72, 118
Edmundson, H. 216
Egerton, D. 203
election methods 236
elections *see* county elections; midterm elections; presidential elections
South Carolina election 51
Eliza Jane (ship) 133
Elmore, F. 224
Emancipation Proclamation 20
Enquirer (Richmond, VA) 42, 44, 45, 228, 246
Era of Good Feelings 4
Erie (ship) 134
Evening Journal (Albany, NY) 183
Evening Post (New York) 245–249
Evening Star (Washington) 215
Ewing, W. H. 85
exchange network 50–51, 245, 246–248
executions 29, 40, 42, 134, 219
Express (New York) 228
extremism in press 212–221

Faneuil Hall 93, 94, 95, 96
Farkas, A. B. 3
Farmer's Cabinet (NH) 119, 147n30
Faxon, L. G. 252, 255
Fayetteville Observer 28, 55, 109, 152n46
Federal Union (Milledgeville, GA) 216
"Federalist, The" 2
Federalist Party 3, 42–44, 52, 58n27
female slaves 128
Fenno, J. 3
Fillmore, M. 77–78, 85–86, 92–95, 121, 148, 151, 170
fire-eaters *see* secessionists
Flag of the Union (Jackson, MS) 27

Fleming, D. G. 122
Florida: Democratic Convention 203, 229; Fugitive Slave Act 26; number of slaves 3; secession 7, *240*; slave trade 128–131
Floyd, J. 43–45
Follett, Foster, and Company 196
Ford, J. S. 83
Fort Sumter 14, 24
Forten, J. 46
Fox, C. J. 126
Frank Leslie's Illustrated Newspaper 183–185
Frantz, A. J. 235–236
Frederick Douglass' Paper 61, 62
Free Soil Party: *Boston Daily Commonwealth* and 91; cartoon *149*; Convention 142; *Evening Post* and 245; Frémont and 7; and Kansas-Nebraska Act 115; Maine 77; Missouri Compromise 116; and Nebraska Bill 110; Pierce and 106, 107; political division and 79, 81, 82, 83–85; Republican Party and 27
Free Trade 51–53, 57
Free Trader (Natchez, MS) 235, 236
freedom of speech 2, 67
Freeman and Messenger 129–130
Freeman's Journal (New York) 173
Freemen's Champion (Prairie City, KS) 73
Freeport Doctrine 191, 228
Fremon, C. 168
Frémont, J. C.: backbone 162; background 167–168; and Fugitive Slave Act 168; *New York Herald* on 182, 213; presidential election 7, 148–149, *149*, 157, 158, 167–176
Fremont Journal 106
Freneau, P. 3
Friend (Philadelphia) 105
Fugitive Slave Act 6, 26, 81, 102, 121, 168
Fugitive Slave Law 64, 91–94, 96, 115, 141–142, 179, 216
Fuhlhage, M. 246

Gabrial, B. 11, 201–209
gag orders 61
Galesburg Democrat 191
Gallagher, G. 15
Gardiner, W. 215
Garrison, W. L. 5, 8, 20, 40–44, 46, 61–67, 213
Gates, H. L. Jr. 125
Gazette of the United States 3
gender 158, 217
Genesee Valley Free Press (New York) 169

George III, King of England 126
Georgia: cartoon *240*; Democratic convention 203, 229; and *The Liberator* 44, 213; and Lincoln 236; Nullification Crisis 53, 55; secession 7, 241; slave population 3; slave trade 126, 127–128, 133
Gholson, J. 45
Glasgow Herald (Scotland) 196
Glasgow Weekly Times (Glasgow, MO) 73
Goldfield, D. 15
Goodwin, D. K. 16
Gordon, N. P. 134
Graham, W. A. 105
Grant, D. 158
Grasshopper Falls 65
"Great Presidential Race of 1856, The" (cartoon) 157
Greeley, H. 4; cartoons *32*, *173*, *205*, *220*; and *Dred Scott v Sandford* 180; and Frémont 167, 174–175; Lincoln-Douglas debates 189, 192; New York press 182–183; and Republican Party 169–170; on slave plots 213–215; and Sumner 217
Green, C. H. 73
Green-Mountain Freedman (Montpelier, VT) 105
Greenville Mountaineer 51, 51n20, 53
Greenville Patriot-Mountaineer 24
Grenville, G. 73
Grenville, W. 126

Hall, A. A. 72
Halstead, M. 30, 202–204, 206–207
Hamilton, A. 2–3
Hammond, J. H. 29–30
Harpers Ferry, Virginia *see* Brown, John
Harper's Weekly 156
Harrison, W. H. 6
Hartford Courant 104–105
Hartford Times 103
Havana 129, 130
Hawthorne, N. 102
Haynes, S. M. 202, 207, 207n40
Henderson, J. D. 75n12
Henry, R. H. 235
Herbert, P. 215, 217
Hickman Courier 252
Higginson, T. W. 98
Hillyer, G. M. 234–235
Hinds County Gazette (Raymond, MS) 30
Hitt, R. 190
Holden, W. W. 31, 194

Holt, M. F. 23n2
Holmes County Farmer (Millersburg, OH) 71
Holzer, H. 190
Horowitz, H. 158
Howard, R. H. 83, 84
Howe, J. W. 96
Howe, S. G. 92, 98
humoral theory 158
Huntzicker, W. E. 11, 179–186

Ibo people 127, 134
Illinois 174, 188–197
Independent (New York) 105, 106
Independent Journal (New York) 2
Independent Press (Abbeville, SC) 108
Independent South (Griffin, GA) 228
Indiana (PA) *True American* 214
Indians *see* Native Americans
Iowa 108, 157
Ivins, S. P. 73

Jackson, A. 4, 19, 26, 42, *54*, 83
Jackson, F. 63
Jackson (MS) 27, 239, 241
Jackson, S. 176
Jay, J. 2
Jefferson, T. 2–3, 21n1, 23–24, 46, 52, 57, 126, 180
Jekyll Island, GA 133
Jersey Independent and Daily Telegraph (Saint Helier, Jersey) 196
John Brown's raid on Harper's Ferry 28–29, 98, 119–122, 201, 212, 217–221, 227–228
Johnson, H. 207
Jones, J. C. 108
Jones, S. 117
Jonesboro 189, 191
Journal (Boston) 245, 246
Journal (New York) 224

Kansas: Buchanan and 157, 192; cartoons *149, 173*; *Dred Scott v Sandford* 179, 183, 190; Frémont and 170; Sumner and 215; *see also* Kansas-Nebraska Act; Lecompton Constitution
Kansas Free State (Lawrence) 118
Kansas Herald of Freedom (Lawrence) 71, 118
Kansas-Nebraska Act 7, 105–110, 114–122; Democratic support for 170; and doughfaces 156, 159, 160; Kansas as antislavery 27; national division 95, 96;

Pierce and 101, 115, 117, 119; popular sovereignty and 7, 64, 65, 81, 115, 169–170, 179, 188, 190, 192, 193, 194, 196, 202–203, 204, 228; press coverage 64–65; unpopularity 169
Kansas Tribune (Lawrence) 117–118
Kansas Weekly Herald (Leavenworth) 72, 73, 74
Keating, T. 215
Keitt, L. 144, *145*, 146, 216, 219, 223
Kelley, R. 72, 74
Kentish Gazette (UK) 196
Kentucky 209
Kentucky Resolution 52, 58n27
Keowee Courier 193–194
Keyes, E. L. 63
Kierkegaard, S. 14
King, W. 104, 105
Knapp, I. 5
Know-Nothings 157, 159, 170, 172, 174, 175, 234–235
Knoxville Standard 73

Lane, J. 208
Lang, G. 16
Lang, K. 16
Laughlin, S. 71
Lawrence (KS) 71, 114, 117–119, 122, 144
Lecompton Constitution 27–28, 61, 65–67, 225–226
Lexington Observer & Reporter 252
Liberator 5, 20, 40–42, 44, 61, 62–64, 66–67, 213
Liberia 44, 131
Lincoln, A. 4; backbone 162; campaign 1860 167; cartoon *185*; *Charleston Mercury* and 229–230; Chicago Republican Convention 204; Cooper Union Speech 217; and Douglas 181, 188–197; election 7, 14, 16, 24, 28, 31, 209, 236; and Frémont 169–170, 175; "House Divided" speech 21n1; Northern vote for 32–33; rebellion 32; on slavery 176; support for ACS 131
Lincoln-Douglas debates 181, 188–197
Lingle, P. 11, 212–221
Lippmann, W. 16
Livengood, N. C. 9, 91–98
Locke brothers 70
London Morning Herald 129
Lone Star (Washington) 87
Loughery, R. W. 83
Louisiana 3, 7, 56, 133, 203, 229, *240*

Louisiana Purchase 64, 77
Louisville Daily Democrat, The 254
Louisville Daily Express, The 254–255
Louisville Daily Journal, The 251
Louisville Journal 107–108, 193
Lovejoy, E. P. 61
Lowell Daily Citizen and News 118, 160

McCardle, W. H. 235
McCombs, M. 16
McGehee, E. F. 237
McMaster, J. A. 173
Macon Weekly Telegraph 119, 121, 143, 148, 246
McPherson, J. 15
Madison, J. 2, 52, 53, 57
Mail (AL) 215
Maine 4, 77
Manifest Destiny 6
Marcy, W. L. 102
marital status of candidate 158–159, 160, 172
Marryat, F. 70
Marshall Republican (TX) 87
Maryland 3, 26, 42, 148, 174, 251–253
masculinity 155–162, 217; barrenness and sexual impotency 159–161
Mason, J. 143–144
Mason-Dixon line *104*, 215, 218
Massachusetts 67, 91, 92, 93, 95, 114–115, 126, 216
Massachusetts Anti-Slavery Society 63
Medill, J. 189, 190, 191
Meigs County Telegraph (Pomeroy, OH) 159n42, 159n43
Memphis Appeal 29, 219
Memphis Daily Appeal 194
Memphis Enquirer 26
Mends, W. 129
Merritt, R. 15, 17–18
Metzger, J. 246
Mexican-American War 6, 9, 25, 62–64, 78–79, 102
Mexico 61, 82, 130
midterm election 1854 101, 107–110
Millersburg Holmes County Republican 193
Milwaukee Daily Sentinel 151
Minkins, S. 92–93
Minooie, M. 7–8, 12, 14–21
Minute Men 230
Mississippi: anti-secession press 234–242; Democratic Conventions 203, 228, 229; Lincoln-Douglas debates 194–195;

Nullification Crisis 56; number of slaves 3; secession 7, 26–27
Mississippian (Jackson) 236–237, 238
Mississippian (Vicksburg) 56
Mississippian and State Gazette (Jackson) 115
Missouri: and Compromise 116–119; editorial feud 72; emancipation 176, 182, 226; presidential election 209; press portrayal 122, 142; secession period 251, 254
Missouri Compromise: division and 30; and Kansas-Nebraska Act 3–4, 64–65, 105, 114, 115–116; prosperity and 78; repeal 7, 27; Taney on 179
Missouri Courier (Hannibal) 116
"Missouri Pro-Slavery Convention" 116
Missouri Republican (St. Louis) 190
mob violence 42, 61, 71, 95–96, 118
Mobile Register 29
Monroe, J. 4, 131
Montgomery Confederation 32
Montgomery Mail 31
Moore, F. Jr. 84, 85
Morning Post (London) 195
Mowry, S. 74
Mullen, A. G. 9, 70–74
Mumford, K. 160

Nashville 26, 71, 72
Nashville Banner 252
Nashville Patriot 194
Nashville Union 26, 71, 72
Nashville Union and American 73
Nashville Union and Patriot 194
Nashville Whig 26
Nat Turner's Revolt 39–46, 213
Natchez (MS) 56, 235
Natchez Weekly Courier 60n56
Natchez Weekly Democrat (later *Weekly Courier*) 56
National Era 61, 62, 66, 93, 106
National Gazette 3
National Register (Austin) 83
nationalism 4
Native Americans 42, 43, 55, 86
nativism 170, 174, 176, 185
natural disasters 73
Navarro Express (Corsicana, TX) 31
Neal, B. 85
Nebraska *see* Kansas-Nebraska Act
Nevins, A. 15, 19
New Berne Daily Progress 28

New England Anti-Slavery Society (later American Anti- Slavery Society) 46
New Era (Newmarket, Ontario) 196
New Hampshire 102–103, 126
New Hampshire Statesman (Concord) 151
New Jersey 148, 174, 209, 230
New Mexico 26, 64, 77–79, 81–82, 84, 85–87
New Orleans 130
New Orleans Bee 32
New Orleans Courier 131
New Orleans Crescent 28
New Orleans Daily Crescent 194, 214
New Orleans Delta 133, 150–151
New Orleans Picayune 28–29
New York 126–128, 130, 132, 157, 185, 192, 237
New York Commercial 246
New York Courier 152, 246
New-York Daily Advertiser 2, 129
New York Daily Times 159, 182–183
New York Day Book 194–195
New York Evening Post 128, 160, 180–181
New York *Herald* 227–228
New York Herald: on Brooks and Sumner 146; on Buchanan 150, 157, 182; on Democratic Convention 202, 204, 207; on Kansas 27, 118, 120–121, 183; on Lincoln 206; on Lincoln-Douglas debates 192, 194; marketing 5; on slave plots 213–214; on slave trade 131–132
New York Journal of Commerce 43, 107, 134
New York Mirror 145
New York Observer and Chronicle 105
New-York Packet 2
New York Times 127, 133–134, 174, 183, 192, 203–206, 209, 224
New York Tribune 182–183; on *Dred Scott v Sandmore* 180, 185; on Frémont 167, 170; on John Brown 219; on Lincoln-Douglas debates 192; on slavery 174–175, 213–214; on Sumner 143, 217
Newbern Spectator 55–56
News (Galveston TX) 87
Nicholson, A. O. P. 112n32
Nicollet, J. 168
Nightingale (ship) 127
Niles Register (Baltimore) 42
Nixon, J. 194
Noble, J. C. 252
North American and United States Gazette (Philadelphia) 120, 146–147

North Carolina: fear of slave revolt 40–42; Nullification crisis 55–56; presidential election 230; secession 14, 25, 32; slave trade 131
North-Carolina Standard (Raleigh) 194
North Carolinian (Fayetteville) 31, 104, 107, 109
North Star (Rochester, NY) 20, 61, 62
Northern Standard (Clarksville) 84, 87
Nueces Valley 85, 86
Nullification Crisis 8, 19, 24, 30, 49–57, 225

Oates, S. B. 45
Observer (London) 132–133, 134
Ohio State Journal (Columbus) 46
Oregon 202
Oregon Trail 168
Orr, J. L. 194, 216, 230
Osawatomie area 119, 122, 175
Osthaus, C. R. 51n14
Ottawa Free Trader 110, 193

Paducah Herald 252
Paine, T. 1
Palmer, T. 27
paper supply 73
Parker, T. 92, 95, 96, 98
Patten, N. 72
Pearce, J. 81, 85–86, 87
Pendleton Messenger 52
Pennington, William 29
Pennsylvania 126, 148, 170, 174
Pennsylvanian 158
"penny press" 5
Pensacola Gazette 115
Peoria Press 116
Perry, B. 24
Philadelphia Public Ledger 169
Phillips, W. 65
Picayune (New Orleans) 246
Pickett, E. 27
Pierce, B. 102
Pierce, F. 101–111; as doughface 155, 159; Kansas-Nebraska Act 105–107, 115, 117, 119; loss of support 148–150; presidency 168–171
Pierson, M. 158
Pittsburg Gazette 217, 219
Pittsburgh Gazette 130
Pittsfield Sun 116, 119, 142
Pleasants, J. H. 40, 42
Poindexter, G. 72, 73

Poinsett, J. 168
political affiliation 70–72
political neutrality 75n12, 83, 192, 251
Polk, J. K. 62, 63, 79, 82, 168
Pollard, H. R. 72, 74
Port Tobacco Times and Charles County Advertiser 253
Portland Advertiser 217
post offices 3, 6, 12, 17
Postal Act 4
Postal Service Act 3
Potter, D. 79, 81
Potter, J. 93
Prairie News (Okolona, MS) 195
Prentice, G. D. 166n61
Prephan, N. 246
presidential elections: 1856 121, 141–152, 156–158, 167–176, 180, 182, 212, 213–214; 1860 14, 15, 16, 24–25, 27, 30–33, 157, 167, 174, 175–176, 182, 185, 189, 194–195, 201–209, 217, 226–227, 229–231, 236–240, 253
Pribanic-Smith, E. 8, 49–57
Protestantism 171, 174
Providence Evening Press 245
Puerto Rico 127

Quincy Herald 191
Quincy Whig 191
Quinn, K. J. 9, 101–111

racial mixing 181
radical Southern Righters 24
railroad expansion 6
Raleigh Register 114–115
Raleigh Standard 31
Raleigh Star 109
Randall, J. G. 14
Randolph Citizen 254
Randolph County Journal 189
Ratner, L. A. 216, 217
Raymond, H. *173*, 74, 182–183, 192, *205*
Red-Lander (San Augustine, TX) 83
Regency, South Carolina clique 224
Republic (Washington) 104–105
Republican (Brandon, MS) 234, 235–236, 238–239
Republican (Lynchburg, VA) 226
Republican Party: blame by South for conflict 14; cartoons *32*, *171*, *173*, 205, *220*; Convention 169, 201, 204; Douglass on 66; formation of 4, 6, 27; funding of press 3; growth of 148,

169–170, 176, 219, 227, 231; and John Brown 28, 228; Kansas-Nebraska Act 96, 116; Lecompton Constitution 225; and masculinity 156, 157, 158–162, 172; naming of 169; Nullification Crisis 52–53, 57; presidential election 7, 24–25, 29, 31, 151, 152, 172, 213, 217, 230; on slavery 181, 214; Sumner on 215–216
Revere, P. 93–94
Revolution 1, 9, 33, 91, 94, 96, 126, 230
Revolutionary press 2
Revolutionary War 52, 57, 84, 237
Rhett, R. B. 5, 201–202, 224–226, 229, 245
Rhett, R. B. Jr. 224–226, 227
Rhode Island 126
Richardson, J. 195
Richardson, W. 87
Richie, T. 42, 44, 45
Richmond Dispatch 196–197, 204, 208–209
Richmond Enquirer: breakdown of compromise 109; on Democratic Convention 203–204, 206, 208; on *Dred Scott v Sandford* 181; on John Brown 29; on Lincoln-Douglas debate 193; Nullification Crisis 56; on Pierce 103; on Republicans 201, 214; on slavery 151–152
Richmond Examiner 226
Richmond South 197
Richmond Whig 216–217
Rio Grande 77, 79, 82
Ripley Bee 121
Rushing, S. K. 16–17
Rust, A. 214, 217

"Sack of Lawrence" 71, 175
Sachsman, D. xi–xiii
St. Cloud Democrat 73n21
St. Louis Herald 253
St. Louis Leader 172
St. Louis Pilot 172
Salem Register 134
San Jacinto 84, 88
Sanford, J. F. A. 179; *see also Dred Scott v. Sandford*
Santa Fe 77, 82, 84, 86
Satterlee, R. C. 74
Savannah Republican 53
Scott, D. *see Dred Scott v. Sandford*
Scott, H. 183, *184*, 186
Scott, W. 63, *104*, 105
Scythes, J. 8, 39–46

secessionists 16, 25, 26, 31–33, 203, 204, 223–225
Semi-Weekly Journal (Galveston, TX) 83, 84, 85
Semi-Weekly Mississippian (Jackson) 120
Semi-Weekly North Carolina Standard (Raleigh) 103
Semi-Weekly Star (TX) 85, 86
Seward, W. H. 21n1, 23–24, 176, 183, 192, 194, 204, *205*
Shannon, M. 235
Shaw, D. 7–8, 12, 14–21, 125–135
Shemberger, M. 12, 251–255
Sheridan, J. B. 190
Sherman, W. 120
Sherman, W. T. 231, 236
Shulman, S. 16
Sibley, J. H. 79
Sibley, M. M. 83
Sims, T. 92, 93–95
slave trade 5, 26, 105, 125–135, 151, 225, 226; press coverage of 127–135
slave uprisings 39–46, 212–214
Smith, G. 115
South America 125, 135
South Carolina: Free Trade 51–52; Nullification Crisis 4, 19, 49–57; numbers of slaves 3; secession 7, 14, 24, 26, 31, 33; slave trade 126; slave uprising 39–44; and tariffs 50; Unionism 51–53; *see also* Charleston
"South Carolina Exposition and Protest" pamphlet 50
South Carolinian (Columbia) 146, 224
South-Western (Shreveport, LA) 106–107
South-Western American (Austin, TX) 85
Southampton County, Virginia 39–44
Southern Rights: *Charleston Mercury* 224–225, 227, 228–229, 231; Democratic Convention 30, 228–229; moderate 24, 29; radical 31
Southern Times and State Gazette (Columbia, SC) 52
Spain, colonial rule 128–129
Speaker of the House of Representatives 29, 79, 194
Spectator 56
Speer, J. 117
Springfield Republican 192
Squatter Sovereign (Atchison, KS) 72, 74
Standard (Raleigh) 182
Star 141
Star State Patriot (Marshall, TX) 87

State Register (Springfield, IL) 216
Staunton Spectator 29
Stephen H. Townsend (ship) 134
Stephens, A. H. 252
Stevens, I. 207
Stevens, T. 29, 159, 174
Stevenson, A. 22n40
Stokes, M. 42
Street, J. 242
Strong, G. T. 156, 162
"submissionism" 40
subscriptions 72
sugar cultivation 56, 125, 128
Sugar Planter (West Baton Rouge) 194
Sumner, C.: attack by Brooks 141–149, 212, 214–217; cartoon *10*; election to Senate 115; Fugitive Slave Law 91, 95, 96
Sumner, E. V. 119
Sumter, T. 52
Sun (Columbus, GA) 249
Sun (New York) 5
Sunbury American 72
Supreme Court 25, 61, 193, 194; *see also* *Dred Scott v. Sandford*; *Worcester v. Georgia*
Swenson Arnold, B. 10–11, 155–162

Tallahassee Sentinel 26
Taney, R. B. 7, 11, 66, 175, 179–181, 183, 185
Tariff of 1828 ("Tariff of Abominations") 8, 49–50
Tariff of 1832 8, 49–50
Tavernier (ship) 134
Taylor, F. M. 254
Taylor, M. J. C. 113n51
Taylor, Z. 6, 77, 78, 79, *80*, 81, 82, 85
technology 6, 18, 19
Teeter, Dwight L. Jr. 216, 217
telegraph 5–6, 28, 144, 185, 190, 217, 227, 236
Telegraph (Macon, GA) 246
Tennessee 14, 16–17, 26, 71, 194, 209, 213
Terry, T. 14–21, 125–135
Texas: coastal slave trade 131; and Compromise of 1850 26, 77–88; Democratic Convention 203, 229; independence from Mexico 130; and New Mexico 64, 79; politics and class 79; secession 7; slavery 126–128; US as protector 84
Texas Democrat (Austin) 83
Texas Monument (La Grange) 87–88
Texas Republican (Marshall) 83

Texas State Gazette (Austin) 84–85, 86, 88
Texian Advocate (Victoria) 85, 86, 87
Thomas, R.W. 193
Thompson, D. P. 112n15
Thompson, R. A. 193–194
Thrubis, E. 12, 245–249
Times-Picayune (New Orleans) 219
Tocqueville, Alexis de 3
Toombs, R. *145*, 147, 219
Topeka Constitution 225
Townsend, J. 230
Trans-Atlantic Slave Trade Database 125, 134
Transcontinental Railroad 7, 105
Traveler (Boston) 226
Travis, J. 39, *41*
Treaty of Guadalupe Hidalgo 25, 63–64, 77, 79, 82
Trent, J. 95
Trenton Banner 26
Tribune (Matagorda, TX) 87
True American (Fayette, KY) 71
True Wesleyan (Boston) 64
Turner, N. 39–46, 213

Unionists: and *Charleston Mercury* 223, 224; and *Evening Post* 245, 248; and John Brown 29; Kentucky 252; and Nullification Crisis 51–53, 56–57; and secession 24–25, 31–32, 239
USS Cyane 128
USS Erie 129
USS Niagara 132
USS North 130
Utah 64, 73, 81

Vallandigham, C. 71
Van Buren, M. 22n40, 108
Van Tuyll, D. R. 8, 12, 23–33, 223–231
VanderVelde, L. 183, 185
Vanity Fair 157, 160
Vermont 108, 127
Vermont Patriot & State Gazette (Montpelier) 143–144
Vicksburg, siege of 235
"Vigilance Association of Columbia" 43
Vigilance Committees 92, 95
Virginia: Compromise 1790 3; Constitution ratification 237; Democratic Convention 197, 207, 228, 229; first slaves 125; Lincoln-Douglas debates 192; new slave code 44–46; Nullification Crisis 55–56; Pierce and 102; presidential election 209; secession 14, 25, 32; slave revolt 39; *see also* Brown, Jesse; Burns, A.; Turner, N.
Virginia Resolutions 52, 58n27
von Frank, A. J. 98

Walden, J. 156, 159
Walker, R. J. 225, 226
Walker, S. 246
Wanderer (ship) 133
War of Independence 2, 93
Warren, G. 252
Washington, DC 26, 64, 78, 127, 132, 168, 214–215, 230
Washington, G. 2, 51, 104
Washington Daily Union (later *Washington Union*) 103, 107, 110, 112n32, 179, 192, 227
Watterson, H. 26
Waugh, J. C. 79
Webster, D. 81, 84, 91, 92–93
Weed, T. 169, 183
Weekly Arizonian (Tubac) 74
Weekly California (MO) *News* 253
Weekly Herald (New York) 115, 148–149, 151
Weekly North Carolina Standard (Raleigh) 109, 214
Weekly Portage Sentinel (Ravenna, OH) 158, 193
Weekly Raleigh Register 146
Western Democrat (Charlotte, NC) 215
Western Monitor (Fayette, MO) 71
Western Reserve Chronicle (Warren, OH) 214
Westport Border Times 120
Wheeling (VA) 230
Whig (Vicksburg) 234, 235, 236–242
Whig Party: collapse of 6, 11, 25, 27, 170; and Constitutional Union Party 202; formation of 4; and Fugitive Slave Law 91; and Kansas-Nebraska Act 116; midterm elections 107–109; Northern 115; Pierce and compromise 105; presidential election 148; press funding 71; Texas 78, 83
White, H. 190
Whiting, A. B. 167–168
Wide Awakes 230
Wilberforce, W. 126
Wilkinson, A. 120
William Clarke (ship) 133
Wilmington Daily Journal 30
Wilmington Herald 106

Wilmington Journal 103
Wilmot, D. 25, 62, 81–82
Wilmot Proviso 25–26, 62–63, *80*, 82
Wilson, H. 147
Windsor Journal 108
Woodville Democrat 60n56
Worcester Spy 42–43
Worcester v. Georgia 55

Wyandotte Constitution 28
Wyoming Republican and Herald (Kingston, PA) 127

Yancey, W. L. 203, 223, 226
Yazoo Democrat 107, 109, 194–195

Zion's Herald (Boston) 64